Mathematical Modeling in Optical Science

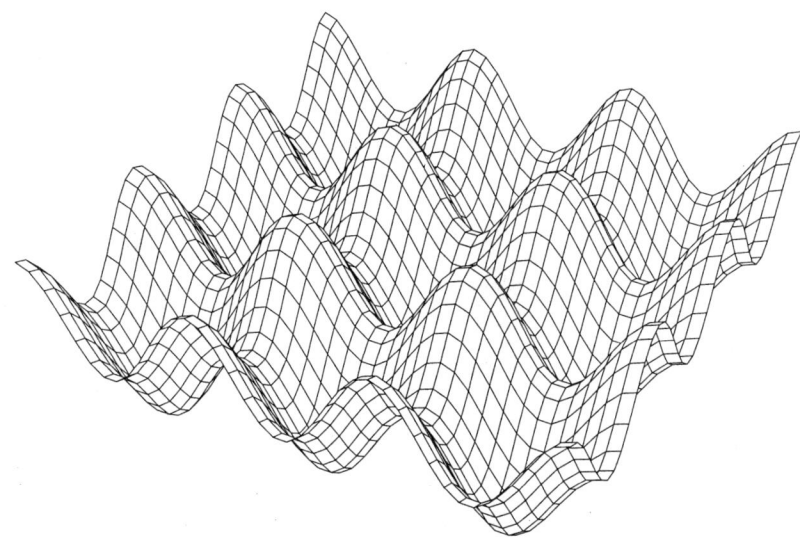

FRONTIERS IN APPLIED MATHEMATICS

The SIAM series on Frontiers in Applied Mathematics publishes monographs dealing with creative work in a substantive field involving applied mathematics or scientific computation. All works focus on emerging or rapidly developing research areas that report on new techniques to solve mainstream problems in science or engineering.

The goal of the series is to promote, through short, inexpensive, expertly written monographs, cutting edge research poised to have a substantial impact on the solutions of problems that advance science and technology. The volumes encompass a broad spectrum of topics important to the applied mathematical areas of education, government, and industry.

EDITORIAL BOARD

H.T. Banks, Editor-in-Chief, North Carolina State University

Richard Albanese, U.S. Air Force Research Laboratory, Brooks AFB

Carlos Castillo Chavez, Cornell University

Doina Cioranescu, Universite Pierre et Marie Curie (Paris VI)

Pat Hagan, Nomura Global Financial Products, New York

Matthias Heinkenschloss, Rice University

Belinda King, Virginia Polytechnic Institute and State University

Jeffrey Sachs, Merck Research Laboratories, Merck and Co., Inc.

Ralph Smith, North Carolina State University

Anna Tsao, Institute for Defense Analyses, Center for Computing Sciences

BOOKS PUBLISHED IN FRONTIERS IN APPLIED MATHEMATICS

Bao, Gang; Cowsar, Lawrence; and Masters, Wen, editors, *Mathematical Modeling in Optical Science*

Banks, H. T.; Buksas, M. W.; and Lin, T., *Electromagnetic Material Interrogation Using Conductive Interfaces and Acoustic Wavefronts*

Oostveen, Job, *Strongly Stabilizable Distributed Parameter Systems*

Griewank, Andreas, *Evaluating Derivatives: Principles and Techniques of Algorithmic Differentiation*

Kelley, C. T., *Iterative Methods for Optimization*

Greenbaum, Anne, *Iterative Methods for Solving Linear Systems*

Kelley, C. T., *Iterative Methods for Linear and Nonlinear Equations*

Bank, Randolph E., *PLTMG: A Software Package for Solving Elliptic Partial Differential Equations. Users' Guide 7.0*

Moré, Jorge J. and Wright, Stephen J., *Optimization Software Guide*

Rüde, Ulrich, *Mathematical and Computational Techniques for Multilevel Adaptive Methods*

Cook, L. Pamela, *Transonic Aerodynamics: Problems in Asymptotic Theory*

Banks, H. T., *Control and Estimation in Distributed Parameter Systems*

Van Loan, Charles, *Computational Frameworks for the Fast Fourier Transform*

Van Huffel, Sabine and Vandewalle, Joos, *The Total Least Squares Problem: Computational Aspects and Analysis*

Castillo, José E., *Mathematical Aspects of Numerical Grid Generation*

Bank, R. E., *PLTMG: A Software Package for Solving Elliptic Partial Differential Equations. Users' Guide 6.0*

McCormick, Stephen F., *Multilevel Adaptive Methods for Partial Differential Equations*

Grossman, Robert, *Symbolic Computation: Applications to Scientific Computing*

Coleman, Thomas F. and Van Loan, Charles, *Handbook for Matrix Computations*

McCormick, Stephen F., *Multigrid Methods*

Buckmaster, John D., *The Mathematics of Combustion*

Ewing, Richard E., *The Mathematics of Reservoir Simulation*

Mathematical Modeling in Optical Science

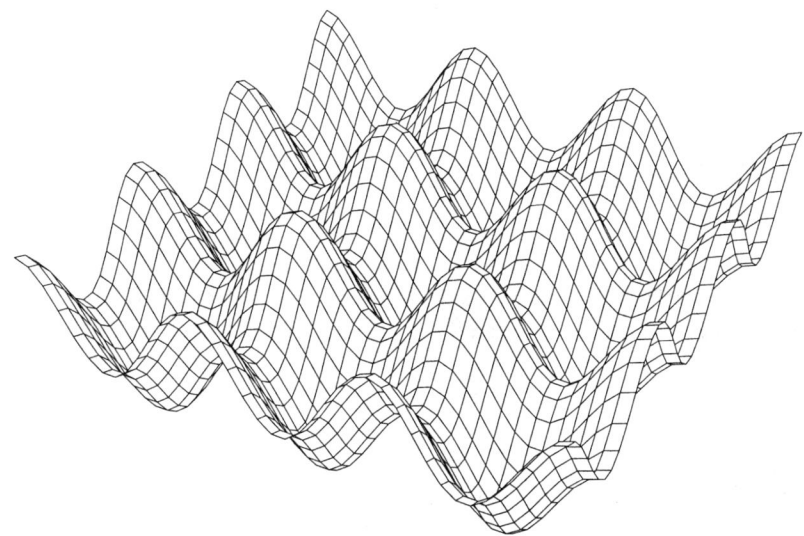

Edited by
Gang Bao
Michigan State University
East Lansing, Michigan

Lawrence Cowsar
Bell Laboratories
Murray Hill, New Jersey

Wen Masters
Office of Naval Research
Arlington, Virginia

Society for Industrial and Applied Mathematics
Philadelphia

Copyright © 2001 by the Society for Industrial and Applied Mathematics.

10 9 8 7 6 5 4 3 2 1

All rights reserved. Printed in the United States of America. No part of this book may be reproduced, stored, or transmitted in any manner without the written permission of the publisher. For information, write the Society for Industrial and Applied Mathematics, 3600 University City Science Center, Philadelphia, PA 19104-2688.

Library of Congress Cataloging-in-Publication Data

Mathematical modeling in optical science / edited by Gang Bao, Lawrence Cowsar, Wen Masters.
      p. cm. -- (Frontiers in applied mathematics)
      Includes bibliographical references and index.
      ISBN 0-89871-475-3
         1. Optics--Mathematics. I. Bao, Gang. II. Cowsar, Lawrence. III. Masters, Wen. IV. Series.

QC372 .M37 2001
535'.01'5118--dc21

00-067125

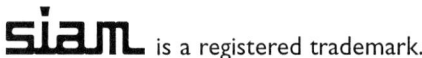 is a registered trademark.

List of Contributors

H. Ammari
Centre de Mathématiques Appliquées
Ecole Polytechnique
91128 Palaiseau, France

G. Bao
Department of Mathematics
Michigan State University
East Lansing, MI 48824-1027

A.-S. Bonnet-Ben Dhia
SMP/URA 853 du Centre National de la Recherche Scientifique
Ecole Nationale Supérieure de Techniques Avancées
32 Bd Victor,
75739 Paris Cédex 15, France

O. P. Bruno
Applied Mathematics
California Institute of Technology
Pasadena, CA 91125

J. A. Cox
Honeywell Research Center
Honeywell, Inc.
3660 Technology Drive,
Minneapolis, MN 55418

D. C. Dobson
Department of Mathematics
Texas A & M University
College Station, TX 77843-3368

P. Joly
INRIA
Domaine de Voluceau-Rocquencourt
BP 105,
78153 LeChesnay Cédex, France

P. Kuchment
Department of Mathematics and Statistics
Wichita State University
Wichita, KS 67260-0033

L. Li
Center for Optical Sciences
University of Arizona
Tucson, AZ 85721
(Present address:
Department of Precision Instruments
Tsinghua University
Beijing 100084, People's Republic of China)

M. S. Mirotznik
Department of Electrical Engineering and Computer Science
The Catholic University of America
Washington, DC 20064

J.-C. Nédélec
Centre de Mathématiques Appliquées
Ecole Polytechnique
91128 Palaiseau, France

D. W. Prather
Department of Electrical and Computer Engineering
University of Delaware
Newark, DE 19716

F. Reitich
School of Mathematics
University of Minnesota
Minneapolis, MN 55455

S. Shi
Department of Electrical and Computer Engineering
University of Delaware
Newark, DE 19716

Contents

Foreword	xi
Preface	xiii
Chapter 1 Overview and Applications of Diffractive Optics Technology J. Allen Cox	1
Chapter 2 Variational Methods for Diffractive Optics Modeling Gang Bao and David C. Dobson	37
Chapter 3 High-Order Boundary Perturbation Methods Oscar P. Bruno and Fernando Reitich	71
Chapter 4 Mathematical Reflections on the Fourier Modal Method in Grating Theory Lifeng Li	111
Chapter 5 Electromagnetic Models for Finite Aperiodic Diffractive Optical Elements Dennis W. Prather, Mark S. Mirotznik, and Shouyuan Shi	141
Chapter 6 Analysis of the Diffraction from Chiral Gratings Habib Ammari and Jean-Claude Nédélec	179
Chapter 7 The Mathematics of Photonic Crystals Peter Kuchment	207
Chapter 8 Mathematical Analysis and Numerical Approximation of Optical Waveguides Anne-Sophie Bonnet-Ben Dhia and Patrick Joly	273
Index	325

Foreword

This volume grew out of a proposal to the SIAM Frontiers in Applied Mathematics series for a conference proceedings volume on optics. The original material was presented during a minisymposium at SIAM's 1997 Annual Meeting at Stanford University. Although the Frontiers series does not publish proceedings, the timeliness and importance of the topics proposed in optical sciences were compelling. Realizing that it might be difficult, if not impossible, to obtain a comprehensive treatment written by a single set of authors, SIAM and the Frontiers editorial board subsequently invited the editors for this volume to carefully organize and edit a collection of chapters, which was then subjected to the usual careful review procedures for the series. The resulting volume, consisting of chapters written by leading research contributors on the topics, does embody the spirit of the SIAM Frontiers series. Indeed, it conforms to the original Frontiers goals and format dating to 1983; see volumes 1 and 2 in the series, wherein distinguished researchers wrote state-of-the-art chapters on related aspects of a topic (for example, on reservoir simulation or seismic exploration).

The present volume provides cutting-edge discourses on areas motivated by emerging technology (communications, information storage, and computing) in optics that provide significant challenges and opportunities for applied mathematicians, physicists, and engineers. The first six chapters focus on diffractive optics, and five of these chapters combine to offer a balanced presentation of current approaches to micro-optics, including computational methods, analytical perturbation techniques, and Fourier modal methods. An additional stimulating chapter treats the relatively new (to applied mathematics) field of chiral media. The final two chapters concern two areas—photonic crystals (the optical analogs of semiconductors) and optical waveguides (such as optical fibers)—that possess great potential for significant research development in applied mathematics.

The thrust of this volume—that micro-optics, in which many classical approximations are not adequate, must be done on a scale that requires careful and precise solutions of Maxwell systems—provides a clear call to applied mathematicians, physicists, and engineers. The editorial board is pleased to have this volume in the Frontiers series and feels that it is especially timely in view of the astounding advances in computing capabilities in the last decade.

H. T. Banks
Center for Research in Scientific Computation
North Carolina State University
Raleigh, NC

Preface

This book addresses some recent developments in mathematical modeling in three areas of optical science: diffractive optics, photonic band gap structures, and waveguides. Particular emphasis is on the formulation of the mathematical models and the design and analysis of new computational approaches. Chapters are organized to present model problems, physical principles, mathematical and computational approaches, and engineering applications corresponding to each of these three areas.

The fundamental importance of optical science is clear. Its various disciplines provide enabling technology applicable to numerous industries, including communication, computing, manufacturing, and data storage. In each of these disciplines, there is an increasing demand for modeling of the relevant physical phenomena. While some members in the applied mathematics community have begun to address a few of the challenging problems in these areas, the barriers to entry are formidable. Much of the existing literature is either highly specialized or lacking focus on the mathematical formulation of the problem.

This book is motivated by a desire to foster communication among the mathematics, physics, and engineering communities on modeling problems in optics. The book's contributors are mathematicians, physicists, and engineers in academia and industry.

While some of the subject matter is classical, the topics in this book are new and represent the latest developments in their respective fields. The model problems discussed in this book are motivated by recent technological developments. For example, in diffractive optics the focus is on micro-optics, i.e., structures of scales comparable to the wavelength of the visible light. Because of the tiny structural scales, wave propagation can no longer be predicted accurately by the classical geometrical optics approximation. Instead, one must solve the Maxwell equations rigorously.

Each of the three topics is presented through a series of survey papers to provide a broad overview focusing on the mathematical models. It is our sincere hope that this book will help researchers and especially graduate students to gain a broad exposure to these important problems. The book should be readable by both the mathematician and the engineer. For the more experienced researcher in either discipline, the book will provide up-to-date results and references. For graduate students and mathematicians less familiar with these areas, the book will provide introductory material to the three areas in optics that offer rich and challenging mathematical problems. With an extensive list of references, it should also enhance the accessibility to specialized literature.

In the areas covered by this book, modeling and simulation have become an important part of the engineering process. The applied mathematics community may contribute significantly to the analysis of these models, as well as to the design and analysis of simulation techniques and automated design tools. Because the applied mathematics community has been rapidly addressing many challenging problems of optical science during the last decade, this book is also intended to introduce researchers in appropriate engineering disciplines to recent mathematical advances in theory, analysis, and computational techniques.

Outline of the Book

Chapters 1–6 concern diffractive optics. Chapter 7 focuses on the mathematics of photonic crystals. Chapter 8 discusses the mathematical analysis and numerical approximation of optical waveguides.

In Chapter 1, Cox gives an account of some recent applications of diffractive optics, particularly in the areas of subwavelength structures, optical imaging, and diffractive optical elements. Through specific examples, he illustrates the great need for new modeling tools in optical science. This section motivates the rest of the chapters on diffractive optics.

In the subsequent four chapters, a balanced description of the most commonly used modeling approachs are given. These approaches are designed for dealing with a variety of modeling situations. Together they provide a complete and up-to-date picture on mathematical modeling of microdiffractive optics.

Chapter 2 by Bao and Dobson is devoted to variational methods. The authors present a variational formulation and well-posedness results for the model. Computationally, various types of finite element methods will be discussed. Recent results on inverse and optimal design problems will also be highlighted.

In Chapter 3, Bruno and Reitich introduce another approach: the method of boundary variations and analytic continuation. This method is based on the deep mathematical observation that electromagnetic fields behave analytically with respect to perturbations of a scattering surface, so they can be represented by convergent power series in a perturbation parameter.

In Chapter 4, Li surveys an engineering approach, namely, the Fourier modal method, or FMM. This method is also known as the BKK method, the rigorous coupled-wave method, the Fourier expansion method, and the Moharam–Gaylord method, among others. He discusses various existing formulations and points out that despite its remarkable popularity and its success in many areas of applications, until recently the FMM had serious problems in modeling metallic gratings in TM polarization. To date there appears to be a lack of a detailed mathematical analysis of the convergence characteristics of the method.

In Chapter 5, Prather, Mirotznik, and Shi discuss the development and application of numerical electromagnetic models as they apply to the analysis of microdiffractive optical elements that are finite in extent and/or aperiodic. The computational methods range from the finite-difference time-domain and frequency-domain techniques to boundary integral techniques. A summary based on generally accepted appraisals for each method will be presented along with a comparison of the computational requirements needed in the analysis of several diffractive optics elements.

Chapter 6 deals with a new research subject in diffractive optics. Ammari and Nédélec introduce the reader to chiral media, which have interesting electromagnetic and bianisotropic properties. Chiral media can be characterized by a set of constitutive relations in which the electric and magnetic fields are coupled. The authors present the model system and mathematical formulation. They discuss the features of this model not shared by the standard system of Maxwell's equations. They also discuss mathematical analysis and computational methods on this new class of problems.

The central topic of Chapter 7 is photonic crystals, also known as photonic band gap structures. A photonic crystal is a periodic dielectric low-loss material for which there exist invervals of frequency for which electromagnetic waves cannot propagate in this medium. In other words, a photonic crystal is an optical analogue of a semiconductor. Because of the great potential for a technological revolution in optics, computers, and other areas, this field has attracted much attention from physicists and engineers. In this chapter, Kuchment provides an up-to-date account of the mathematical formulation, analysis, and optimal design of photonic crystals. A broad range of topics within this subject is covered, with a focus on those for which rigorous mathematical results have been obtained. These include properties of the periodic Maxwell operator, conditions for the existence of band gaps in photonic crystals, defects in photonic crystals and conditions for the existence of defect modes with frequencies in band gaps, and recent developments in numerical techniques for the calculation of photonic structures. Several conjectures are also included, which point to challenging problems that need to be solved.

Chapter 8 is devoted to optical waveguides. Bonnet-Ben Dhia and Joly present a rigorous mathematical analysis and computational method for waveguide problems in which the full set of Maxwell equations must be solved. Optical fibers with circularly symmetric refractive index profiles are an important example of optical waveguides in which a number of mathematically simpler reduced models may be applied such as those derived under the assumption that the modes are "weakly guided." The reader interested in that work will find ample information elsewhere in the literature, for instance, in Marcuse's classic book, *Theory of Dielectric Optical Waveguides* (2nd ed., Academic Press, 1991) or Snyder and Love's *Optical Waveguide Theory*, supplemented by Dunford and Schwartz's *Linear Operators*.

Acknowledgments

In the process of producing the book, we have received generous help from a number of individuals. In particular, we thank all of the contributors for contributing the papers in a timely fashion, yet without sacrificing the quality. We also thank the reviewers for their many constructive (often critical) reviews of each one of the papers. Professor H. T. Banks, the editor-in-chief of the SIAM "Frontiers in

Applied Mathematics" series, deserves special thanks for his encouragement and guidance. The editors are grateful to the anonymous reviewers of our original book proposal for their invaluable inputs. Finally, the highly professional and helpful assistance provided by the SIAM staff is graciously appreciated.

GANG BAO
LAWRENCE COWSAR
WEN MASTERS

Chapter 1

Overview and Applications of Diffractive Optics Technology

J. Allen Cox

1.1 Introduction

Diffractive optical elements represent a generalization of diffraction gratings that provide powerful light manipulation properties. As the technology to produce them with high efficiency and low cost has developed, their power has been used advantageously in a variety of optical systems. The first efforts to insert diffractive elements took the form of volume holograms as combiners in head-up displays for aircraft cockpits in the late 1960s [1]. However, both the success and the limitation of volume holograms can be summarized by the observation that, 30 years later, head-up displays in military aircraft are still one of the very few applications where holograms are used in imaging systems. The primary limitations to more widespread use arise from environmental susceptibility of the holographic media and expensive, specialized laser equipment and facilities needed for recording.

A second class of diffractive optical elements was discovered in the early 1960s [2, 3, 4] and takes the form of phase-coherent surface-relief gratings. This class of diffractive elements evolved rather slowly over the following 20 years due mostly to fabrication limitations. However, as advancements were made in various forms of precise micromachining techniques, this technology came to dominate the field of diffractive optics and has been applied to a wide variety of both imaging and nonimaging optical systems. These surface-relief diffractive structures are known by a number of names, such as kinoforms, Fresnel phase lenses, binary optics, and, more generically, diffractive optics (used exclusively here), and are the subject of this chapter.

The availability of efficient diffractive phase elements and micro-optical components has had a significant impact not only on design of imaging systems but also on the development of entirely new approaches to implementing some optical functions. For imaging applications, diffractive elements used in combination with refractive elements permit the designer to "dial in" the dispersive power of an element made of simple glass, thus providing an easy means to either correct or enhance chromatic aberration. Similar flexibility can be applied to correct other aberrations and to athermalize an optical system. New approaches for artificial dielectrics, polarization control, and extremely narrow-band, highly reflective mirrors have emerged with the use of low-order diffractive gratings. A variety of unique roles for diffractive optics and micro-optics becomes evident when one considers current developments in optoelectronics that integrate laser diodes, optics, and receivers in miniaturized packages.

The objective of this chapter is not to provide a survey of the mathematical basis for diffractive optics—several aspects of this topic can be found in subsequent chapters—but rather is first to provide an overview of the current status in diffractive optics technology and then to illustrate three applications that require rigorous modeling tools. This chapter is written primarily from an engineering standpoint and has a decidedly nonmathematical nature. It is curious to note that the need for increasingly sophisticated mathematical modeling tools is well correlated with technological limitations in the fabrication of minimum feature size in the period, or Fresnel zone, of diffractive elements. Initially, the ratio of wavelength to minimum zone width was rather small, and simple approximations to the mathematical diffraction problem, usually expressed as the Fourier or Fraunhofer approximation, were quite adequate. As the ability to fabricate ever smaller features progressed with improved lithographic, holographic, and micromachining techniques, the need arose for reliable models based on the rigorous electromagnetic theory of gratings derived directly from Maxwell's equations. This has been a very active area of research in recent years, and several numerical approaches have been developed. In Section 1.3 we describe three applications that illustrate this point. In the final section we attempt to identify problem areas and possible future directions in the technology that applied mathematicians may find interesting and challenging.

1.2 Technology Overview

1.2.1 General Description

A diffractive optical element brings about an optical transformation by acting only on the phase of an incident wavefront. It consists of a set of surface-relief contours that have 2π phase modulation in both width and depth. Figure 1.1 illustrates in cross-sectional view how such an element can be formed from a conventional refractive element in a manner analogous to a Fresnel lens. However, it must be emphasized that a traditional Fresnel lens is not phase coherent because the transition points in the collapsed element do not occur at regular 2π phase intervals.

Chapter 1. Overview and Applications of Diffractive Optics Technology

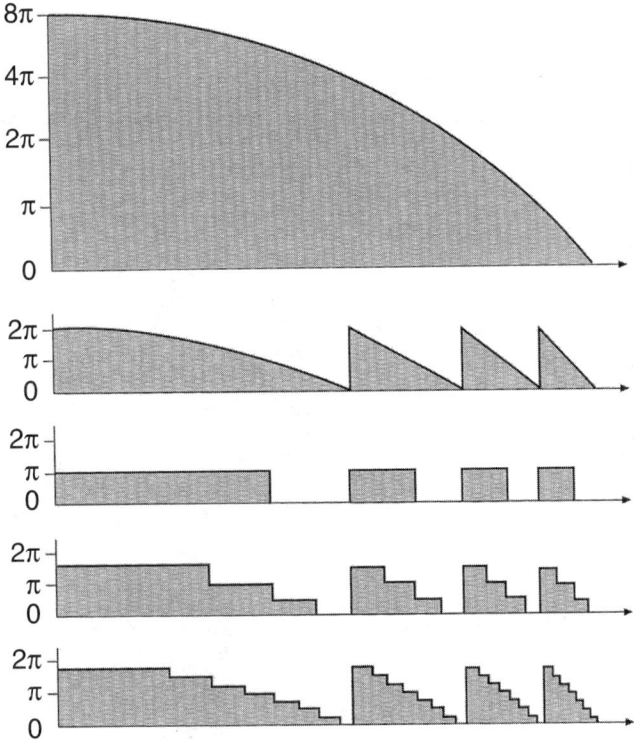

Figure 1.1: Cross-sectional view of the transition from a refractive lens to a diffractive element and its discrete phase approximation.

The generality of the optical transformations that can be implemented is one important reason for the wide impact diffractive optics has had on optical systems. This generality derives from the fact that the phase map describing the transformation is specified mathematically in the design process. The mathematical specification can be quite complex, thus permitting the use of complicated transformations (i.e., "optical elements") in a system. In many cases, diffractive optics provides a particularly simple implementation of a concept which otherwise is difficult or impossible to do with conventional optics. Examples of functions or transformations commonly performed by diffractive surfaces are

- aspheric aberration correction,
- chromatic aberration correction,
- higher order aberration correction,
- tilt and decenter functions,
- special functions such as "circle-to-line" conversion and "spot array" generation (Dammann gratings).

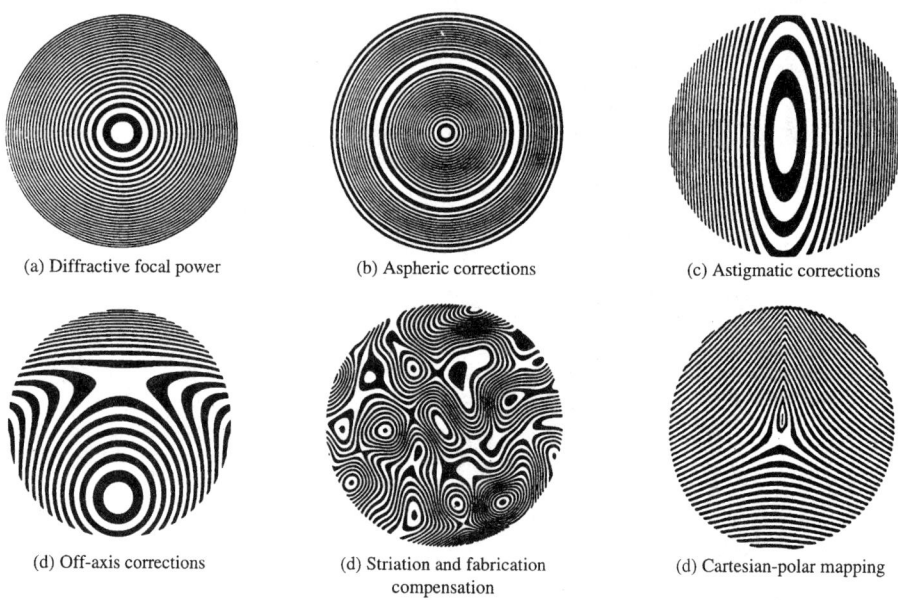

Figure 1.2: Examples of phase contours representing optical transformations.

Some special cases are illustrated in Figure 1.2 in a planar view of the phase contours forming the transformation function. A phase contour is a region where a 2π phase change is added to the wavefront relative to the neighboring contours; such contours are generally referred to as Fresnel zones. In practice, the utility of diffractive elements is generally most effective with complicated optical systems when a hybrid approach is taken in which simple spherical elements embody the optical power and the diffractive elements implement the "fine tuning" corrections.

Diffractive optics has benefited from, and owes much of its rapid growth to, precision micromachining and high-resolution pattern generation techniques originally developed to support other industries, such as microelectronics. Generally, continuous surface profiles are preferred for best performance, and several techniques, such as diamond turning, laser writing, grayscale lithography, and direct-write electron beam lithography, are being developed for this purpose. Discrete phase structures are preferred in some cases for phase retardation, artificial index modulation, and resonant waveguide gratings, and are in other cases a matter of practical compromises, so methods using UV lithography and dry etching will continue to play an important role. In all cases tight control of process errors is necessary to approach theoretical performance limits, and applications will continue to be very demanding in this respect. For reasons to be described below, diamond turning has become the most important machining method for imaging applications.

The term "binary optical elements" refers to diffractive elements having staircase phase profiles with 2^n phase steps fabricated in n processing cycles [5]. The lower three profiles in Figure 1.2 show binary profiles having two, four, and eight

Micromachining

- Single Point Diamond Turning
  - capable of large areas
  - applied to IR and plastic mat'ls
  - sequential process limits scale-up potential
  - limited to coarse surface features and circular symmetry

- "Binary" Optics
  - standard semiconductor processing
  - minimum processing/phase levels
  - limited scale-up for large elements
  - high sensitivity to process errors
  - very fine features

- Isobathic Process
  - standard semiconductor processing
  - maximum processing/phase levels
  - no symmetry limits
  - low sensitivity to process errors
  - very fine features

Replication

- Polymer Embossing
  Dry Photopolymer Embossing/Surphex
  - well patented (DuPont)
  - diffraction-limited imagery demo'd (1993)
  - high fidelity, low shrinkage
  - replicates high aspect, submicron features
  - acrylic and polycarbonate substrates only
  - environmental susceptibility (?)

  Cast and Cure
  - optical epoxy on glass demo'd
  - good image quality
  - environmental durability (?)

- Polymer Injection Molding
  - larger range of standard mat'ls (~6)
  - very large scale-up potential
  - shrinkage, warpage, birefringence issues
  - requires tight process control
  - poor environ. stability of standard mat'ls
  - new polymers promising

Figure 1.3: Summary of manufacturing methods for diffractive optics.

phase levels for the same diffractive element. The isobathic process is another method to produce staircase phase profiles that generates m phase levels in $m-1$ processing cycles. Both methods of fabrication involve multiple and iteratively sequential steps of photolithography and pattern transfer using wet-chemical or dry-plasma etch procedures. In general, several masks or templates are used to build up the desired surface-relief profile. The resulting structure is a discrete staircase approximation to an ideal continuous phase profile.

Many of the fabrication methods discussed here are labor intensive, time consuming, or inherently sequential and do not easily permit scale-up to large production volumes. In such cases, these methods are used to fabricate master elements, and replication techniques are employed to achieve high-volume, low-cost producibility. Interest in replication methods is growing quickly as applications mature into products, particularly in the visible region. Replication methods include embossing, cast-and-cure, and injection molding. Shrinkage, warpage, and environmental degradation affect fidelity, image quality, and product life of the replicas, and control of these factors represents an intense area of research in replication technology. Figure 1.3 summarizes the main features of the most commonly used micromachining and replication methods. An excellent recent survey of the entire scope of diffractive optics technology can be found in Lee [6].

1.2.2 Phase Function and Diffraction Efficiency

The most basic constraint limiting the diffractive transformation is set by the minimum feature size of the lithography and processing errors in fabrication. Minimum feature size in a diffractive element is usually expressed in terms of the minimum Fresnel zone, or a fraction thereof, and is generally a linear function of wavelength and optical power. For example, in a Fresnel phase lens which brings on-axis, collimated, monochromatic light to diffraction-limited focus, the phase function is embodied in circular zones with phase function $\phi(r)$ having radial dependence,

$$\phi(r) = \text{Mod}((2\pi/\lambda)\{f - \sqrt{(f^2 - r^2)}\}, 2\pi), \tag{1.1}$$

where λ is the wavelength and f is the focal length. Each zone contributes a phase change to the wavefront of exactly 2π relative to its neighboring contours, and from this it can be shown that the outer radius of the mth contour is given by

$$r_m^2 = 2mf\lambda - (m\lambda)^2. \tag{1.2}$$

And, finally, from this follows that the Fresnel zone with the smallest width ("minimum Fresnel zone width" $\Delta r_{\min}$) is

$$\Delta r_{\min} \approx 2\lambda f/, \tag{1.3}$$

where $f/$ is the focal ratio, or "f number," of the element defined as the ratio of the focal length to the diameter of the element. The power of the element is inversely proportional to $f/$.

Raytracing is by far the most common method to calculate the phase function required to bring about a given optical transformation. As a reminder, the method of raytracing, corresponding to the approximation of geometrical optics, is derived from the rigorous theory embodied in Maxwell's equations for the electromagnetic field in the limit of vanishing wavelength and of neglecting polarization (scalar field). In this limit one derives the eikonal equation describing the phase propagation of the field (cf. Born and Wolf [7, Chap. 3]),

$$\text{grad}(S) \cdot \text{grad}(S) = n(x)^2,$$

where $S(x)$ represents the spatial phase of the field (also called the optical path length) and $n(x)$ is the refractive index of the medium in which the wave propagates. The surface $S(x) = $ constant is the geometrical wavefront; an optical "ray" is just the local normal at one point on the wavefront. Raytracing codes calculate the function $S(x)$ by propagating a large number of rays from a given point over a range of angles through the optical system to a specified surface, and the value of the calculated optical path length along the path is associated with the intersection point on the surface. If a sufficiently large number of rays are used, an accurate numerical estimate of the function $S(x)$ is obtained. Any desired optical transformation on that surface can be calculated simply as the difference between the calculated $S(x)$ and the desired phase on that surface. In particular, the phase function $\phi(x)$ needed to specify a diffractive element is simply

$$\phi(x) = \text{Mod}((2\pi/\lambda)\Delta S(x), 2\pi), \tag{1.4}$$

where $\Delta S(x)$ is the optical path difference on the surface.

Chapter 1. Overview and Applications of Diffractive Optics Technology

The discussion above relates to the wavefront transformational properties of the diffractive element. The diffractive nature of such elements naturally gives rise to multiple orders or channels into which the incident light can be directed by the diffractive element. The fraction of the incident power directed into a specific order is called the diffraction efficiency of that order. The real practical utility of diffractive elements derives from the possibility of achieving perfect efficiency in one order by "blazing" a surface profile to implement the phase function.

The physical depth of the surface relief profile is determined from the phase profile in (1.1) by the following expressions:

$$d(r) = \begin{cases} [\phi(r)/2\pi]\lambda/(n-1) & \text{transmissive element with refractive index } n, \\ -[\phi(r)/4\pi]\lambda & \text{reflective element.} \end{cases} \quad (1.5)$$

Being a grating in its basis structure, a diffractive element supports a number of diffractive orders—transmissive, reflective, positive, and negative—into which incident light may be directed. The diffraction efficiency of a given order is the radiant power propagating in that order divided by the total radiant power propagating in all orders. In general, rigorous electromagnetic theory based on Maxwell's equations must be applied to accurately predict diffraction efficiency for an arbitrary surface relief profile as a function of wavelength, polarization, and angle of the incident wave. However, in virtually all cases of interest to visual imaging systems the minimum Fresnel zone (grating period) is much larger than the relevant wavelengths, and the Fraunhofer approximation [8] accurately describes the transformation of the incident light into the various propagating orders supported by the diffractive element. In this approximation, diffraction efficiency of the mth order is simply the squared modulus of the mth Fourier coefficient of the phase profile. For a strictly periodic phase grating of period Λ, the diffraction efficiency of order m can be shown to be [5, 9]

$$\eta_m = \left[\frac{\sin\{\pi(\phi/2\pi - m)\}}{\{\pi(\phi/2\pi - m)\}} \right]^2. \quad (1.6)$$

It is clear that imposing the phase condition on the grating, $\phi = 2\pi m$, through the proper fabrication of the surface depth profile in (1.5) yields a diffraction efficiency in the mth order of 100%. Most commonly, $m = 1$ and the element is said to work in first order. Note that the efficiency is independent of the period in this approximation but does depend on wavelength. As long as the surface relief profile satisfies the phase condition in (1.5) for a specific wavelength and periodicity is enforced by making the profile repeat modulo $2\pi m$, the diffractive element will have very high efficiency in mth order at that wavelength. It is this feature that enables high efficiency in diffractive elements having more general phase contours illustrated in Figure 1.2, which are not strictly periodic but only quasi-periodic. Further, by fixing the depth profile in (1.5) to yield maximum efficiency at one reference wavelength λ_0, (1.5) and (1.6) can be combined to obtain an expression for the efficiency at any other wavelength, as illustrated in Figure 1.4.

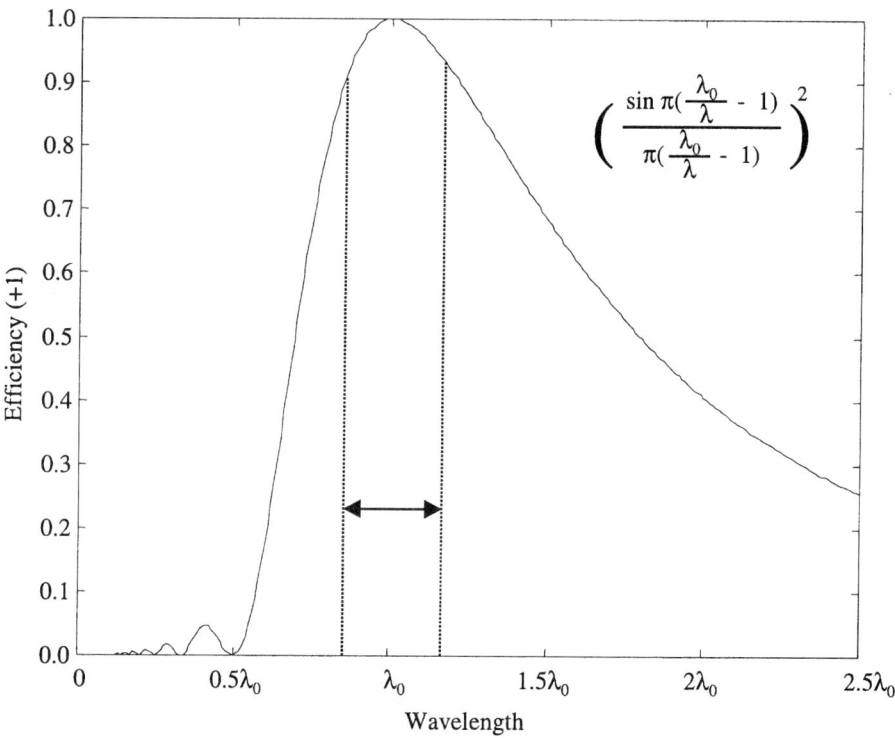

Figure 1.4: Spectral variation of diffraction efficiency for an ideal diffractive element working in first order and optimized for wavelength λ_0.

Single point diamond machining can be used to fabricate a continuous surface profile defined by (1.5) to achieve high efficiency provided that the phase contours have circular symmetry, that the minimum Fresnel zone is much larger than the tool radius, and that the substrate material is compatible with the machining process. When these conditions are not satisfied, a multilevel discrete phase approximation to the continuous profile can be fabricated by lithographic and dry etch techniques as illustrated in Figure 1.1. In this case, the efficiency in mth order for N phase levels can be shown to be [5, 6]

$$\eta_m = \left[\frac{\sin\{\pi m/N\}}{\{\pi m/N\}}\right]^2. \tag{1.7}$$

We show in Figure 1.5 the efficiency in first order as a function of N. Most noteworthy is the rapid increase such that efficiencies greater than 95% are possible in principle with at least eight phase levels. Since every real surface profile will have some degree of fabrication errors, we give an indication of the sensitivity of diffraction efficiency to mask alignment error used in making a multilevel structure. Figure 1.6 shows first-order efficiency for a four-level structure as a function of Fresnel zone width for an ideal structure having 81% efficiency, for a structure

Chapter 1. Overview and Applications of Diffractive Optics Technology 9

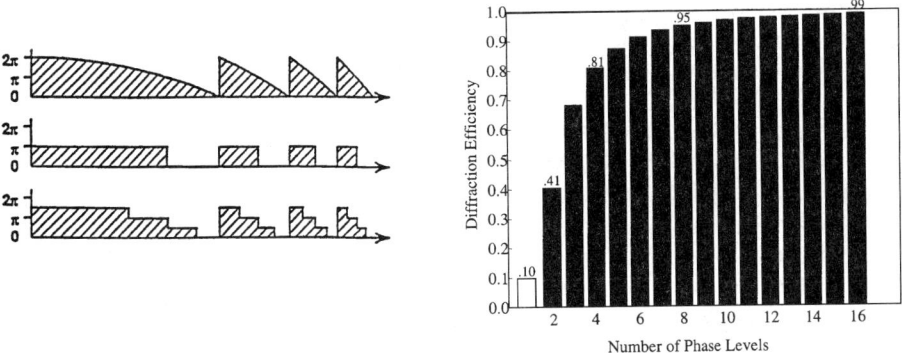

Figure 1.5: Diffraction efficiency for an element working in first order showing the rapid increase with number of discrete phase levels. The ideal continuous profile has an efficiency of 100%. The calculation is done in the Fraunhofer (Fourier) approximation.

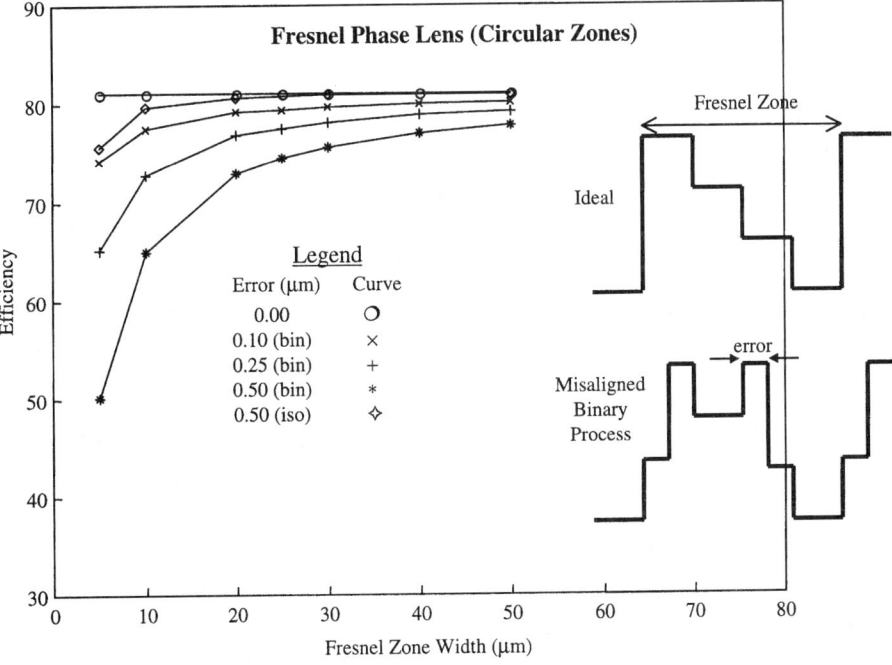

Figure 1.6: Sensitivity of diffraction efficiency to mask alignment error for a four-level diffractive element working in first order.

made using the binary optics procedure assuming mask alignment errors of 0.1 μm, 0.25 μm, and 0.5 μm, and for a structure made using the isobathic process with a mask alignment error of 0.5 μm. The insert on the lower right illustrates the effect of mask alignment error on the surface relief structure in the binary optics procedure. For binary optics structures the efficiency can be seen to be quite sensitive to mask alignment error and to decrease sharply for Fresnel zones smaller than 20 μm even with an alignment error of 0.25 μm. The isobathic procedure, however, is remarkably insensitive to fabrication errors in general, and this provides the strongest motivation for its use in making master diffractive elements for replication despite its additional cost.

Regardless of the fabrication method, some amount of machining error or artifact is always present to cause a deviation from the ideal profile, and in practice an upper limit on diffraction efficiency is approximately 98%. It should be clear that if the actual fabricated profile can be measured, then the distribution of light among all orders can be calculated accurately and taken into account in the overall image quality of the optical system.

1.2.3 Image Quality in Multiorder Systems

We now turn to the issue of image quality in an imaging system having at least one diffractive element with light distributed over multiple diffractive orders (see [10]). Here we refer to the representation of image quality strictly on an optical basis in terms of quantitative measures such as the point spread function and the modulation transfer function. We emphasized earlier in this section that a diffractive element supports multiple orders into which incident light can be channeled and that, although it can be blazed to achieve high efficiency in one order, there is in practice always some light directed in higher orders. Any light occurring in higher orders can lead to degraded image quality, and here we briefly describe and illustrate how this happens.

The optical representation of image quality in noncoherent systems is discussed by a number of authors. Born and Wolf's classic text [7] provide detailed definitions of various aberrations in image quality as viewed from the standpoint of both geometrical optics (Chapters 4 and 5) and diffraction theory (Chapter 9). Other authors, such as Goodman [8] and Gaskill [11], take an approach based on linear systems theory, which is widely embraced by the engineering community. Regardless of one's favored viewpoint, all discussions of image quality ultimately stem from the point source response of the optical system. Its importance lies in the fact that light or radiation emitted from any scene or extended object can be considered as a linear superposition of infinitely many point sources comprising the object. The point source response of an optical system is simply the intensity distribution that results at the image plane from a point source at the object. The point source response function is generally dependent on wavelength, location of the source in the field, and the design of the optical system itself.

Although the point source response contains all information about the image quality of an optical system, perhaps a more practical measure is the modulation transfer function (MTF), which can be derived from the point response. First, recognize that as the image of a point in object space, the point source response intensity distribution represents a blurring, or smearing, of the point in the transformation from object to image space. Such blurring leads to a loss of contrast in the image of an extended object and is the basis of the MTF measure of image. First, modulation is defined conceptually in terms of the luminance L ("brightness") of light ($L_{\max}$) and dark ($L_{\min}$) bars in a periodic square wave pattern,

$$\text{modulation } M_o = (L_{\max} - L_{\min})/(L_{\max} + L_{\min}),$$

and, as the ratio of the difference to the sum of the two luminances, modulation is a measure of contrast. When an optical system forms an image of this pattern, the blurring action of the point source response reduces the sharpness of definition of the light and dark bars in the pattern, causing a reduction in the difference $L_{\max} - L_{\min}$ and hence in the modulation. Clearly, the loss in contrast depends on the spatial frequency of the periodic pattern, and the ratio of modulation of the image, M_i, to modulation of the object, M_o, as a function of spatial frequency f is called modulation transfer of the optical system

$$MTF(f) = M_i/M_0.$$

Gaskill [11] explains the mathematical procedure to convert point source response into MTF. Basically, the point source response is converted into a linear source response by integration of one spatial variable, and the MTF is the absolute value of the Fourier transform of the linear source response.

We illustrate the effects of higher diffractive orders on image quality with a specific imaging system used for a head-mounted display (HMD) called INIGHTS. This system has been used extensively at Honeywell to verify the theory experimentally [10]. The baseline optical design for the system contains only conventional refractive elements. A hybrid design was generated by replacing one refractive element with a plano-concave element having a diffractive element on the planar surface to do aspheric correction. Eight phase levels were machined into the diffractive element, and it was designed to operate at +1 order.

Image quality of the system was determined both by photographing the point source response and by measuring the MTF. One can observe the point source response directly with a microscope, and Figure 1.7(a) shows the spot diagrams predicted for the hybrid INIGHTS system. One can see the −7 order and +9 order present as diffuse distributions centered on the sharply focused +1 order, as expected for an eight-level element. This diagram corresponds quite closely to actual visual observations of the system point source response.

Figure 1.7(b) shows the MTF data for the baseline and hybrid systems measured on-axis at 543.5 nm with a 10 mm pupil. At very low spatial frequencies, below ∼4 cy/mm, the MTF of the hybrid system is a few percent less than the baseline MTF. This is due to the diffraction inefficiency of ∼8% in the hybrid element; i.e., approximately 8% of the light passing through the hybrid system is diffracted into higher orders and appears to the eye as a faint diffuse background. At spatial frequencies above ∼6 cy/mm and out to the noise floor occurring at ∼40 cy/mm, the MTF of the hybrid system is significantly larger than the baseline. Here, the

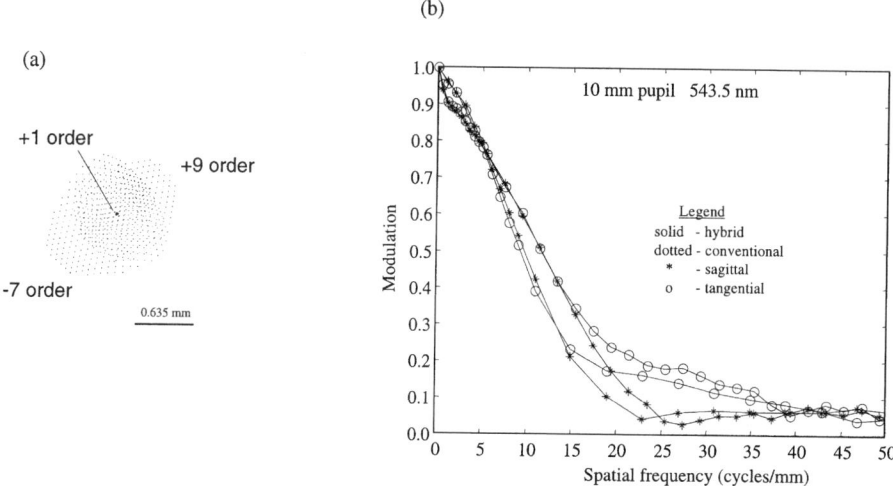

Figure 1.7: Summary of the method to calculate MTF for a multiorder hybrid optical system.

diffractive optical element (DOE) is performing the desired function of reducing aberrations and improving image quality.

Because light in the higher orders appears as a diffuse distribution, it is not unreasonable to expect its effect on MTF to resemble that of scattered light. However, unlike the situation with scattered light in which predictions of MTF usually are based on a statistical approach, the light distributed in the higher diffractive orders is quite deterministic, and hence its effect on MTF can be predicted with high confidence. Buralli, Morris, and Rogers [9, 12] and Cox, Fritz, and Werner [10] have described a procedure for calculating the MTF of a hybrid system as a straightforward extension of the usual theory by incorporating the efficiency of each order as a weight for the order's MTF.

1.3 Challenging Applications

Having already summarized diffractive optics technology used in current engineering practice, we now wish to turn to topics at the leading edge in research that represent challenges in both modeling and fabrication. Typically, the challenges arise because the grating period or minimum feature size is on the order of or less than the operating wavelength. In addition, polarization effects can be significant and are often exploited. Many of the mathematical aspects of the topics discussed here are treated in subsequent sections and chapters of this book. Here we have selected three applications to illustrate the improved performance and extended functionality possible with new diffractive structures and to provide motivation for pursuing these challenges.

As a bridge with the previous overview, we first choose a novel application of diffractive optics to photonics and describe a dual diffractive element integrated with a vertical-cavity surface-emitting laser (VCSEL) used in fiber-optic data communications. The diffractive elements have minimum Fresnel zones of $\sim 4\lambda$ and thus are not accurately described by the Fourier approximation for calculating diffraction efficiency. Nevertheless the phase functions can be specified with conventional design tools based on raytracing. We then move to two examples that can only be described at the level of Maxwell's equations and thus designed and analyzed with rigorous numerical models. The first application deals with a high-aspect metallic grating having a continuously adjustable period fabricated with the microelectromechanical systems' (MEMS) LIGA technology. This device was developed as a tunable infrared transmittance filter. Finally, we consider recent developments in grating-enhanced resonances to show significant potential both for photonic devices such as VCSELs and for exploiting nonlinear phenomena, and we present a summary of recent work of guided-mode grating filters for application to lasers.

1.3.1 VCSEL Beam Monitor and Focusing Element

The VCSEL is rapidly replacing semiconductor edge-emitting laser and light-emitting diode sources in a wide variety of applications, such as fiber-optic data communications and sensors [13, 14]. With further development it is expected to dominate other applications, such as printing and optical data storage, currently considered the domain of edge-emitting lasers. Diffractive optical elements play an important role for many VCSEL-based applications, and here we introduce one particular case for data communication modules.

There are several features and characteristics inherent to the VCSEL structure that are responsible for its favorable standing. First, as implied in the name, light from a VCSEL is emitted normal to the surface of the substrate on which it is grown, while edge-emitting laser diodes direct light parallel to the surface through faces cleaved at the edge of the chip. The VCSEL aperture is circular and has small diameter, typically 5–15 μm, while the edge-emitting aperture is a highly elongated rectangle typically 1–2 μm by 50–100 μm. Consequences of these different features are that VCSELs have circular low-divergence output beams and 10–100 times smaller lasing threshold currents for the same output power, and they can be tested at the wafer level prior to dicing into individual devices. Further benefits are more efficient coupling into optical fibers, higher modulation rates, and simpler drive electronics. Thus, with these benefits it should be clear why VCSELs have become so dominant in data communications.

Here we illustrate the use of DOEs both for monitoring laser output power and for coupling the beam into a fiber. DOEs are well suited for this application because they can be integrated naturally into a cover glass plate already present and because the narrow laser wavelength permits significant optical power in the DOE without the accompanying problem of spectral dispersion. Furthermore, the DOE used for monitoring beam power operates in reflection, while the DOE focusing on the fiber operates in transmission. As suggested in the diffraction efficiency expressions in (1.6), (1.7) in the previous section, the phase profile for high efficiency in one mode leads to low efficiency in the other mode, and thus it is possible to use a reflective DOE and a transmissive DOE in series and still achieve good overall efficiency.

As discussed below, the DOEs needed for this application require small features (Fresnel zones) on the order of a few wavelengths, and thus the Fourier, or Fraunhofer, approximation can yield an error of ~10% for diffraction efficiency in the smallest zones of the elements. At such values of minimum Fresnel zone, polarization also has a small effect on efficiency and an even smaller effect on image quality (a slight amount of astigmatism is introduced). From a design standpoint, however, these small deviations are relatively insignificant for this application. Design codes based on raytracing still give an accurate prescription of the Fresnel zone radii defining the optical transformation of the DOEs. Furthermore, for this application precise specification of the diffraction efficiency is not important for reasons discussed below. So, this application represents a case where the effects of small Fresnel zone size exist to a small degree and are observable if examined closely, but the effects are sufficiently small and the tolerances are sufficiently loose that the raytracing approximation is adequate for designing the elements.

Figure 1.8 presents a cross-sectional view of a VCSEL mounted in a standard TO-46 header/can package. The VCSEL is centered on top of an annular silicon detector that monitors the radiant power emitted by the VCSEL. Two diffractive elements are machined on opposite sides of the glass cover plate. The power monitor DOE, located on the inner surface of the cover glass, operates in reflection and focuses ~2% of the emitted radiation into an annular ring centered on the VCSEL. Approximately 88% of the transmitted light is in the zero order and is available for focusing on a fiber by a second DOE located on the back (outer) surface of the cover glass.

There are several reasons, perhaps not immediately obvious, why diffractive elements are very well suited for both the power monitoring and the focusing functions in this application. First, VCSELs used in data communications generally are highly multimode, and not only is the radiant power distributed among the many modes, but also the modal distribution varies within each pulse as the drive current rises and falls. Thus, in order to monitor accurately the total radiant power in the pulse it is necessary to sample the entire wavefront during the whole pulse. Thus, the reflective DOE is designed to span the complete angular aperture of the VCSEL and to map the VCSEL point source into a ring focus at the plane of the power monitor detector underneath the VCSEL substrate. The Fresnel zone pattern that accomplishes this function is exhibited on the left side of Figure 1.8. The magnitude of the reflected signal equals the diffraction efficiency of the reflective DOE times the Fresnel reflectance of the cover plate glass (~4%).

The transmissive DOE operates on the zero-order light transmitted by the reflective DOE and brings it to focus on the core of the optical fiber. The Fresnel zone pattern required for this operation is shown on the right side of Figure 1.8 and is a straightforward design exercise. Less straightforward requirements are eye safety and the need for high-speed operation. High modulation rates (> gigabit/second) in VCSELs generally require driving the device into saturation with maximum current, and hence maximum output power [15]. High speed thus usually implies more power available than actually needed to achieve maximum bit error rates, and this

Chapter 1. Overview and Applications of Diffractive Optics Technology

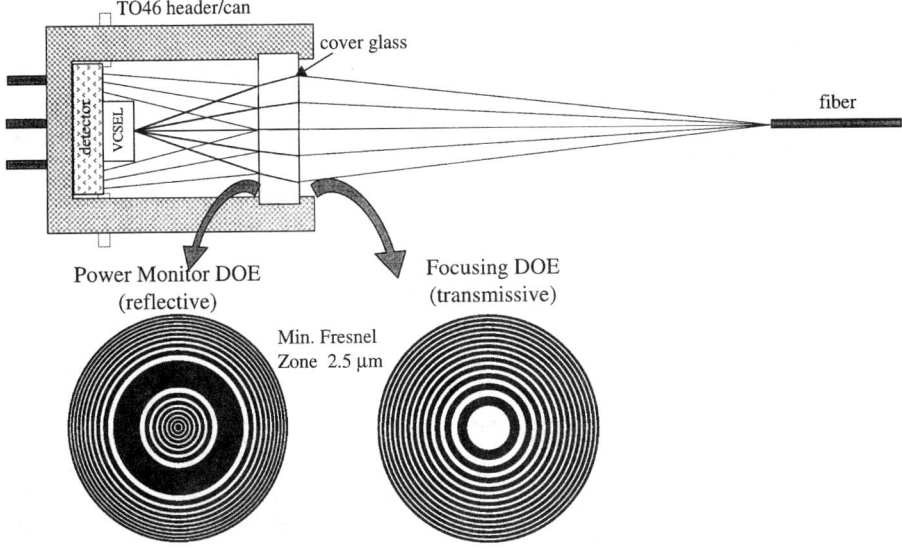

Figure 1.8: Cross-sectional view of VCSEL data communication module with power monitor and end-view of two diffractive optical elements.

in turn raises the issue of eye safety in the event of failed connection. For this reason it is useful to be able to focus only part of the radiant power into fiber and to disperse the remainder over a wide angle, thus providing both high-speed operation and a higher eye safety margin. This combination of otherwise conflicting requirements is easily implemented with a DOE by selecting the efficiency appropriately.

From the above considerations, as well as a limit on minimum feature size in the DOEs, the final design specified a blaze of three phase levels in the reflective DOE (power monitor) and four phase levels in the transmissive DOE (fiber coupler), giving theoretical diffraction efficiencies of 68% and 81%, respectively. Because the minimum Fresnel zone width in both DOEs is $\sim$2.5 microns, direct-write electron-beam lithography has to be used in order to fabricate the phase profiles with minimum loss in efficiency from process errors.

To verify the power monitor concept, lower efficiency, one-sided, two-level DOEs were fabricated using standard contact lithography of the reflective and transmissive power monitor elements. This was feasible since $\sim$1.25 micron linewidths are attainable with contact lithography and no mask alignment is required for two phase levels. Although both the reflective and transmission DOEs have only $\sim$40% diffraction efficiency, we were able to test several design variations. Preliminary examination of the transmissive elements indicated that the spot size was $\sim$30 microns (as designed) and at approximately the correct distance from the DOE. The performance of the reflective power monitor DOE is illustrated in Figure 1.9 and showed a well-defined ring with uniform brightness and ring diameters consistent with the design. The average source-to-DOE separation was 1.760 mm. Thus, diffractive optics yields an attractive solution for this application that offers

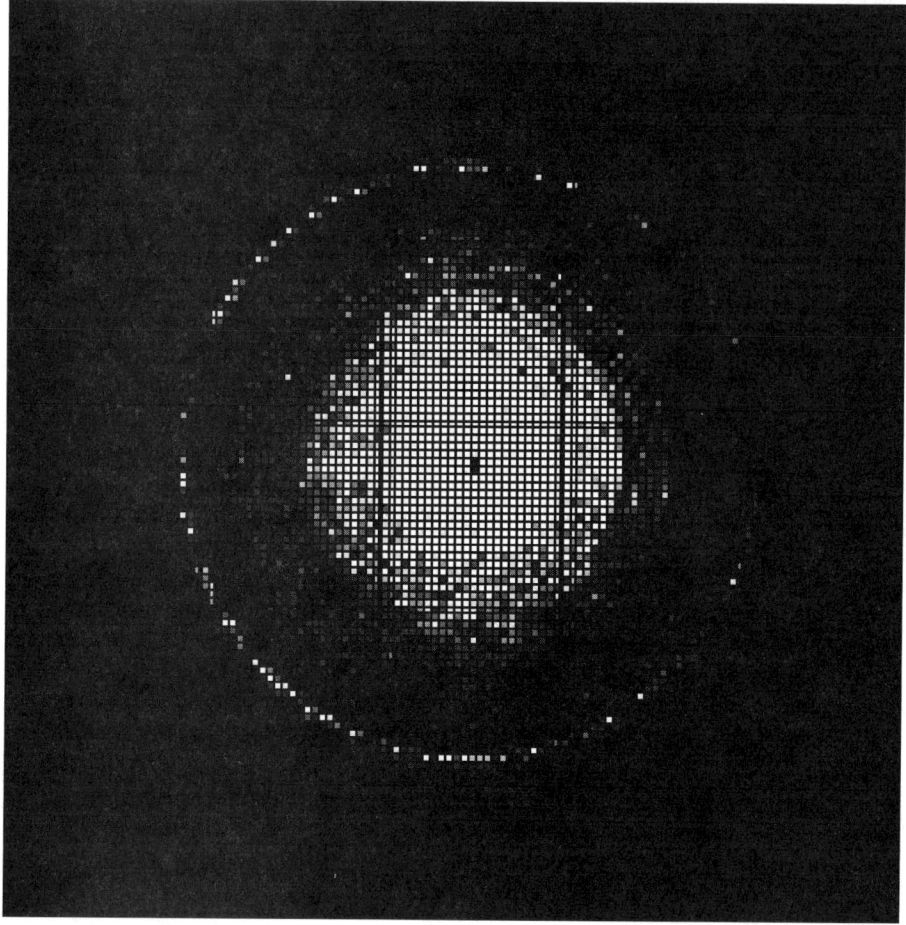

Figure 1.9: Measured image of annular ring formed by reflective power monitor DOE under VCSEL illumination.

excellent performance with cost-effective producibility and conforms with special features of VCSELs.

1.3.2 LIGA Tunable Grating Filter

As originally conceived, the term *MEMS technology* implied the merging of micromachining techniques with microelectronic fabrication methods to provide a new approach for mass production of miniature electronic and magnetic sensors and actuators. Concurrently, micromachining methods have been adapted primarily from the microelectronics industry and applied to produce both diffractive and microoptical elements. Here we describe one such device, a tunable infrared (IR) filter, and describe its optical properties [16, 17]. It represents a good example of the

Chapter 1. Overview and Applications of Diffractive Optics Technology

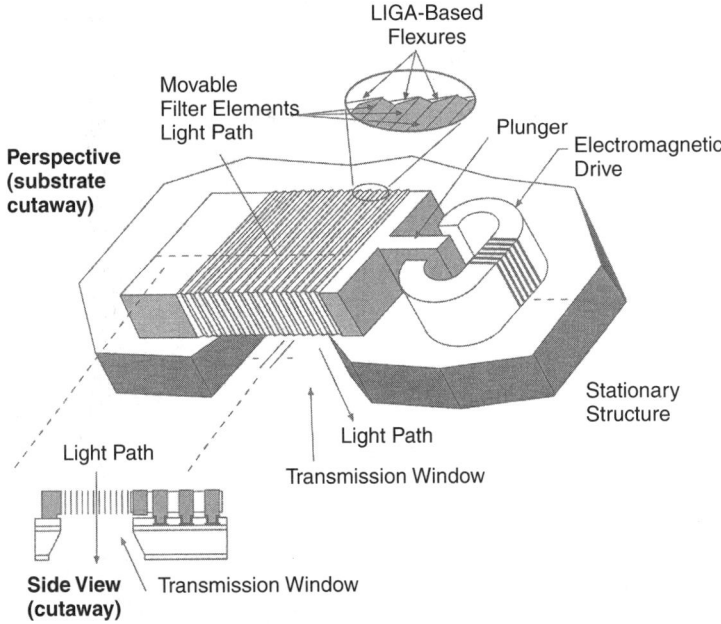

Figure 1.10: LIGA-based tunable IR filter showing vertical parallel-plate filter structure and linear-magnetic-drive actuator.

confluence of MEMS and diffractive optic technologies in that its components are fabricated with LIGA techniques developed for MEMS, while its design and performance prediction rely on rigorous Maxwell solver codes developed for diffractive optics. (LIGA derives from the German LIthographie, Galvanoformung, und Abformung, denoting the use of x-ray lithography in thick poly (methyl methacrylate) (PMMA) to define structures with very high aspect ratio, following by nickel electroplating [galvanoformung], and subsequent replication in PMMA [abformung] to continue the process.)

The tunable IR filter device consists of two components: a flexible optical filter structure itself and a linear magnetic actuator. A conceptual drawing is shown in Figure 1.10. The filter is made up of a linear array of parallel nickel plates forming a linear diffraction grating and is joined at the ends by spring flexures to give mechanical flexibility. The grating period and the spacing between plates is adjusted by a linear electromagnetic actuator. The filter structure is fixed to the substrate at one end and free to move linearly within the actuator cavity at the other end. The linear actuator consists of an electromagnet with an air gap. A plunger is attached to the free end of the grating structure and is driven into the gap by the activated electromagnet. Friction with the substrate is avoided by magnetically levitating the plunger. Figure 1.11 shows a scanning electron microscope (SEM) image of a parallel array of metal plates (top view) comprising the spring flexures and their interface to the grating plates for a flexible grating structure. The high

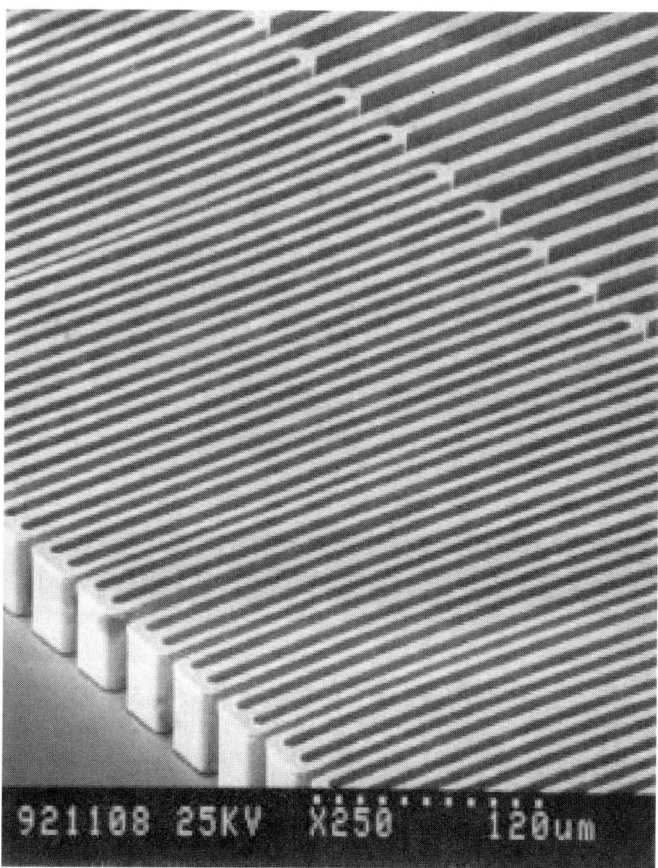

Figure 1.11: Detailed view of filter spring flexures joined to grating plates at the side of the active filter area.

degree of edge acuity and smoothness of the vertical sidewalls typical of the LIGA are evident in the image.

The device acts as a low-pass spectral filter to linearly polarized radiation with the electric field parallel to the metal plates. Thus, incident, unpolarized radiation must be prepolarized, for example, with a wiregrid aligned orthogonally with respect to the metal plates. The cutoff property of the structure derives from the fact that each unit cell of the grating is also a planar waveguide that only supports propagating modes in parallel polarization with wavelength less than twice the gap between plates. Wavelengths greater than this value cannot pass through the structure and are mostly reflected; hence,

$$\lambda_{\text{cutoff}} = 2 \cdot \text{gap}. \tag{1.8}$$

Figure 1.12, taken from reference [16], illustrates these points by showing the predicted transmittance for the two basic polarization states, $E_{\|}$ and $H_{\|}$ ($= E\perp$

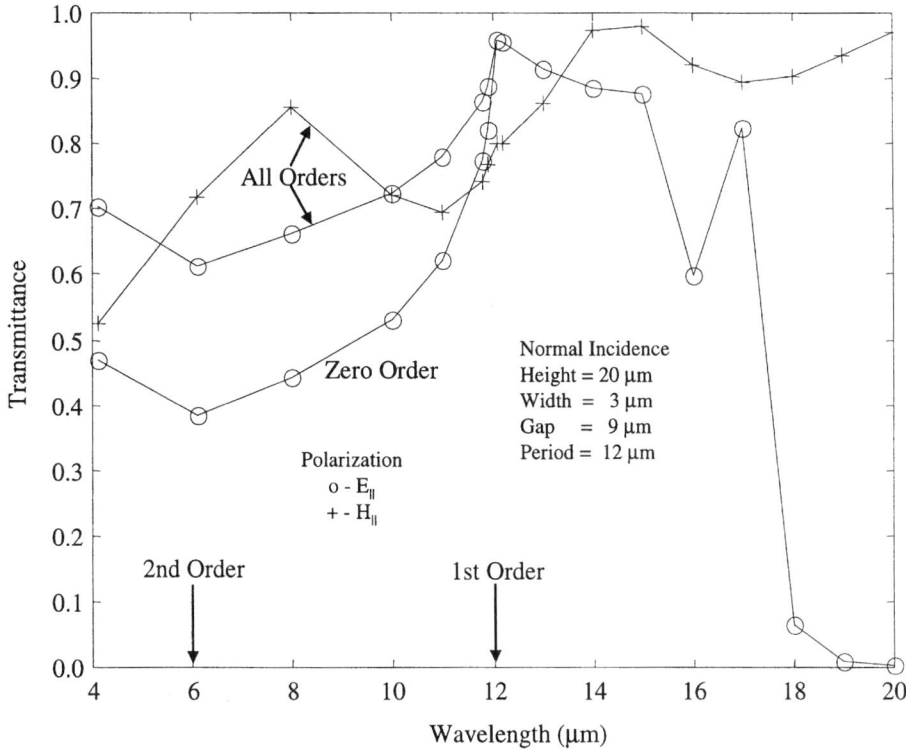

Figure 1.12: Model prediction of LIGA grating transmittance for $E_\parallel$ and $H_\parallel$ polarization

at normal incidence), and summed over all propagating orders. The existence of a cutoff in $E_\parallel$ polarization at twice the gap value (18 μm) is evident in contrast to the high transmittance at all wavelengths for $H_\parallel$ ($E\perp$) polarization. A small transmittance tail exists beyond cutoff; this is a consequence of the finite height of the grating walls, which here nearly equals the wavelength. Note that averaging the transmittance of the two polarizations over the lower range of wavelengths gives a value approximately equal to the clear aperture fill factor of the grating (75%). Two arrows on the abscissa mark the wavelengths at which the first and second diffractive orders begin to propagate, and over this range we have plotted the transmittance in zero order. The difference between the zero-order curve and that summed over all orders represents the radiation propagating in higher diffractive orders and thus highlights the grating nature of the filter.

Model

The task of predicting performance of LIGA gratings in the infrared is made difficult both because the geometric dimensions (period and height) are comparable to the wavelength and because of the very high conductivity of the metal lines. A reliable

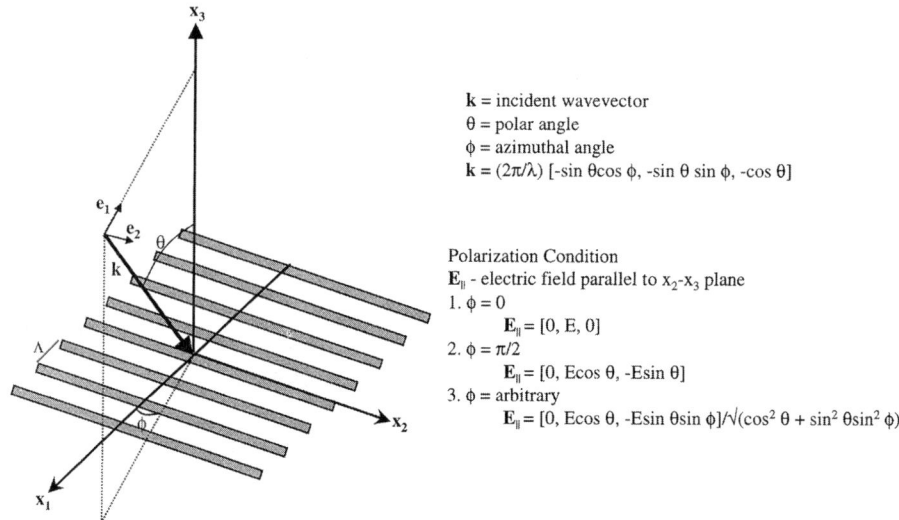

Figure 1.13: Grating geometry and incident wave definitions.

model must be based on the rigorous electromagnetic theory of gratings derived directly from Maxwell's equations. This has been a very active area of research in recent years, and several numerical approaches have been developed. In our work we use Dobson's variational method [18] implemented in a finite element code (MAXFELM) [19]. The problem can be stated as follows: we are given an arbitrary biperiodic interface separating two homogeneous half-spaces with a plane wave incident at an arbitrary angle on the interface from the upper half-space, and we have to calculate the scattered radiation in both half-spaces (i.e., both reflected and transmitted radiation in all diffracted orders).

The variational method has proven to be well suited for treating the diffraction problem for LIGA structures. It is one of the few methods for which statements of existence, uniqueness, convergence, and rate of convergence have been proved [20], and hence we know that the solutions of the model are close approximations of physical reality.

We implement the variational method by first applying the Floquet–Bloch theorem to confine the problem in the lateral dimensions to just one unit cell of the periodic interface. We then confine it to a finite box in the transverse dimension by developing "transparent" boundary conditions for the field over the ends of the box in terms of the usual outgoing radiation condition. A detailed description of this procedure can be found in references [18–20]. The problem can be discretized in various finite element spaces using standard Galerkin-type schemes. In the MAXFELM model used here, piecewise trilinear elements on a uniform grid are used. The geometry treated in our model is shown in Figure 1.13; various grating parameters and incident plane wavevector definitions are also included.

Chapter 1. Overview and Applications of Diffractive Optics Technology

Summary of Optical Performance

Application of the rigorous Maxwell solver to the LIGA grating problem has predicted optical performance of the structure in remarkable agreement with experiments. In addition to the general filter properties, various subtle features observed in the measurements are evident in the predicted results and provide convincing confirmation of the value of rigorous modeling. Here we simply summarize the main points and conclusions learned from the previous study reported in [16, 17] wherein measured values of optical constants for nickel [21] were used. Note that a full three-dimensional vector model is required to treat the arbitrary incident angles encountered in this problem even though the grating itself is basically two-dimensional.

"Organ pipe" resonances. One or more sharp dips in transmittance are frequently observed in the cutoff region. An example of one such dip can be seen at $\sim$16 μm in Figure 1.12. These features are associated with a resonance of the grating cavity and are well known in acoustics as organ pipe resonances [22]. The number of resonances in the transition region increases with the height of the grating walls.

Sensitivity to incident angle. Spectral transmittance curves exhibit markedly different sensitivity to incident polar angle θ depending on the plane of incidence ($\phi = 0$ or $\phi = 90$). For the case $\phi = 0$, the plane of incidence is perpendicular to the grating lines, and the cutoff wavelength is independent of the incident angle, but the transmittance at all wavelengths below cutoff decreases significantly with increasing angle. Conversely, with the incident plane parallel to the grating lines ($\phi = 90$), there is significant shift in the cutoff with angle and relatively insignificant change in transmittance. The shape of the transmittance curve in the latter case is independent of angle. The behavior of transmittance below cutoff can be understood in both cases in terms of the apparent fill factor of the grating as a function of polar angle. The different behavior in cutoff wavelength is explained in terms of the effective wavelength of the incident wave when $\phi = 90$.

Sensitivity to wall height. In addition to affecting the number of resonances, the height of the grating walls also has a significant effect on the sharpness of the transmittance cutoff. When the height is smaller than the cutoff wavelength, there is a significant tail beyond cutoff and a gradual transition to high transmittance. As the height increases, the tail is attenuated and the transition sharpened such that when the height is approximately three times the cutoff wavelength, the tail is insignificant and a sharp leading edge exists.

Wall taper. Providing the top edge of the grating plates with a taper gives some apparent improvement in reducing angular sensitivity in the $\phi = 0$ plane. However, it only holds for small angles of incidence, and in fact most of the improvement can be attributed to elimination of organ pipe resonances. Thus, there does not appear to be significant advantage to use of the wall taper.

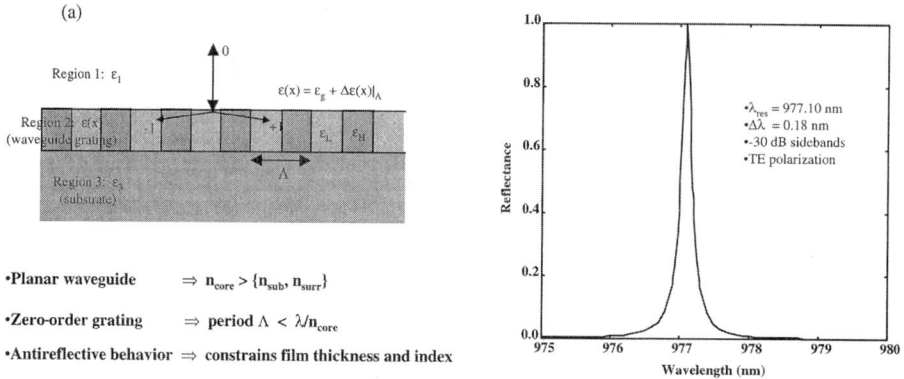

Figure 1.14: Main features of the GMGRF concept (a) and illustration of representative reflectance properties.

1.3.3 Guided-Mode Grating Resonance Filters

One of the most interesting new developments in diffractive optics involves the integration of a zero-order grating with a planar waveguide to create a resonance. Taking the form of a planar dielectric layer with the grating providing a periodic modulation of the dielectric constant in one or more of the layers, such structures have been demonstrated to yield ultranarrow bandwidth filters for a selected center wavelength and polarization with ~100% inband reflectance and ~30 dB sideband suppression [23], [25]–[30]. With such extraordinary potential performance, it is not surprising that these "resonant reflectors" have attracted attention for many applications. In particular, one natural application is a laser mirror, and we shall describe our effort to incorporate a guided-mode grating resonance filter (GMGRF) into a VCSEL. We start, however, with a brief description of the structure and a design methodology developed by Wang and Magnusson [25]–[27]. As will be emphasized, an accurate prediction of the optical properties of GMGRFs requires the use of a rigorous Maxwell solver, and here there are several open issues that should be addressed by the applied mathematics community.

Background and Overview

A conceptual structure representing a GMGRF is illustrated in Figure 1.14a. Region 2, consisting of a planar thin film, separates two homogeneous half-spaces. The upper half-space is the "surrounding" and designated region 1; the lower half-space is the "substrate," called region 3. Electromagnetic radiation ("light") in the form of a polarized plane wave can be incident on region 2 from either half-space. Region 2 is constrained to have two special properties: (1) it must satisfy the requirement of a planar waveguide and have an average refractive index greater than the refractive indices of both half-spaces; and (2) it must have a periodically modulated dielectric function (note that for nonconductive media the dielectric function equals the square of the refractive index).

Thus, in addition to being the core of a waveguide that supports guided modes, region 2 is also a grating. For a given incident plane wave of wavelength λ, angle, and polarization it is possible to find a grating period Λ such that a first diffractive order of the grating couples to a guided mode of the waveguide. However, because the refractive index of the waveguide core is periodically modulated, any excited guided mode also undergoes diffraction, and by reciprocity, one possible diffracted order must lie along the vector of the incident wave. By arranging the grating to support only the zero propagating order, energy of the guided mode diffracted out of the core can only lie along the direction of the incident wave, and through this coupling a resonance is established which can lead in principle to 100% reflectance in a very narrow spectral bandwidth, as illustrated in Figure 1.14b. The resonant wavelength is determined primarily by the grating period, and the bandwidth primarily by the modulation of refractive index in the grating. Furthermore, for wavelengths outside the resonance region, the structure appears "homogenized" in its dielectric properties, and thus it may be considered approximately as a simple thin film structure with reflectance properties described by well-known thin film expressions (cf. Born and Wolf [7, section 1.6]). In particular, it is possible to achieve antireflection conditions in the thin film structure away from the resonant wavelength.

Hessel and Oliner [23] were the first to study the guided-mode grating resonance in detail, and in particular they clearly distinguished between such resonances and the more commonly known Wood's anomalies which occur at the Rayleigh points in a grating (where a diffractive order just passes from evanescent to propagating). Hessel and Oliner also cleared up the confusion about the two types of anomalies by showing that guided-mode resonance commonly, but not necessarily, occurs at a Rayleigh point. Nevière [30] revisited the problem from the standpoint of "leaky" guided waves that is appealing both physically and mathematically. Magnusson and his students [24, 25, 26, 29] have been at the forefront for the past few years in the area of guided-mode grating structures in dielectric films and have not only developed a practical design methodology but have also pointed out many practical uses and applications.

Design Methodology

Wang and Magnusson [26] have described a detailed iterative method to design GMGRFs in dielectric multilayer thin film structures. Their method is summarized in this section and illustrated with a specific structure that we have designed, fabricated, and tested.

We suggested in the previous section that zero-order gratings are most effective in coupling incident radiation to a guided mode. Although it is possible to achieve a resonance with a higher order grating, the existence of more than one channel for light to propagate makes it difficult to get high reflectance, and in practice we constrain the design only to zero-order gratings. Intuitively, the coupling between

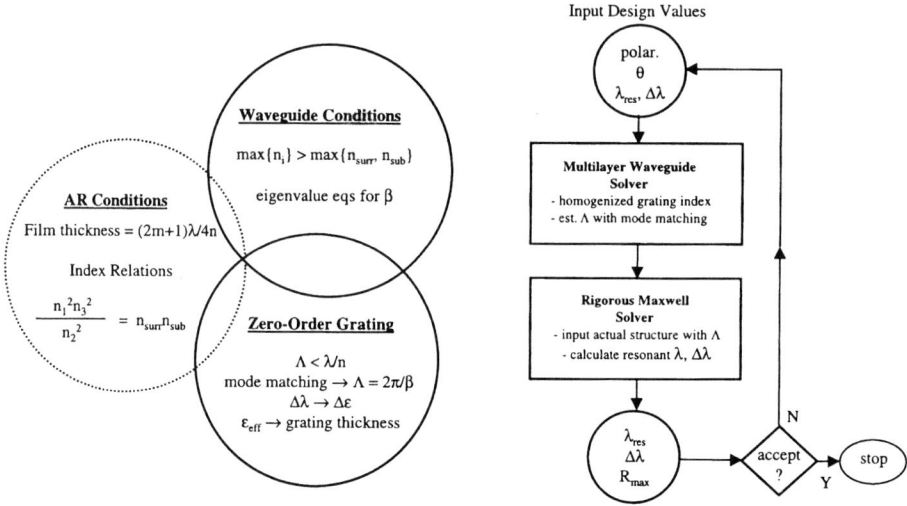

Figure 1.15: Elements of the design methodology for GMGRFs.

the grating and waveguide is realized by equating the first-order wavevector of the grating to a wavevector of a guided mode, viz.,

$$\beta = k(n_0 \sin\theta_0 \pm \lambda/\Lambda), \text{ mode matching condition}, \qquad (1.9)$$

where

$\beta =$ guided mode eigenvalue = axial component of guided mode wavevector,

$\lambda =$ wavelength of incident wave in vacuum,

$k =$ incident wavenumber $= 2\pi/\lambda$,

$\theta_0 =$ angle of incident plane wave in region 1 with respect to surface normal,

$n_0 =$ refractive index of region 1,

$\Lambda =$ grating period.

Design requirements are specified in terms of the resonant wavelength, λ_{res}, at which maximum reflectance occurs, the spectral bandwidth of the resonance $\Delta\lambda$, as well as the polarization and angle of the incident wave, and possibly the "out-of-band" reflectance away from resonance.

Figure 1.15 summarizes the overall conditions, design parameters, and sequence in the design process. On the left side of the figure, various necessary conditions that must be imposed to achieve resonance are enclosed in two solid circles representing the waveguide and zero-order grating; antireflective (AR) conditions that are not necessary for resonance but are often convenient are enclosed in a dotted circle. A convenient starting point in the design process is with the waveguide and AR conditions. Given refractive indices for the surrounding (region 1) and the substrate (region 3), a thin film material must be chosen with refractive index greater than both. Regardless of the use of the AR conditions, it is a good idea to start with a film thickness that is a quarter-wavelength, or a multiple thereof. If AR properties are desired, then a multiple thin film structure is generally required for the waveguide.

The zero-order grating must exist in one of the thin films to bring about the coupling, and from a practical fabricational standpoint the topmost film is usually the best choice. Wang and Magnusson [25] point out that, to a very good approximation over several orders of magnitude, spectral bandwidth $\Delta\lambda$ is directly proportional to the modulation in the waveguide's dielectric function $\Delta\varepsilon$, which in effect is the modulation of the film containing the grating. The dielectric modulation is adjusted most easily by the fill factor of the grating, and we recommend [25] for guidelines to relate $\Delta\lambda$ and $\Delta\varepsilon$.

With initial estimates of ε and $\Delta\varepsilon$ selected to impose the waveguide condition and spectral bandwidth, respectively, an effective value for the dielectric function of the grating film is needed to proceed with calculation of waveguide eigenvalues and thin film reflectance properties. Since the grating supports only zero order, the period is smaller than the wavelength, and the first-order effective medium approximation can be used to estimate the effective dielectric constant of the grating film. The value depends on polarization. In the most common implementation, the grating is fabricated as a surface-relief profile in a thin film of thickness t_g and dielectric constant ε_g. If $s(x)$ represents the surface profile, with maximum value t_g and period Λ, that is, the interface between the film material and surrounding material of dielectric constants ε_g and $\varepsilon_{\text{surr}}$, respectively, the effective dielectric constant for the grating film is given by

$$\varepsilon_{\text{eff}} = [1/\Lambda t_g] \int_0^\Lambda [\varepsilon_g s(x) + \varepsilon_{\text{surr}}(t_g - s(x))]\, dx \qquad \text{(TE polarization)}, \qquad (1.10)$$

$$\varepsilon_{\text{eff}}^{-1} = [1/\Lambda t_g] \int_0^\Lambda [\varepsilon_g s(x) + \varepsilon_{\text{surr}}(t_g - s(x))]^{-1}\, dx \qquad \text{(TM polarization)}. \qquad (1.11)$$

This value of ε_{eff} is then used in ordinary thin film design expressions for the AR stack, as illustrated in Figure 1.15 and explained in detail in thin film design texts (cf., for example, [31]) and in the calculation of the waveguide eigenvalues. The calculation of the waveguide eigenvalues makes use of standard expressions for planar waveguides and can be found for a simple structure in Marcuse [32] and for a multilayer structure in Wang and Magnusson [25]. It should be noted that the eigenvalues are also polarization dependent. With the eigenvalues calculated for the effective structure, one can then estimate the required grating period using (1.9). The design is not complete at this point because the expressions (1.9)–(1.11) are only approximately true, being rigorous in the limit $\Delta\varepsilon \to 0$ [24, 25, 26]. To complete the design, one must calculate the performance of the structure using a rigorous Maxwell solver code to get accurate values for the resonant wavelength, spectral bandwidth, and spectral reflectance. These values are then compared with the desired values, and if they exceed specified tolerances, the entire procedure is iterated until a satisfactory structure is found. In practice, this procedure is found to converge quite quickly because, as pointed out by Wang and Magnusson [25], (1.9) is remarkably accurate even for relatively large values of dielectric modulation.

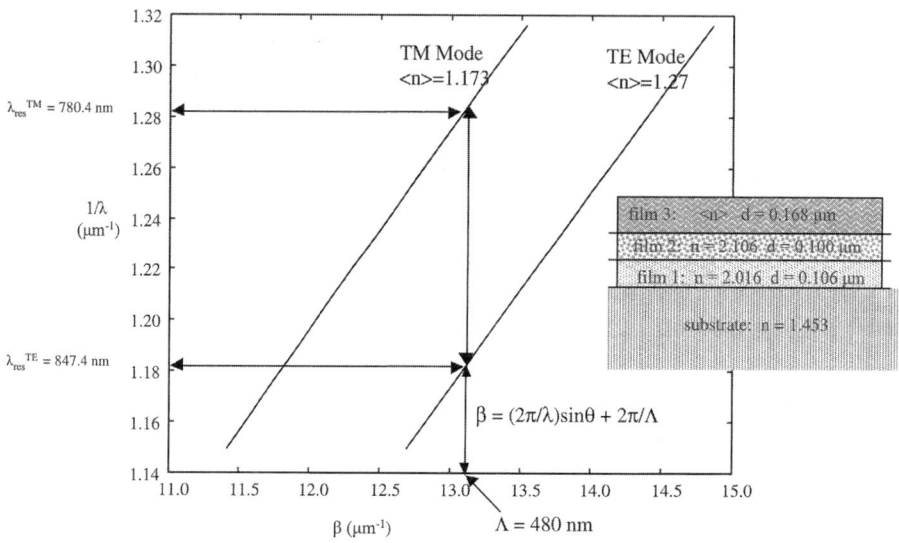

Figure 1.16: Parametric curves derived for effective waveguide structure used to initiate mode-matching condition.

Design Example

We illustrate the design methodolgy for a GMGRF intended for a mirror in an external cavity laser diode. The primary design drivers were a resonant wavelength at ∼850 nm under normal incidence and a broad bandwidth to permit easy measurement and characterization. A three-film multilayer waveguide was selected to facilitate imposing both the resonance condition and the out-of-band AR condition. Fused quartz was chosen for the substrate and was coated on the backside with a standard broadband AR coating to minimize the effects of Fresnel reflection from that surface. The top film contains the grating and is sputtered SiO_2 (refractive index = 1.485) for ease in etching the profile. The profile itself was etched as a simple binary structure with 50% fill factor in order to maximize the modulation and hence the spectral bandwidth. Films 2 and 3 are sputtered blended oxides with refractive indices 2.106 and 2.016, respectively. The thickness of the three films was a quarter wavelength at 850 nm.

With the assumptions cited above, the effective refractive index of the grating film was calculated using equations (1.10), (1.11) to be 1.27 for TE and 1.17 for TM polarization. Figure 1.16 shows the homogenized structure and a plot of reciprocal wavelength versus guided-mode eigenvalue in both polarizations for the multilayer waveguide structure. It can be seen that for a given period the resonant wavelengths in each polarization are well separated. From the parametric design standpoint, the graph in Figure 1.16 may enter from either axis. If the resonant wavelength is of primary interest, one enters from the vertical axis with the appropriate value of $1/\lambda$, finds the corresponding value of β, and converts it into a grating period Λ. For example, if we wish a resonant wavelength of 850 nm in TE polarization at

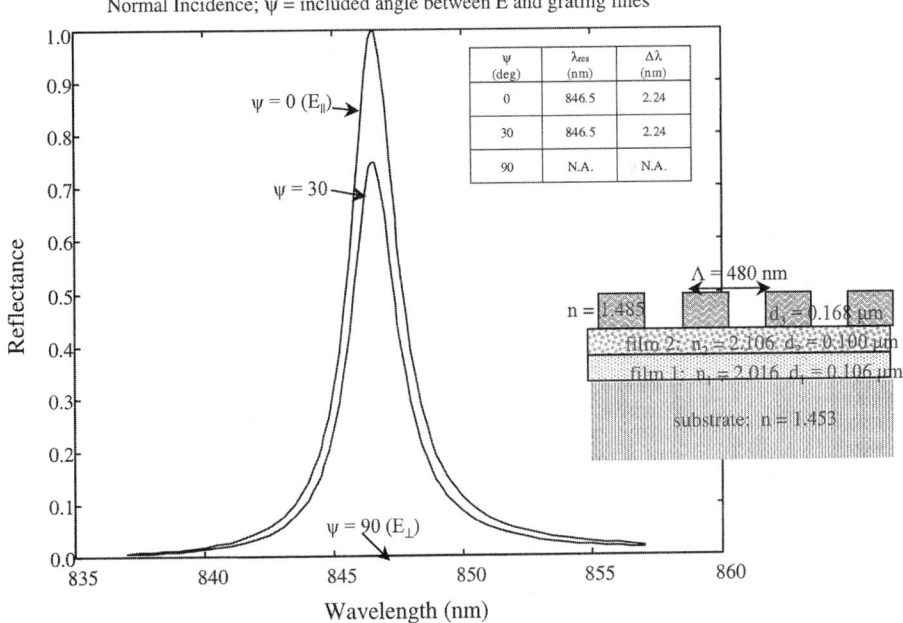

Figure 1.17: Predicted resonant filter reflectance for three polarizations.

normal incidence, we find from interpolation that the corresponding guided-mode eigenvalue is 13.0422 μm^{-1}, and the required grating period to couple to this mode is 481.8 μm. For this particular case, it was convenient to fix the period at 480 nm in order to take advantage of a "snap-to-grid" feature in our direct-write electron beam lithographic machine used to fabricate the grating. We thus entered the graph on the horizontal axis and predicted an approximate resonant wavelength of 847.4 nm.

The results of applying Dobson's rigorous Maxwell solver (MAXFELM) [19] to the exact structure is shown in Figure 1.17. Here it is seen that the "actual" resonant wavelength occurs at 846.5 nm (0.9 nm smaller than the approximate prediction) and the spectral bandwidth is 2.24 nm. Notice also that far from resonance the AR nature of the multilayer film does indeed provide very low reflectance. The polarization sensitivity of the design is explicitly illustrated for two cases with the incident polarization vector oriented 30 degrees and 90 degrees to the grating lines.

The grating structure illustrated in the inset in Figure 1.17 was fabricated patterning with direct-write electron beam lithography and machining the surface profile with reactive ion etching. A high magnification (25,000X) image of the grating lines was formed by a scanning electron microscope and is shown in Figure 1.18. The period of the grating is uniform and the fill factor is close to the design value of 50%. However, there is clearly significant roughness in the grating lines, leading to random nonuniformity in the grating profile. The measured reflectance of this grating is shown in Figure 1.19. The general shape and location of the resonant

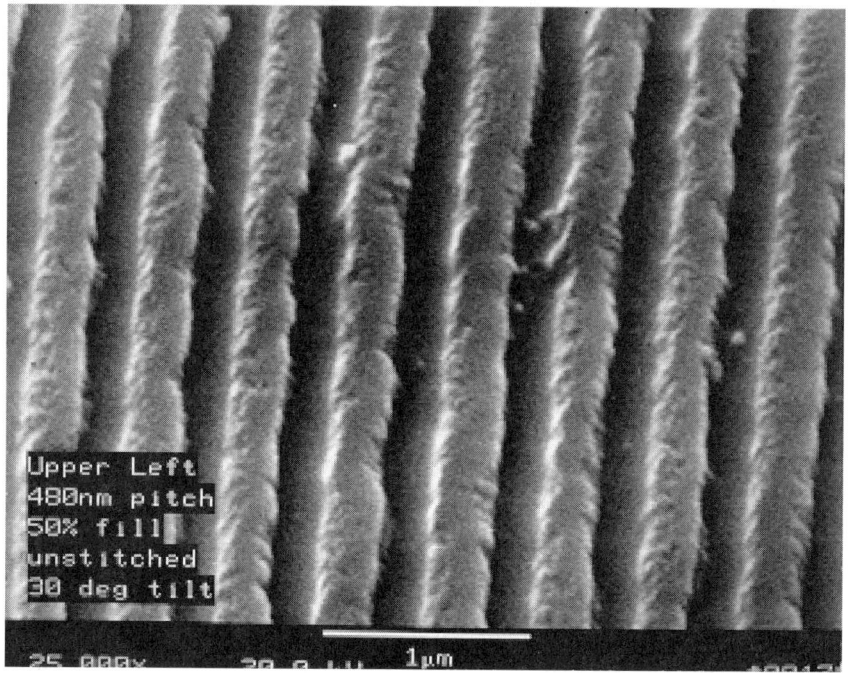

Figure 1.18: SEM photo of prototype GMGRF designed for 850 nm.

wavelength are close to the prediction in Figure 1.17, but the peak reflectivity is reduced by ~10% and the bandwidth is approximately half the predicted value. Both results can be attributed to loss caused by the roughness of the grating lines, a topic to be examined more closely in the next section. The measured resonant wavelength occurs at ~847.1 nm, or 0.5 nm greater than predicted. This can be attributed to two factors: (1) the incident beam had to be tilted slightly (0.8°) in a plane containing the grating lines in order to bring the reflected beam onto a detector, which causes a slight positive shift in the resonant wavelength, and (2) as can be seen from Figure 1.18, the fill factor of the grating profile is not exactly 50%, thus causing the effective refractive index of the grating film to be slightly different than assumed and hence leading to a slightly different guided-mode eigenvalue.

Peng and Morris [27] have reported ~90% peak reflectance at 633 nm in a resonant filter having a grating fabricated holographically in photoresist. Magnusson's group has recently achieved 98% peak reflectance in a GMGRF using the same holographic process for the grating [29]. In both cases the maximum reflectance was limited by scatter attributed to roughness in the grating.

Application to VCSEL Mirrors

In section 1.3.1 we briefly described several of the beneficial features of VCSELs and explained their relevance to a number of existing and emerging applications.

Chapter 1. Overview and Applications of Diffractive Optics Technology

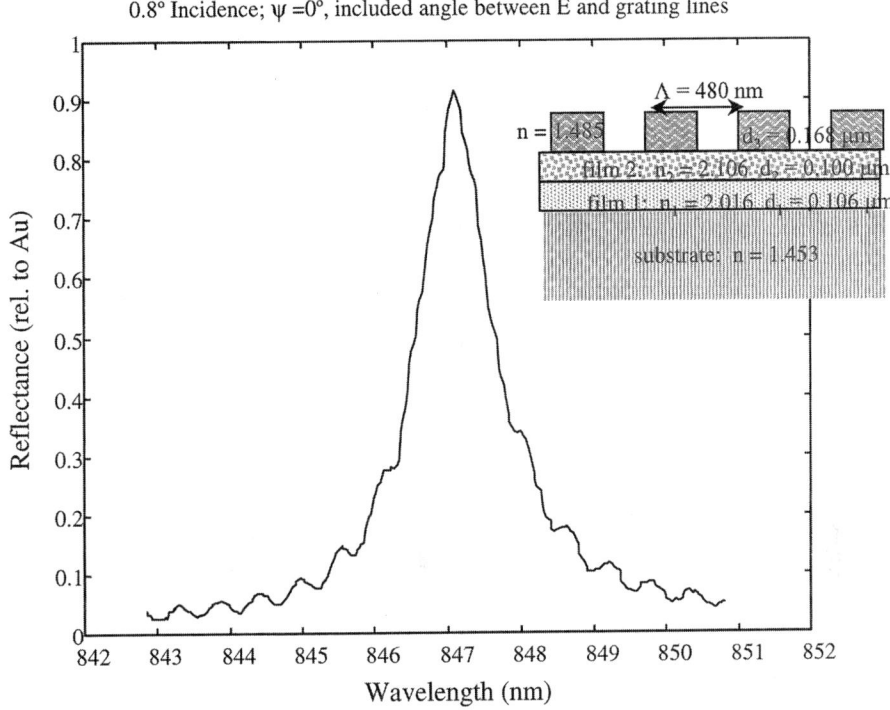

Figure 1.19: Measured GMGRF reflectance.

Although we did not discuss the actual structure of the device itself, in fact one of the most critical components of a VCSEL are the cavity mirrors formed from multiple thin film layers of GaAlAs semiconductor material [33, 34]. The most successful approach for present high-performance VCSELs uses a distributed Bragg reflector (DBR) to form a zero-order Fabry–Perot cavity. Typically, high reflectivity (>99.5%), low sidebands (−30dB), and narrow bandwidth require 30–40 periods extending 3–4 μm in thickness. The gain region in a VCSEL is comprised of a few quantum wells, each typically only ∼100 Å thick, and high-reflectance mirrors are required to balance the small gain-distance product in the active region. (One real advantage of the very thin active region is much lower ([∼10×] threshold current compared to edge-emitting lasers.) Furthermore, the active gain region is pumped by injecting current directly through the mirrors. Thus, the DBR mirror structure must also be electrically conductive. Apparently, the GMGRF should be well suited to satisfying the optical requirements for VCSEL mirrors, and in this regard GMGRFs have attracted attention for their potential to do modal engineering of VCSELs. However, there are a number of issues that must be addressed before the GMGRF will be suitable for a VCSEL, not the least of which is sensitivity to loss and the need for compatibility with current injection.

In addition to sensitivity to loss, there are several other issues that are expected to challenge the application of GMGRFs to VCSELs. The high refractive index of III–V materials leads to small grating periods; e.g., for a wavelength of 850 nm a period of 480 nm is required for fused quartz, while a period of 250 nm is needed for GaAs. This further compounds the fabrication difficulty. Direct-write, electron-beam lithography does not scale up easily to large volume manufacturing. Phase mask lithography and holographic lithography represent viable alternatives with better potential for large-scale production. Potentially more serious are temperature-induced changes in the resonance and stress mismatch in the dielectric and III–V materials. Further, it should be remembered that the entire VCSEL structure is a relatively thin resonant cavity, and in the end a design of any hybrid structure incorporating a GMGRF will likely have to treat the entire structure as one integral device. This in turn will require more sophisticated and capable models.

1.4 Summary

We have presented a fairly lengthy survey of diffractive optics technology as it is currently practiced, and we have described three applications considered to be at the leading edge of the technology. The variety of enabling roles for diffractive optics becomes evident when one considers current developments in optoelectronics to integrate VCSELs, optics, and receivers in miniaturized packages for optical interconnects, switches, and sensors. Regarding the status of mathematical models, it should be clear from the examples that current models of gratings agree quite well with measurements in those cases where the finite boundary of the grating can be neglected. This was illustrated for scalar theory in our first example of the VCSEL power monitor and focusing elements and for the full vector theory in the last two examples. However, the success of modeling exhibited in these examples should not be taken to imply that the field is closed, with few challenges remaining to applied mathematicians. In fact, the field continues to be an active and relevant area of research, and we conclude this chapter with a brief survey of current topics and comments on possible future directions for applied mathematics in diffractive optics.

Modeling, in the form of both the direct problem and the inverse problem of optimal design, continues to be important as the practical significance of new devices, especially low-order gratings and gratings with high aspect ratio as illustrated in the last two examples above, are developed and integrated into optical systems. Both classes of the problem are attracting wide attention and are generating ever more efficient numerical codes. Earlier activity in the direct problem dealt with formulating approaches to solve Maxwell's equations describing the full vector field for periodic diffraction grating. There is a very long list of references in this area, with many contributions from both optical scientists and applied mathematicians, and many of the original authors and references appear in the following chapters. More recently, activity in this area has been directed at either diffractive or refractive micro-optic elements of finite extent. Approaches taken for this have included the finite-difference time-domain (FDTD) method [35], the finite element method [36], and the boundary element method [37]. These approaches are by no means fully satisfactory for all applications. For example, as noted by Prather, Mirotznik, and Maik [37], the boundary element method can bring a significant computational burden. Currently, the FDTD method seems to be popular for a wide variety of both diffractive and micro-optical problems having finite aperture, but it seems that other approaches are worth examining if only to give approximate, analytical results. One straightforward approach could consist of modeling a finite-aperture grating in the surface of a substrate by matching exact solutions for an infinite grating and for a smooth interface separating two half-spaces across the aperture boundary.

In addition, new approximation methods are valuable not only to get quick estimates of performance but also to strengthen understanding and intuition. For the optimal design problem, much current activity is directed at subwavelength structures and diffractive elements implemented with artificial dielectrics synthesized from such structures. Effective-medium theory is applied most commonly in this context, with approaches for improved accuracy taking the form of either higher order terms [38] or zero-order theory with precompensation to the phase function [39]. Generally, the justification for these approaches relies mostly on experimental verification and physical intuition. LaLanne et al. [38], however, do point to other references for mathematical justification, and while there is no special reason to doubt the mathematical legitimacy of these methods, there is room for work to place them on a mathematically rigorous foundation.

Especially challenging problems in modeling lie in the application of periodic structures to nonlinear media and to active media with gain (lasers). In the case of nonlinear media, Reinisch, Nevière, and Popov [40, 41] have explored the application of guided-mode grating structures to significantly enhance coupling to nonlinear media and shown that second harmonic generation can be enhanced by two, and perhaps three, orders of magnitude. Bao and Dobson [42] have extended their finite element model to this effect and obtained some preliminary results. Intuitively, the physical principle of this phenomenon is similar to the guided-mode grating resonance discussed in section 1.3.3, but the coupling of the evanescent first-order to the second-order susceptibility of the waveguide core makes the mathematical model significantly more complicated. The practical implications of enhanced second harmonic generation to sensing warrant more detailed investigation of this problem.

With regard to the problem of periodic structures integrated active media with optical gain (e.g., lasers), we have already suggested in the previous section some areas for study in modeling guided-mode resonant filters applied to VCSELs. The application of the GMGRF to VCSELs is particularly appealing as a means to enable modal engineering for control of polarization, wavelength, and wavefront. However, the overall scope of the modeling problem for VCSELs is extremely complex, involving the intersection of Maxwell's equations (optical resonator defined by multilayer thin film mirrors) with quantum mechanics (electrical transport in semiconductor PN junctions with optical gain provided by epitaxially grown quantum wells) with the diffusion equation (thermal effects due to localized heating).

Moloney's group at the University of Arizona has been active in developing various approaches to modeling the entire problem [43].

One of the most active new areas in "diffractive" optics deals with the behavior of electromagnetic waves in three-dimensionally periodic dielectric structures. Such structures are conceived as artificial dielectrics comprised of zero-order three-dimensional gratings. The formal development of wave propagation in these structures has shown properties with very close analogy to propagation of electrons in real crystals, and thus the term "photonic crystals" has been coined to describe such periodic dielectric structures. Reference [44] contains a number of papers that review various properties, experimental results, and applications of photonics crystals. The existence of a band gap—a region of frequencies for which wave propagation is forbidden—is one particular property of photonic crystals with especially important practical applications. In particular, just as with semiconductors, it is possible to introduce defects, or impurities, in the periodic lattice and thereby create isolated states in the band gap which support metastable states (waves). An example of a one-dimensional photonic band gap crystal [45] was fabricated and demonstrated to possess a strongly resonant microcavity with very high Q and can serve as a very efficient waveguide coupler. Very recently, and of great significance, Scherer's group has demonstrated the first laser in a two-dimensional photonic band gap structure [46].

Dobson [47] has extended his finite element model of diffractive elements to two-dimensional photonic crystals and shown this approach to be quite efficient in predicting the properties of an infinite crystal. For the design of the actual lasers demonstrated by Scherer's group, the FDTD approach is needed to account for the finite extent of the device. And thus we come back to the point emphasized at the beginning of this section: improved modeling tools for diffractive and periodic structures continue to be needed and will be motivated by continued technology developments such as photonic crystals.

References

[1] D. W. Swift, *The application of HOEs to head-up displays*, in Proc. IERE Holographic Systems, Components, and Applications Conference, Cambridge, UK, 1987, pp. 93–97.

[2] K. Miyamoto, *The phase Fresnel lens*, J. Opt. Soc. Amer., 51 (1961), pp. 17–20.

[3] L. B. Lesem, P. M. Hirsch, and J. A. Jordan, Jr., *The kinoform: A new wavefront reconstruction device*, IBM J. Res. Develop., 13 (1969), pp. 150–155.

[4] L. D'Auria, J. P. Huignard, A. M. Roy, and E. Spitz, *Photolithographic fabrication of thin film lenses*, Optics Communications, 5 (1972), pp. 232–235.

[5] G. J. Swanson, *Binary Optics Technology: The Theory and Design of Multi-Level Diffractive Optical Elements*, Technical Report 854, MIT Lincoln Laboratory, August 1989 (available from National Technical Information Service).

[6] S. H. Lee, ed., *Diffractive and Miniaturized Optics*, SPIE Critical Reviews of Optical Science and Technology CR49, SPIE, Bellingham, WA, 1993.

[7] M. Born and E. Wolf, *Principles of Optics*, 6th ed., Cambridge University Press, New York, 1980.

[8] J. W. Goodman, *Introduction to Fourier Optics*, McGraw-Hill, San Francisco, 1968.

[9] D. A. Buralli, G. M. Morris, and J. R. Rogers, *Optical performance of holographic kinoforms*, Appl. Optim., 28 (1989), pp. 976–982.

[10] J. A. Cox, T. A. Fritz, and T. Werner, *Application and demonstration of diffractive optics for head-mounted displays*, Proc. SPIE, 2218 (1994), pp. 32–40.

[11] J. D. Gaskill, *Linear Systems, Fourier Transforms, and Optics*, Wiley, New York, 1978.

[12] D. A. Buralli and G. M. Morris, *Effects of diffraction efficiency on the modulation transfer function of diffractive lenses*, Appl. Optim., 31 (1992), pp. 4389–4396.

[13] B. Fritz, J. A. Cox, T. Werner, and J. Gieske, *Diffractive optic power monitor for use with a VCSEL*, in Diffractive Optics and Micro-Optics 10, OSA Technical Digest Series, Optical Society of America, Washington DC, 1998, pp. 206–208.

[14] R. A. Morgan, J. A. Lehman, and M. K. Hibbs-Brenner, *Vertical Cavity Surface Emitting Laser Arrays: Come of Age*, Invited paper, published in SPIE Vol. 2683-04, OE LASE 96; Photonics West: Fabrication, Testing, and Reliability of Semiconductor Lasers (SPIE, Bellingham, WA, 1996).

[15] J. A. Lehman, R. A. Morgan, M. K. Hibbs-Brenner, and D. Carlson, *High-frequency modulation characteristics of hybrid dielectric/AlGaAs single-mode VCSELs*, Electron. Lett., 31 (1995), pp. 1251–1252.

[16] J. A. Cox, J. D. Zook, T. D. Ohnstein, and D. C. Dobson, *Optical performance of high-aspect LIGA gratings*, Opt. Eng., 36 (1997), pp. 1367–1373.

[17] J. A. Cox, D. C. Dobson, T. D. Ohnstein, and J. D. Zook, *Optical performance of high-aspect LIGA gratings II*, Opt. Eng., 37 (1998), pp. 2878–2884.

[18] D. C. Dobson, *A variational method for electromagnetic diffraction in biperiodic structures*, RAIRO Modél. Math. Anal. Numér., 28 (1994), pp. 419–439.

[19] D. Dobson and J. A. Cox, *Mathematical modeling for diffractive optics*, in Diffractive and Miniaturized Optics, Critical Reviews, SPIE CR49, Sing Lee, ed., 1994, pp. 32–53.

[20] G. Bao, D. Dobson, and J. A. Cox, *Mathematical issues in rigorous electromagnetic grating theory*, J. Opt. Soc. Amer. A, 12 (1995), pp. 1029–1042.

[21] E. D. Palik, ed., *Handbook of Optical Constants of Solids I*, Academic Press, Orlando, FL, 1985, pp. 313–323.

[22] P. M. Morse, *Vibration and Sound*, 2nd ed., McGraw-Hill, New York, 1948, pp. 255–258.

[23] A. Hessel and A. A. Oliner, *A new theory of Wood's anomalies on optical gratings*, Appl. Optim., 4 (1965), pp. 1275–1297.

[24] R. Magnusson and S. Wang, *New principle for optical filters*, Appl. Phys. Lett., 61 (1992), pp. 1022–1024.

[25] S. Wang and R. Magnusson, *Theory and applications of guided-mode resonance filters*, Appl. Optim., 32 (1993), pp. 2606–2613.

[26] S. Wang and R. Magnusson, *Multilayer waveguide-grating filters*, Appl. Optim., 34 (1995), pp. 2414–2420.

[27] S. Peng and G. M. Morris, *Sub-nanometer linewidth resonant grating filters*, in Diffractive Optics and Micro-Optics, OSA Technical Digest Series 5, Optical Society of America, Washington DC, 1996, pp. 257–260.

[28] S. Peng and G. M. Morris, *Experimental demonstration of resonant anomalies in diffraction from two-dimensional gratings*, Optics Letters, 21 (1996), pp. 549–551.

[29] Z. S. Liu, S. Tibuleac, D. Shin, P. P. Young, and R. Magnusson, *High-efficiency guided-mode resonance laser mirror*, postdeadline paper in Diffractive Optics and Micro-Optics, OSA Technical Digest Series 5, Optical Society of America, Washington DC, 1996.

[30] M. Nevière, *The homogeneous problem*, in Electromagnetic Theory of Gratings, Chapter 5, Springer-Verlag, New York, 1980.

[31] A. Musset and A. Thelen, *Multilayer antireflection coatings*, in Progress in Optics, Vol. VIII, E. Wolf, ed., American Elsevier, New York, 1970, Chapter IV.

[32] D. Marcuse, *Theory of Dielectric Optical Waveguides*, 2nd ed., Academic Press, New York, 1991.

[33] J. A. Cox, R. A. Morgan, R. Wilke, and C. Ford, *Guided-mode grating resonant filters for VCSEL applications*, in Diffractive and Holographic Device Technologies and Applications V, Ivan Cindrich and Sing H. Lee, eds., Proceedings of SPIE 3291 (1998), pp. 70–76.

[34] R. A. Morgan, J. A. Cox, R. Wilke, and C. Ford, *Application of guided-mode grating resonant filters to VCSELs*, in Diffractive Optics and Micro-Optics, OSA Technical Digest Series 10, Optical Society of America, Washington DC, 1998, pp. 18–20.

[35] A. Taflove, *Computational Electrodynamics: The Finite-Difference Time-Domain Method*, Artech-House, Boston, MA, 1995.

[36] B. N. Lichtenberg and N. Gallagher, *Numerical modeling of diffractive devices using the finite element method*, Opt. Eng., 33 (1994), pp. 3518–3526.

[37] D. W. Prather, M. S. Mirotznik, and J. N. Mait, *Boundary integral methods applied to the analysis of diffractive optical elements*, J. Opt. Soc. Amer. A, 14 (1999), pp. 34–43.

[38] Ph. LaLanne, S. Astilean, P. Chavel, E. Cambril, and H. Launois, *Design and fabrication of blazed binary diffractive elements with sampling periods smaller than the structural cutoff*, J. Opt. Soc. Amer. A, 16 (1999), pp. 1143–1156.

[39] J. N. Mait, D. W. Prather, and M. S. Mirotznik, *Design binary subwavelength diffractive lenses by use of zeroth-order effective-medium theory*, J. Opt. Soc. Amer. A, 16 (1999), pp. 1157–1167.

[40] R. Reinisch and M. Nevière, *Gratings as electromagnetic field amplifiers for second-harmonic generation*, in Scattering in Volumes and Surfaces, M. Nieto-Verperinas and J. C. Dainty, eds., North-Holland, Amsterdam, 1990, pp. 269–285.

[41] E. Popov and M. Nevière, *Surface-enhanced second-harmonic generation in nonlinear corrugated dielectrics: New theoretical approaches*, J. Opt. Soc. Amer. B, 11 (1994), pp. 1555–1564.

[42] G. Bao and D. C. Dobson, *Second harmonic generation in nonlinear optical films*, J. Math. Phys., 35 (1994), pp. 1622–1633.

[43] Z. Yang, R. A. Indik, and J. V. Moloney, *Self-consistent approach to thermal effects in vertical-cavity surface-emitting lasers*, J. Opt. Soc. Amer. A, 12 (1995), pp. 1993–2004.

[44] J. Opt. Soc. Amer. B, *Special Feature Issue on Photonic Bandgap Structures*, 10 (1993), p. 283.

[45] P. R. Villeneuve, S. Fan, J. D. Joannopoulos, K.-Y. Lim, G. S. Petrich, L. A. Kolodziejski, and R. Reif, *Air-bridge microcavities*, Appl. Phys. Lett., 67 (1995), pp. 167–169.

[46] O. Painter, R. K. Lee, A. Sherer, A. Yariv, J. P. O'Brien, P. D. Dapkus, and I. Kim, *Two-dimensional photonic band-gap defect mode laser*, Science, 284 (1999), pp. 1819–1821.

[47] D. C. Dobson, *An efficient method for band structure calculations in 2D photonic crystals*, J. Comp. Phys., 149 (1999), pp. 363–376.

Chapter 2

Variational Methods for Diffractive Optics Modeling

Gang Bao and David C. Dobson

2.1 Introduction

The field of diffractive optics has developed over the past few decades in concurrence with the semiconductor industry. Many of the same micromachining techniques developed for producing semiconductor devices can also be applied to create intricate structural features on optical substrates. The propagation of light waves in devices with such small features is typically dominated by diffraction, rather than by geometrical optics as in classical "smooth" optical elements. This creates a challenge for optical engineers, since traditional raytracing methods for analyzing optical systems can no longer be accurately applied when diffractive optical elements are present. Increasingly, sophisticated computational methods are being applied to simulate the behavior of these devices and to aid in their design.

There are, of course, many important approaches to solving and analyzing diffraction problems [32, 33, 50, 63]. The reader is referred to Chapters 1, 3, 4, and 5 for a sample. This paper focuses on variational methods in the particular case of spatially periodic diffractive structures, also known as *diffraction gratings*. We discuss weak formulations and finite element approximations of the time-harmonic Maxwell's equations in such structures. We also discuss the inverse problem of determining shapes from scattered fields and describe some recent work aimed at finding optimal shape designs for surface-relief gratings. The underlying mathematical models covered here are, of course, not restricted to optics; by changing the length scale one can study, for example, classical electromagnetic fields in microwave structures, antenna arrays, and crystalline materials, as well as analogous structures in acoustics.

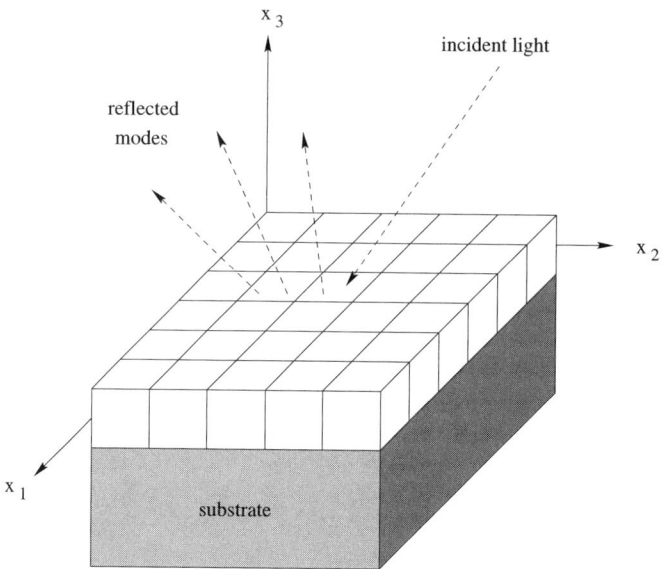

Figure 2.1: Basic geometrical configuration. Each periodic cell contains a copy of a given structure with arbitrary, spatially varying refractive index. The periodic pattern extends infinitely in the (x_1, x_2) plane, and the substrate extends infinitely in the $-x_3$ direction.

The classic book by Petit [66] provides a good introduction to diffraction grating problems and a description of various numerical approaches. Earlier results and references on the mathematical analysis of diffraction gratings may be found in [71]. The more recent books by Friedman [47, 48, 49] contain descriptions of several mathematical problems that arise in diffractive optics modeling in industry.

This chapter is structured as follows. In the next section, the model diffraction problem is introduced. A variational formulation is then presented which is fundamental to the rest of the paper. Results on the well-posedness of the model problem are also discussed. Section 2.3 is devoted to computation and numerical analysis of finite element methods for solving the diffraction problem. Inverse and optimal design problems are examined in section 2.4. Finally, we conclude the paper with a sample of various related research problems.

2.2 The General Diffraction Problem

The basic problem we consider is that of predicting the scattered modes that arise when an electromagnetic wave is incident on some periodic structure. Roughly speaking, any incident wave of interest can be decomposed into a sum of plane waves, and because the problem is linear, it suffices to consider only one incident plane wave at a time. A schematic picture of the basic problem is shown in Figure 2.1.

Chapter 2. Variational Methods for Diffractive Optics Modeling

The most useful and interesting feature of scattering by infinite periodic structures is that at large distances from the scatterer, only a *finite* number of modes propagate. These modes appear as plane waves propagating in directions that can be easily calculated, given the period of the structure and the frequency of the incident wave. Physically, one can think of this as being caused by interference between waves scattered from neighboring cells. Mathematically, it is even easier to see: by separation of variables one finds that the field can be expanded as an infinite sum of plane waves. For example, for a reflected field u of two variables from a 2π-periodic structure, this expansion is

$$u(x_1, x_3) = \sum_{n=-\infty}^{\infty} a_n e^{i(nx_1 + \beta_n x_3)}, \qquad (2.1)$$

where the a_n are unknown coefficients. This is sometimes called the *Rayleigh expansion*. In the case of a medium with real refractive index $k > 0$, the coefficients β_n in the sum (2.1) are defined by

$$\beta_n = \begin{cases} \sqrt{k^2 - n^2} & \text{if } k \geq |n|, \\ i\sqrt{n^2 - k^2} & \text{if } k < |n|. \end{cases}$$

Since β_n is real for only a finite number of indices n, we see from (2.1) that only a finite number of plane waves in the sum propagate, with the remaining *evanescent* modes decaying exponentially as $x_3 \to \infty$.

The quantities of most interest to optical engineers are the energies in each propagating mode as a proportion of the total incident energy. The ratio of the energy in a given mode to the incident energy is called the *efficiency* of that mode. One typically wishes to calculate the efficiency in each diffracted mode for a given structure, for a given incident beam or range of incident beams.

2.2.1 Maxwell's Equations in a Periodic Medium

The standard model for classical electromagnetic wave propagation is provided by Maxwell's equations. Most applications in optics deal with relatively narrow bands of radiation, so that studying fixed-frequency problems is useful. In this context, assuming that all fields have time dependence $e^{-i\omega t}$, the time-harmonic Maxwell equations take the following form:

$$\nabla \times E - i\omega\mu H = 0, \qquad (2.2)$$
$$\nabla \times H + i\omega\epsilon E = 0. \qquad (2.3)$$

The equations hold over all of $\mathbb{R}^3$. Here E and H denote the complex-valued electric and magnetic field vectors, respectively. The real constant μ is the magnetic permeability, assumed to be constant since most optical devices are made from nonmagnetic materials. The physical structure is then described by the dielectric coefficient $\epsilon(x)$, $x = (x_1, x_2, x_3)$. Since the structure is periodic in the x_1 and x_2

directions, there must be two constants L_1 and L_2, the axis lengths of each cell, such that

$$\epsilon(x_1 + n_1 L_1, x_2 + n_2 L_2, x_3) = \epsilon(x_1, x_2, x_3)$$

for all $x \in \mathbb{R}^3$ and all integers n_1, n_2. The structure is assumed to have finite extent in the x_3 direction; that is, there exists a number $b > 0$ such that

$$\epsilon(x) = \epsilon_1 \quad \text{for } x_3 \geq b,$$
$$\epsilon(x) = \epsilon_2 \quad \text{for } x_3 \leq -b.$$

Here ϵ_1 and ϵ_2 are the constant dielectric coefficients of the transmission medium and the substrate, respectively. The transmission medium is assumed to be non-conductive, so ϵ_1 is real. The case Im $\epsilon_2 > 0$ accounts for a substrate that absorbs energy [27].

Assume that a plane wave $(E_I, H_I) = (s, p)e^{iq \cdot x}$ is incident on the structure from above. Here $q = (\alpha_1, \alpha_2, -\beta)$ is the incidence vector. To ensure that the incident field satisfies Maxwell's equations, we take s, p to be constant vectors that satisfy

$$s = \frac{1}{\omega \epsilon_1}(p \times q), \quad q \cdot q = \omega^2 \epsilon_1 \mu, \quad p \cdot q = 0.$$

Because of the well-known Floquet–Bloch theorem, it is natural to consider so-called *quasi-periodic* solutions E, H. Define $\alpha = (\alpha_1, \alpha_2, 0)$ and let

$$E_\alpha(x) = e^{-i\alpha \cdot x} E(x),$$
$$H_\alpha(x) = e^{-i\alpha \cdot x} H(x).$$

Assuming that solutions to (2.2)–(2.3) are unique, the fields E_α, H_α must be spatially periodic in (x_1, x_2), with the same periodicity as $\epsilon(x)$.

Substituting $E = e^{i\alpha \cdot x} E_\alpha$ and $H = e^{i\alpha \cdot x} H_\alpha$ into the Maxwell equations, one obtains a new set of equations for E_α, H_α:

$$\nabla_\alpha \times E_\alpha - i\omega \mu H_\alpha = 0, \tag{2.4}$$
$$\nabla_\alpha \times H_\alpha + i\omega \epsilon E_\alpha = 0, \tag{2.5}$$

where $\nabla_\alpha = \nabla + i\alpha$. Since all quantities in (2.4)–(2.5) are periodic in (x_1, x_2), the equations can be taken to hold on *one* fundamental cell, say $(0, L_1) \times (0, L_2) \times (-\infty, \infty)$, with periodic boundary conditions in the x_1 and x_2 directions. To complete the statement of the problem, a radiation condition is enforced, which states that all reflected and transmitted waves are outgoing and remain bounded as $|x_3| \to \infty$. In a weak sense, (2.4)–(2.5) are equivalent to the decoupled system

$$\nabla_\alpha \times \left(\frac{1}{\epsilon \mu} \nabla_\alpha \times H_\alpha\right) - \omega^2 H_\alpha = 0, \tag{2.6}$$
$$\nabla_\alpha \times H_\alpha + i\omega \epsilon E_\alpha = 0. \tag{2.7}$$

Chapter 2. Variational Methods for Diffractive Optics Modeling

In the next section we will consider the variational form of this problem over a simple Sobolev space. In this context, a coercivity consideration leads naturally to

$$\nabla_\alpha \times \left(\frac{1}{\epsilon\mu}\nabla_\alpha \times H_\alpha\right) - \nabla_\alpha\left(\frac{1}{\epsilon_c\mu}\nabla_\alpha \cdot H_\alpha\right) - \omega^2 H_\alpha = 0, \tag{2.8}$$

$$\nabla_\alpha \times H_\alpha + i\omega\epsilon E_\alpha = 0, \tag{2.9}$$

where ϵ_c is a constant chosen to satisfy $\inf \operatorname{Re}\frac{1}{\epsilon(x)} \geq \frac{3}{4\epsilon_c}$. It can be shown [19] that the system (2.8)–(2.9) admits a unique weak solution and that the scattering problems (2.6)–(2.7) and (2.8)–(2.9) are actually equivalent, except possibly at a discrete set of parameter values. The formulation (2.8)–(2.9) may be useful because it allows well-posed approximations in nodal finite element spaces.

2.2.2 Boundary Conditions and the Variational Formulation

In this section we derive a variational formulation of problem (2.8)–(2.9). The presentation follows that of [19]. The fundamental step of the variational formulation is to reduce the domain of (2.4)–(2.5) from the periodic fundamental cell with infinite extent in the x_3-direction to a bounded periodic "box" with finite extent in the x_3-direction. To do so requires the derivation of appropriate boundary conditions; this is done by establishing artificial plane boundaries $\{x_3 = b\}$ and $\{x_3 = -b\}$ and matching the fundamental solution and the Fourier expansion of the field on these boundaries. One then obtains nonlocal operators, which map the traces of the field components on the artificial boundaries to their derivatives. This gives rise to exact "transparent" boundary conditions. Artificial boundary conditions of other types, for example, absorbing boundary conditions, have been developed extensively. Further references may be found in [44, 45, 52].

Since H_α is periodic in x_1 and x_2, we have the Fourier series expansion

$$H_\alpha(x) = \sum_{n \in Z^2} H_\alpha^{(n)}(x_3) e^{-i\alpha_n \cdot x}, \tag{2.10}$$

where $n = (n_1, n_2)$, $Z = \{0, \pm 1, \pm 2, \ldots\}$,

$$H_\alpha^{(n)}(x_3) = \frac{1}{L_1 L_2} \int_0^{L_1} \int_0^{L_2} H_\alpha(x) e^{i\alpha_n \cdot x} dx_1 dx_2,$$

and

$$\alpha_n = (2\pi n_1/L_1, 2\pi n_2/L_2, 0).$$

Let $\Omega_0 = \{x \in \mathbf{R}^3 : -b < x_3 < b\}$, $\Omega_1 = \{x \in \mathbf{R}^3 : x_3 > b\}$, $\Omega_2 = \{x \in \mathbf{R}^3 : x_3 < -b\}$. Denote

$$\Gamma_1 = \{x \in \mathbf{R}^3 : x_3 = b\}, \quad \Gamma_2 = \{x \in \mathbf{R}^3 : x_3 = -b\}.$$

Define for $j = 1, 2$ the coefficients

$$\beta_j^n(\alpha) = e^{i\gamma_j^n/2} \left|\omega^2 \epsilon_j \mu - |\alpha_n - \alpha|^2\right|^{1/2}, \quad n \in Z^2, \tag{2.11}$$

where
$$\gamma_j^n = arg(\omega^2 \epsilon_j \mu - |\alpha_n - \alpha|^2), \quad 0 \leq \gamma_j^n < 2\pi.$$

We assume that $\omega^2 \epsilon_j \mu \neq |\alpha_n - \alpha|^2$ for all $n \in Z^2$, $j = 1, 2$. This condition excludes "resonance." Note that β_j^n is real for at most finitely many n.

In particular, for real ϵ_2, we have the following equivalent form of (2.11):

$$\beta_j^n(\alpha) = \begin{cases} \sqrt{\omega^2 \epsilon_j \mu - |\alpha_n - \alpha|^2}, & \omega^2 \epsilon_j \mu > |\alpha_n - \alpha|^2, \\ i\sqrt{|\alpha_n - \alpha|^2 - \omega^2 \epsilon_j \mu}, & \omega^2 \epsilon_j \mu < |\alpha_n - \alpha|^2. \end{cases} \quad (2.12)$$

Observe that inside Ω_1 and Ω_2 the dielectric coefficients ϵ are constants, i.e, $\epsilon = \epsilon_j$ in Ω_j, $j = 1, 2$. Maxwell's equations then imply that

$$(\Delta_\alpha + \omega^2 \epsilon_j \mu) H_\alpha = 0, \quad (2.13)$$

where $\Delta_\alpha = \Delta + 2i\alpha \cdot \nabla - |\alpha|^2$. The method of separation of variables implies that H_α can be expressed as a sum of plane waves in Ω_j:

$$H_\alpha|_{\Omega_j} = \sum_{n \in Z^2} a_j^n e^{\pm i \beta_j^n(\alpha) x_3 - i\alpha_n \cdot x}, \quad j = 1, 2, \quad (2.14)$$

where the a_j^n are constant complex-valued vectors.

Since β_j^n is real for at most finitely many n, there are only a finite number of propagating plane waves in the sum (2.14). The remaining waves must either decay exponentially or become unbounded as $|x_3| \to \infty$. We impose the radiation condition that H_α is composed of bounded outgoing plane waves in Ω_1 and Ω_2, plus the incident incoming wave in Ω_1.

From (2.10) and (2.14) we deduce

$$H_\alpha^{(n)}(x_3) = \begin{cases} H_\alpha^{(n)}(b) e^{i\beta_1^n(\alpha)(x_3 - b)}, & n \neq 0, \text{ in } \Omega_1, \\ H_\alpha^{(0)}(b) e^{i\beta_1(x_3 - b)} + p e^{-i\beta_1 x_3} - p e^{i\beta_1(x_3 - 2b)}, & n = 0, \text{ in } \Omega_1, \\ H_\alpha^{(n)}(-b) e^{-i\beta_2^n(\alpha)(x_3 + b)} & \text{ in } \Omega_2, \end{cases} \quad (2.15)$$

where p is the polarization vector associated with the incident wave H_0. From (2.15) we can then calculate the derivative of $H_\alpha^{(n)}(x_3)$ with respect to ν, the unit normal, on Ω_0:

$$\left. \frac{\partial H_\alpha^{(n)}}{\partial \nu} \right|_{\Gamma_j} = \begin{cases} i\beta_1^n(\alpha) H_\alpha^{(n)}(b), & n \neq 0, \text{ on } \Gamma_1, \\ i\beta_1 H_\alpha^{(0)}(b) - 2i\beta_1 p e^{-i\beta_1 b}, & n = 0, \text{ on } \Gamma_1, \\ i\beta_2^n(\alpha) H_\alpha^{(n)}(-b) & \text{ on } \Gamma_2. \end{cases} \quad (2.16)$$

Thus from (2.14) and (2.16),

$$\left. \frac{\partial H_\alpha}{\partial \nu} \right|_{\Gamma_1} = \sum_{n \in Z} i\beta_1^n(\alpha) H_\alpha^{(n)}(b) e^{-i\alpha_n \cdot x} - 2i\beta_1 p e^{-i\beta_1 b}, \quad (2.17)$$

$$\left. \frac{\partial H_\alpha}{\partial \nu} \right|_{\Gamma_2} = \sum_{n \in Z} i\beta_2^n(\alpha) H_\alpha^{(n)}(-b) e^{-i\alpha_n \cdot x}, \quad (2.18)$$

where the unit vector $\nu = (0, 0, 1)$ on Γ_1 and $(0, 0, -1)$ on Γ_2.

For positive integer m, define $H_p^m(\Omega)$ to be the closure of

$$C_p^\infty(R^2 \times [-b,b]) = \{u \in C^\infty(R^2 \times [-b,b]) : u(x_1+L_1, x_2+L_2, x_3) = u(x_1, x_2, x_3)\},$$

with respect to the norm $\|\cdot\|_{m,\Omega}$ of the usual L^2-based Sobolev space $H^m(\Omega)$. Denote by $H_p^s(\Gamma_j)(s \in R)$ the L-periodic Sobolev space on Γ_j of order s. We refer to Adams [6] for a complete account of Sobolev spaces. In what follows, for convenience, we shall drop the subscript p.

For functions $f \in H^{1/2}(\Gamma_j)^3$, define the operator T_j^α by

$$(T_j^\alpha f)(x_1, x_2) = \sum_{n \in \Lambda} i\beta_j^n(\alpha) f^{(n)} e^{-i\alpha_n \cdot x}, \tag{2.19}$$

where $f^{(n)} = \frac{1}{L_1 L_2} \int_0^{L_1} \int_0^{L_2} f(x) e^{i\alpha_n \cdot x}$, and equality is taken in the sense of distributions.

Similarly, one could derive the following tangential transparent boundary conditions:

$$\nu \times (\nabla_\alpha \times (H_\alpha - H_{I,\alpha})) = B_1(P(H_\alpha - H_{I,\alpha})) \text{ on } \Gamma_1, \tag{2.20}$$
$$\nu \times (\nabla_\alpha \times H_\alpha) = B_2(P(H_\alpha)) \text{ on } \Gamma_2, \tag{2.21}$$

where the operator B_j is defined by

$$B_j f = -i \sum_{n \in Z^2} \frac{1}{\beta_j^{(n)}} \{(\beta_j^{(n)})^2 (f_1^{(n)}, f_2^{(n)}, 0) + ((\alpha + \alpha_n) \cdot f^{(n)})(\alpha + \alpha_n)\} e^{i\alpha_n \cdot x}, \tag{2.22}$$

where P is the projection onto the plane orthogonal to ν, i.e.,

$$Pf = -\nu \times (\nu \times f).$$

Therefore, the scattering problem can be formulated as follows:

$$\nabla_\alpha \times \left(\frac{1}{\epsilon} \nabla_\alpha \times H_\alpha\right) - \nabla_\alpha \left(\frac{1}{\epsilon_c} \nabla_\alpha \cdot H_\alpha\right) - \omega^2 H_\alpha = 0 \text{ in } \Omega_0,$$
$$\nu \times (\nabla_\alpha \times (H_\alpha - H_{I,\alpha})) = B_1(P(H_\alpha - H_{I,\alpha})) \text{ on } \Gamma_1,$$
$$\nu \times (\nabla_\alpha \times H_\alpha) = B_2(P(H_\alpha)) \text{ on } \Gamma_2,$$
$$\left(T_1^\alpha - \frac{\partial}{\partial \nu}\right) H_{\alpha,3} = 2i\beta_1 p_3 e^{-i\beta_1 b} \text{ on } \Gamma_1,$$
$$\left(T_2^\alpha - \frac{\partial}{\partial \nu}\right) H_{\alpha,3} = 0 \text{ on } \Gamma_2.$$

Applying some simple vector identities along with integration by parts formulae, it follows that the scattering problem has an equivalent variational form: find $H_\alpha \in H^1(\Omega_0)^3$ such that

$$B(H_\alpha, F) = R(F) \quad \text{for all } F \in H^1(\Omega_0)^3, \tag{2.23}$$

where

$$B(H,F) = \int_\Omega \frac{1}{\epsilon} \nabla_\alpha \times H \cdot \overline{\nabla_\alpha \times F} + \int_\Omega \frac{1}{\epsilon_c} \nabla_\alpha \cdot H \, \overline{\nabla_\alpha \cdot F} + \int_{\Gamma_1} \epsilon_1^{-1} B_1(P(H)) \cdot \overline{F}$$

$$+ \int_{\Gamma_2} \epsilon_2^{-1} B_2(P(H)) \cdot \overline{F} - \int_{\partial\Omega} \epsilon_c^{-1} \nabla_{\alpha t} \cdot H \, \overline{\nu \cdot F} - \int_{\Gamma_1} \epsilon_c^{-1} T_1(H_3) \overline{F_3}$$

$$- \int_{\Gamma_2} \epsilon_c^{-1} T_2(H_3) \overline{F_3} - \omega^2 \int_\Omega H \cdot \overline{F}$$

and

$$R(F) = \int_{\Gamma_1} (\nu \times \nabla_\alpha \times H_I - B_1 P(H_I)) \cdot \overline{F} + \int_{\Gamma_1} 2i\beta_1 \epsilon_c^{-1} p_3 e^{-i\beta_1 b} \overline{F_3} \, .$$

2.2.3 Two-Dimensional Geometries

In many cases of interest, the periodic structure is constant in one direction, say, the x_2 direction. That is, $\epsilon(x_1, x_2, x_3) = \epsilon(x_1, x_3)$. This covers the case of "classical" diffraction gratings, certain kinds of polarizers, and many other devices.

The full Maxwell equations can be simplified substantially in this situation. Several different cases arise, depending on the direction and polarization of the incident plane wave:

1. *Transverse electric (TE) polarization:* The incidence vector is orthogonal to the x_2-axis, and the E field is parallel to x_2;

2. *Transverse magnetic (TM) polarization:* The incidence vector is orthogonal to the x_2-axis and the H field is parallel to x_2;

3. *Conical diffraction:* The incidence vector is not orthogonal to x_2.

In the first case, Maxwell's equations reduce to a simple scalar Helmholtz equation

$$(\triangle_\alpha + \omega^2 \epsilon \mu) u_\alpha = 0, \qquad (2.24)$$

where $\triangle_\alpha = \partial_{x_1}^2 + \partial_{x_3}^2 + 2i\alpha_1 \partial_{x_1} - |\alpha_1|^2$ and u_α represents the component of the E_α field in the x_2 direction. In the second case, Maxwell's equations also reduce to a simple scalar model, this time of the form

$$\nabla_\alpha \cdot \left(\frac{1}{\epsilon} \nabla_\alpha u_\alpha\right) + \omega^2 \mu u_\alpha = 0, \qquad (2.25)$$

where u_α now represents the x_2 component of the H_α field. The regularity of solutions u_α of this model is reduced relative to the Helmholtz equation case (2.24). In both cases, simple scalar transparent boundary conditions analogous to (2.17), (2.18) may be easily derived. For the third case, the full vector equations need to be retained, but the problem can be solved over a "two-dimensional" domain. Variational formulations can also be easily derived in a manner analogous to that for the full Maxwell equations described in section 2.2.2.

2.2.4 Well-Posedness of the Variational Problem

Here we give a brief summary of recent results and approaches dealing with issues of well-posedness of the grating problem. For Maxwell's equations in a biperiodic structure that separates two homogeneous materials and is piecewise C^2, the existence and uniqueness of the solutions were established in [42] by an integral equation approach. Using jump conditions, the authors reduced the problem to an equivalent system of integral equations and then applied Fredholm theory. In particular, it was shown that there exists a unique solution at all but a countable set of frequencies. The result generalized the earlier work of Chen and Friedman [34]; see also Nédélec and Starling [65].

The variational approach can also be applied to study well-posedness. It has the advantage of dealing with extremely general diffractive structures and materials. The basic idea is to establish coercivity for the bilinear form of the variational formulation and then apply the Lax–Milgram lemma and the Fredholm alternative. Existence and uniqueness results were proved in TE polarization [39], in TM polarization [12], and finally for biperiodic structures [19]. The general result may be stated as follows.

THEOREM 2.1. *For all but a countable sequence of frequencies ω_j, $|\omega_j| \to +\infty$, the diffraction problem has a unique solution.*

Proof. We prove the result in the simplest TE case. For TE polarization, the scattering problem can be formulated as follows: Find $u_\alpha \in H^1(\Omega)$ such that

$$(\Delta_\alpha + \omega^2 k^2)u_\alpha = 0 \quad \text{in } \Omega, \tag{2.26}$$

$$\left(T_1^\alpha - \frac{\partial}{\partial \nu}\right) u_\alpha = 2i\beta_1 e^{-i\beta_1 b} \quad \text{on } \Gamma_1, \tag{2.27}$$

$$\left(T_2^\alpha - \frac{\partial}{\partial \nu}\right) u_\alpha = 0 \quad \text{on } \Gamma_2. \tag{2.28}$$

In the proof, we denote $k^2 = \omega^2 k^2$ to illustrate the explicit dependence on the frequency parameter ω. One can then write down an equivalent weak form: Find $u_\alpha \in H^1(\Omega)$ such that

$$B(u_\alpha, \phi) = -\int_{\Gamma_1} 2i\beta_1 e^{-i\beta_1 b}\overline{\phi} \quad \text{for all } \phi \in H^1(\Omega). \tag{2.29}$$

Here the bilinear form is defined by

$$B(w_1, w_2) = \int_\Omega \nabla w_1 \cdot \nabla \overline{w_2} - \int_\Omega (\omega^2 k^2 - \alpha^2) w_1 \overline{w_2} - 2i\alpha \int_\Omega (\partial_{x_1} w_1)\overline{w_2}$$
$$- \int_{\Gamma_1} (T_1^\alpha w_1)\overline{w_2} - \int_{\Gamma_2} (T_2^\alpha w_1)\overline{w_2}, \tag{2.30}$$

where $\int_{\Gamma_j}$ represents the dual pairing of $H^{-\frac{1}{2}}(\Gamma_j)$ with $H^{\frac{1}{2}}(\Gamma_j)$.

For simplicity, from now on, we shall remove the subscript and superscript and denote u_α, T_j^α by u, T_j, respectively.

Write $B(w_1, w_2) = B_1(w_1, w_2) + \omega^2 B_2(w_1, w_2)$, where

$$B_1(w_1, w_2) = \int_\Omega \nabla w_1 \cdot \nabla \overline{w_2} + 2\int_\Omega \alpha^2 w_1 \overline{w_2} - 2i\alpha \int_\Omega (\partial_{x_1} w_1)\overline{w_2}$$
$$- \int_{\Gamma_1} (T_1 w_1)\overline{w_2} - \int_{\Gamma_2} (T_2 w_1)\overline{w_2},$$
$$B_2(w_1, w_2) = -\int_\Omega (k^2 + k_1^2 \sin^2\theta) w_1 \overline{w_2}.$$

It follows from integration by parts that

$$B_1(u, u) = \int_\Omega |\nabla u|^2 + 2\int_\Omega \alpha^2 |u|^2 - 2\alpha \int_\Omega \mathrm{Im}(u\,\overline{\partial_{x_1} u})$$
$$- \int_{\Gamma_1} (T_1 u)\overline{u} - \int_{\Gamma_2} (T_2 u)\overline{u}.$$

Next denote $k^2 = \epsilon\mu$ by $\sigma' + i\sigma''$. Clearly, $\sigma' > 0$ and $\sigma'' \geq 0$. Also, denote k_2^2 by $\sigma_2' + i\sigma_2''$, where $\sigma_2' > 0$ and $\sigma_2'' \geq 0$. Thus

$$\mathrm{Re}\{B_1(u,u)\} = \int_\Omega |\nabla u|^2 + 2\int_\Omega \alpha^2 |u|^2 - 2\alpha \int_\Omega \mathrm{Im}(u\,\overline{\partial_{x_1} u})$$
$$- \mathrm{Re}\left\{\int_{\Gamma_1} (T_1 u)\overline{u} + \int_{\Gamma_2} (T_2 u)\overline{u}\right\}$$
$$\geq \int_\Omega \frac{1}{2}|\nabla u|^2 - \mathrm{Re}\left\{\int_{\Gamma_1} (T_1 u)\overline{u} + \int_{\Gamma_2} (T_2 u)\overline{u}\right\}, \qquad (2.31)$$

$$-\mathrm{Im}\{B_1(u,u)\} = \mathrm{Im}\left\{\int_{\Gamma_1} (T_1 u)\overline{u} + \int_{\Gamma_2} (T_2 u)\overline{u}\right\}. \qquad (2.32)$$

Further,

$$-\int_{\Gamma_1} (T_1 u)\overline{u} = -\sum 2\pi i \beta_1^n |u^{(n)}|^2$$
$$= \sum 2\pi \mathrm{Im}\beta_1^n |u^{(n)}|^2 - i\sum 2\pi \mathrm{Re}\beta_1^n |u^{(n)}|^2, \qquad (2.33)$$

and it is easy to see that

$$-\int_{\Gamma_2} (T_2 u)\overline{u} = -\sum 2i\pi \beta_2^n |u^{(n)}(-b)|^2$$
$$= \sum_n 2\pi |\beta_2^n| |u^{(n)}(-b)|^2 q_n, \qquad (2.34)$$

where $q_n = q_n' - iq_n''$ with $q_n' = -\sigma_2''\cos(\gamma_2^n/2) + \sigma_2'\sin(\gamma_2^n/2)$ and $q_n'' = \sigma_2'\cos(\gamma_2^n/2) + \sigma_2''\sin(\gamma_2^n/2)$.

Recall that

$$\gamma_2^n = \arg(\omega^2 \mathrm{Re}(k_2^2) - (n+\alpha)^2 + i\omega^2 \mathrm{Im}(k_2^2))$$

and $0 \leq \gamma_2^n < 2\pi$. Then it follows that $q_n'' > 0$ for all n and the set $\Lambda = \{n; q_n' < 0\}$ is finite. It is also easy to verify that $|q_n''| > |q_n'|$ for $n \in \Lambda$. Moreover, for fixed $\omega \notin \mathcal{B} = \{\omega; \beta_j^n = 0, \ j = 1, 2\}$, we have

$$|\beta_j^n| \geq C(1 + |n|^2)^{1/2}, \qquad j = 1, 2 . \tag{2.35}$$

Combining the above estimates (2.31)–(2.35), we have

$$|B_1(u,u)| \geq C \left[\int_\Omega |\nabla u|^2 + \|u\|^2_{H^{1/2}(\Gamma_1)} \right.$$
$$\left. + \sum_{n \in \Lambda}(|q_n''| - |q_n'|)|u^{(n)}(-b)|^2 + \sum_{n \notin \Lambda} |q_n''||u^{(n)}(-b)|^2 \right]$$
$$\geq C \left[\int_\Omega |\nabla u|^2 + \|u\|^2_{H^{1/2}(\Gamma_1)} + \|u\|^2_{H^{1/2}(\Gamma_2)} \right]$$
$$\geq C\|u\|_{L^2(\Omega)} ,$$

where the last inequality may be obtained by applying standard elliptical estimates. Therefore, we have shown that

$$|B_1(u,u)| \geq C\|u\|^2_{H^1(\Omega)} , \tag{2.36}$$

i.e., that B_1 is a bounded coercive bilinear form over $H^1(\Omega)$. The Lax–Milgram lemma then gives the existence of a bounded invertible map $A_1 = A_1(\omega) : H^1(\Omega) \to (H^1(\Omega))'$ such that $\langle A_1 u, v \rangle = B_1(u,v)$, where $'$ represents the dual space. Moreover, A_1^{-1} is bounded. Notice that the operator $A_2 : H^1(\Omega) \to (H^1(\Omega))'$ defined by $\langle A_2 u, v \rangle = B_2(u,v)$ is compact and independent of ω.

Holding $\omega_0 \notin \mathcal{B}$ fixed, consider the operator $A(\omega_0, \omega) = A_1(\omega_0) + \omega^2 A_2$. Since A_1 is bounded invertible and A_2 is compact, we see that $A(\omega_0, \omega)^{-1}$ exists by Fredholm theory for all $\omega \notin \mathcal{E}(\omega_0)$, where $\mathcal{E}(\omega_0)$ is some discrete set. It is clear that

$$\|A_1(\omega) - A_1(\omega_0)\| \to 0 \quad \text{as } \omega \to \omega_0.$$

Thus, since $\|A(\omega, \omega) - A(\omega_0, \omega)\| = \|A_1(\omega) - A_1(\omega_0)\|$ is small for $|\omega - \omega_0|$ sufficiently small, it follows from the stability of bounded invertibility (see, e.g., Kato [57, Chap. 4]) that $A(\omega, \omega)^{-1}$ exists and is bounded for $|\omega - \omega_0|$ sufficiently small, $\omega \notin \mathcal{E}(\omega_0)$. Since $\omega_0 > 0$ can be an arbitrary real number, we have shown that $A(\omega, \omega)^{-1}$ exists for all but a discrete set of points. $\square$

Abboud and Nédélec [3] independently developed a variational formulation for the Maxwell equations in a nonperiodic bounded inhomogeneous medium. They were interested in the more general problem where the magnetic permeability is nonconstant. Their approach was further extended by Abboud [1, 2] to the periodic case.

In general, the result in Theorem 2.1 is the best possible. There are examples that indeed exhibit the existence of singular frequencies (the sequence $\{\omega_j\}$). We refer to [26], in which explicit examples are constructed and nonuniqueness is shown at the singular frequencies in the TM case. It was also shown that in general the sequence in Theorem 2.1 is unbounded. It is interesting to note that, in addition to Theorem 2.1, Abboud [1] showed the following.

LEMMA 2.2. *If* Im $\epsilon_1 > 0$ *or* Im $\epsilon_2 > 0$, *then the diffraction problem has a unique solution.*

Thus, for media that are absorbing in either Ω_1 or Ω_2, the diffraction problem always has a unique solution.

Continuous dependence of solutions on material parameters was studied by Kirsch [58] in the two-dimensional TE case with gratings that separate a dielectric medium from a perfectly reflecting medium (conductor). The model equation takes the form

$$(\Delta + k^2)u = 0, \tag{2.37}$$
$$u|_S = 0, \tag{2.38}$$

together with the radiation condition, where S is the grating profile. It was shown that the solution depends on k and the angle of incidence θ analytically, provided that $(\alpha_n + k\cos\theta)^2 \neq k^2$ for every integer n, where k^2 is real. However, the dependence of solutions on the grating profile turns out to be a bit more complicated. In fact, using a variational approach, Kirsch [58] showed that

$$\|u_f - u_g\|_{H^1} \leq C\|f - g\|_{C^1}, \tag{2.39}$$

where the constant C is independent of u_f, u_g and f, g. Here u_f and u_g are the solutions of (2.37)–(2.38) generated by grating profiles f and g, respectively, for a fixed incident wave.

2.3 Finite Element Methods

In this section we present some recent convergence results for the finite element method in solving the diffraction problem in various geometries and polarization modes. The method of finite elements offers effective means for solving differential equations numerically. Its mathematical foundation combines the variational approach and the approximation theory. The reader is referred to [35, 30] for the basic theory and additional references.

We consider the simplest TE case. For each $h \in (0,1)$, the domain Ω_0 is discretized with a quasi-uniform mesh of size h. Let $\{S^h : h \in (0,1)\}$ denote a family of finite-dimensional subspaces of H^1, for example, the continuous piecewise linear functions.

We define the finite element approximation of the solution u_α of (2.29) by the following equation: for each $v_h \in S^h$,

$$a(u_h, v_h) = (f, v_h). \tag{2.40}$$

Chapter 2. Variational Methods for Diffractive Optics Modeling   49

Thus the problem is reduced to finite dimensions. Solving the resulting matrix equation gives rise to a finite element approximation of the solution. In fact, this provides the basic idea for solving our model equation: one first chooses a basis of S^h, $\{\phi_1, \phi_2, \ldots, \phi_k\}$, which is a finite set according to the definition of S^h. Substituting the expression of

$$u_h = c_1\phi_1 + c_2\phi_2 + \cdots + c_k\phi_k \qquad (2.41)$$

into the equation (2.16), by choosing $v_h = \phi_i$, $i = 1, \ldots, k$, one gets a system of linear equations. Solving this system for $\{c_j\}$ then leads to an approximation of u_α in S^h.

2.3.1 Convergence Analysis

We analyze two types of errors in the finite element approximation. We consider first the discretization of the continuous problem. The goal is to show that u_h, the solution to (2.40), is a good approximation to u_α. We then analyze truncations of nonlocal boundary operators T_j. The fact that these boundary operators are nonlocal follows immediately from the infinite series expansion. In practice, it is essential to obtain error estimates when truncations of these operators take place. For TE polarization, the following well-posedness results for the discretized problem and error estimates were established in [10].

Concerning the discretization error, the following result holds.

THEOREM 2.3. *There exists a constant $0 < h_0 \leq 1$ such that for any h, $0 < h < h_0$, the discretized problem (2.16) attains a unique solution u^h, and*

$$\|u_\alpha - u_h\|_{L^2(\Omega)} \leq Ch^2, \qquad (2.42)$$

$$\|u_\alpha - u_h\|_{H^1(\Omega)} \leq Ch^1, \qquad (2.43)$$

where the constant C is independent of h. Further, the error estimates are optimal.

Define u_N^h, the truncated finite element approximation to the solution u_α, by the equation

$$a^N(u_N^h, v_h) = (f, v_h) \quad \text{for all } v_h \in S^h, \qquad (2.44)$$

where a^N is given by replacing T_j in (2.11) with

$$(T_j^N f)(x_1) = \sum_{|n|<N} i\beta_j^n(\alpha) f^{(n)} e^{i\alpha_n x_1}. \qquad (2.45)$$

In [10], we showed that the discretized variational problem is well posed for N (the number of terms used to represent the boundary operators) sufficiently large and h sufficiently small. In other words, the convergence properties of Theorem 2.3 remain valid when N is sufficiently large. Similar convergence results were obtained in [53] for waveguide problems.

In the TM case, the situation becomes more complicated. Suppose that the diffraction problem has a unique solution $u_\alpha \in H^1(\Omega)$. The next result concerns the finite element approximation.

THEOREM 2.4. *For any given $\delta > 0$, there exists $h_0 = h_0(\delta)$ such that for $0 < h < h_0$,*

$$\|u_\alpha - u^h\|_{L^2(\Omega)} \leq \delta \|u_\alpha - u^h\|_{H^1(\Omega)} . \tag{2.46}$$

Moreover, if $f \in L^2(\Omega)$ in (2.10), there exists an $h_1 = h_1(\delta)$ such that for all $0 < h < h_1$,

$$\|u_\alpha - u^h\|_{H^1(\Omega)} \leq \delta \|f\|_{L^2(\Omega)} . \tag{2.47}$$

The estimates establish the existence and uniqueness for the corresponding finite element approximation. Actually, because of the finite dimensionality of S^h, the uniqueness implies the existence. Moreover, since h_0, h_1 are independent of u_α, the estimates are uniform with respect to u_α.

Remark. In the TE case, since discontinuous coefficients only occur in the lower order terms, the solution is in H^2. It follows that improved error estimates are possible. On the other hand, no improved regularity on the solution is available in the TM case. The singularities caused by the discontinuous coefficients in the principle part of the operator can spread more destructively. As a result, the solution is only in H^1. Thus one can only expect uniform convergence results similar to the estimates in the above theorems [12]. Our numerical experiments support that although the finite element method does converge in the TM case, the convergence is slower than in the TE case. Convergence results similar to the TM case may be proved for the biperiodic diffraction problem [13].

2.3.2 Implementation Issues

As in all finite element methods, discretization of the original variational problem results in a large, sparse matrix equation. Matrix sub-blocks corresponding to the boundary operators are full, but small relative to the size of the matrix problem. These terms are calculated by truncating the Fourier series representations of the boundary operators, as in (2.45). The discretization can of course be done in many ways depending on the type of grid constructed and the types of elements used. A key question, which can affect the choice of discretization, is, How will the matrix equation be solved? In general, the matrix is complex, indefinite, and sometimes non-Hermitian. These properties preclude the use of many of the most effective iterative solvers available. Effective alternatives do exist, such as GMRES and Orthomin. These solvers generally require a good preconditioner to be effective for diffractive optics problems.

A particular preconditioner, which can be very effective particularly for "low-contrast" structures, can be constructed as follows. The idea is to average the refractive index of the grating structure in the direction(s) of periodicity, to produce

Chapter 2. Variational Methods for Diffractive Optics Modeling   51

a "layered" problem. Thus for TE polarization we solve the problem

$$(\Delta + \tilde{a})u = f \quad \text{in } \Omega, \tag{2.48}$$

$$\left(T_1^\alpha - \frac{\partial}{\partial x_3}\right) u = g_1 \quad \text{on } \Gamma_1, \tag{2.49}$$

$$\left(T_2^\alpha - \frac{\partial}{\partial x_3}\right) u = g_2 \quad \text{on } \Gamma_2, \tag{2.50}$$

where

$$\tilde{a}(x_3) = \frac{1}{2\pi} \int_0^{2\pi} a(t, x_3) dt,$$

via the sequence of one-dimensional problems

$$\left(\frac{d}{dx_3} - n^2 + \tilde{a}\right) \hat{u}_n = \hat{f}_n \quad \text{in } (-b, b),$$

$$\frac{d\hat{u}_n}{dx_3}(b) = i\beta_1^n - \hat{g}_{1n},$$

$$\frac{d\hat{u}_n}{dx_3}(-b) = i\beta_2^n - \hat{g}_{2n},$$

where $\hat{u}_n(x_3)$, $n = 0, \pm 1, \pm 2, \ldots$, are the Fourier components of u in the x_1 variable. Computationally this problem can be solved very rapidly by the fast Fourier transform (FFT). For example, if the problem is discretized with bilinear finite elements on an $N_1 \times N_2$ rectangular grid, a complete solve takes $\mathcal{O}(N_2 N_1 \log N_1)$ time. Using the resulting solution operator as a preconditioner yields a very efficient solution procedure for some difficult problems.

Similar preconditioners can be applied to the TM polarization case and to the full Maxwell equations. The main difference is that the dielectric coefficient must be averaged by a somewhat more complicated homogenization procedure, which results in a tensor-valued approximate coefficient. Convergence is not as fast as for the TE case, but nevertheless this gives a useful approximation.

Finite element methods such as those described above have been applied to analyzing problems of current technological interest. A typical example involving micro-electrical-mechanical structures (MEMS) was described in [37, 38].

2.3.3 Least-Squares Finite Element Methods

More recently, we have developed least-squares finite element methods for solving the diffraction problem [23, 15]. Here, we present the formulation and convergence result in the TM case.

Least-squares finite element methods have received much attention recently (see [25] and the references cited therein). Significant progress has been made in algorithm designs as well as convergence analysis for a large class of partial differential equation models. The least-squares finite element method presented here takes jump conditions at the interfaces into consideration. As with general least-squares finite element methods, the resulting bilinear form is symmetric, coercive, and continuous. Also, it has the flexibility of choosing different finite element spaces for different variables. Unlike the mixed finite element method, the inf-sup condition need not be satisfied, thus improving the stability of the method. The resulting discrete system includes a symmetric and positive definite matrix, so it is easily treated by various existing preconditioning techniques, e.g., multigrid methods [29, 64]. In contrast, the system from the standard finite element may not even be symmetric. The method also has the following additional advantages: Since the interface jump conditions are enforced through the least-squares functional, different finite element spaces may be used easily on either side of the interface. The interface can be a curved surface. With sufficiently smooth interfaces, significantly better estimates can be expected. Note that, even with a smooth interface, standard finite element methods may not have good convergence results. The reason is that the solution, which is possibly very smooth on both sides of the interface, may not be smooth globally.

In [23], we developed an interface least-squares finite element method for TM polarization. In order to introduce the approach, it is crucial to view the grating problem as an interface problem. Let a plane wave $u_I = e^{i\alpha x_1 - i\beta_1 x_3}$ be incident on the grating surface S_1 from the top, where $\alpha = \omega\sqrt{\epsilon_1}\sin\theta$, $\beta_1 = \omega\sqrt{\epsilon_1}\cos\theta$, and $\theta \in (-\pi/2, \pi/2)$ is the angle of incidence. Assume that some possibly nonhomogeneous material lies in between S_1 and S_2. Let Ω_1, Ω_2, and Ω_0 denote the regions above S_1, below S_2, and between S_1 and S_2, respectively. The first step is to rewrite the interface problem as a first-order system of u, ϕ and to formulate it as a least-squares problem. Let $\phi^j = a_j \nabla_\alpha u_\alpha^j$, $j = 0, 1, 2$. We have

$$\nabla_\alpha \cdot \phi^j + u_\alpha^j = 0 \text{ in } \Omega_j, \quad j = 0, 1, 2,$$
$$a_j \nabla_\alpha u_\alpha^j - \phi^j = 0 \text{ in } \Omega_j, \quad j = 0, 1, 2,$$
$$[u_\alpha] = 0 \text{ on } S_j, \quad j = 1, 2,$$
$$[n \cdot \phi] = 0 \text{ on } S_j, \quad j = 1, 2,$$
$$a_1 T_1 u_\alpha^1 - n_1 \cdot \phi^1 = f \text{ on } \Gamma_1,$$
$$a_2 T_2 u_\alpha^2 - n_2 \cdot \phi^2 = 0 \text{ on } \Gamma_2,$$

where $f = 2i a_1 \beta_1 \, e^{-i\beta_1 b}$.

For an integer l, define $\tilde{H}^l(\Omega) = \{u \in L^2(\Omega) : u^j = u|_{\Omega_j} \in H_p^l(\Omega_j), \, j = 0, 1, 2\}$. Let $V = \tilde{H}^1(\Omega)$. Denote $W = \{\phi \in (L^2(\Omega))^2 : \phi^j = \phi|_{\Omega_j} = H_p(\text{div}, \Omega_j), \, j = 0, 1, 2\}$. Define the least-squares functional $J(v, \psi; f)$ by

$$\sum_{j=0}^{2}(\|\nabla_\alpha \cdot \psi^j + v^j\|_{L^2(\Omega_j)}^2 + \|a_j \nabla_\alpha v^j - \psi^j\|_{L^2(\Omega_j)}^2)$$
$$+ \sum_{j=1}^{2}(\|[v]\|_{H^{1/2}(S_j)}^2 + \|[n \cdot \psi]\|_{H^{-1/2}(S_j)}^2)$$
$$+ \|a_1 T_1 v^1 - n_1 \cdot \psi^1 - f\|_{H^{-1/2}(\Gamma_1)}^2 + \|a_2 T_2 v^2 - n_2 \cdot \psi^2\|_{H^{-1/2}(\Gamma_2)}^2.$$

Chapter 2. Variational Methods for Diffractive Optics Modeling   53

Note that the jump conditions are built into the functional. The L^2, $H^{-1/2}$, $H^{1/2}$ norms are standard and may be computed by the Fourier transforms. The least-squares minimization problem is to find $u \in V$, $\phi \in W$, such that

$$J(u, \phi; f) = \inf_{v \in V, \psi \in W} J(v, \psi; f).$$

Taking the variation of J with respect to v, ψ, we obtain an equivalent variational problem: find $u \in V$, $\phi \in W$ such that

$$B(u, \phi; v, \psi) = \langle f, a_1 T_1 v^1 - n_1 \cdot \phi^1 \rangle_{-\frac{1}{2}, \Gamma_1} \text{ for all } v \in V,\ \psi \in W,$$

where

$$B(u, \phi; v, \psi) = \sum_{j=0}^{2} \{ (\nabla_\alpha \cdot \phi^j + u^j,\ \nabla_\alpha \cdot \psi^j + v^j)_{0, \Omega_j} + (a_j \nabla_\alpha u^j - \phi^j,\ a_j \nabla_\alpha v^j - \psi^j)_{0, \Omega_j} \}$$

$$+ \sum_{j=1}^{2} \{ \langle [u], [v] \rangle_{\frac{1}{2}, S_j} + \langle [n \cdot \phi], [n \cdot \psi] \rangle_{-\frac{1}{2}, S_j} + \langle a_j T_j u^j - n_j \cdot \phi^j,\ a_j T_j v^j - n_j \cdot \psi^j \rangle_{-\frac{1}{2}, \Gamma_j} \}.$$

Here $\langle \cdot \rangle_{-\frac{1}{2}, \Gamma_j}$, $\langle \cdot \rangle_{\frac{1}{2}, \Gamma_j}$, $\langle \cdot \rangle_{-\frac{1}{2}, S_j}$, $\langle \cdot \rangle_{\frac{1}{2}, S_j}$ are inner products in the specified spaces that are defined by Fourier transforms (series), and $(\cdot)_{0, \Omega_j}$ denotes the L^2 inner product.

Let V_h, W_h be finite-dimensional subspaces of V, W, respectively. Here h is the mesh size from partitioning Ω into subdomains. A least-squares finite element approximation (u_h, ϕ_h) of (u, ϕ) may be obtained by solving

$$B(u_h, \phi_h; v_h, \psi_h) = \langle f, a_1 T_1 v_h^1 - n_1 \cdot \phi_h^1 \rangle_{-\frac{1}{2}, \Gamma_1} \quad \text{for all } v_h \in V_h,\ \psi \in W_h,$$

which gives rise to a system of linear equations. An important property of the least-squares finite element formulation is that the resulting system is symmetric, positive definite, and thus can be solved efficiently by various existing preconditioning techniques.

Assume that the interfaces S_1, S_2 are sufficiently smooth, say, at least of $C^{1,1}$, so that the solution has the desired regularity. Assume also that V_h and W_h satisfy the following approximation properties and inverse properties.

In the following, consider the convergence properties in the L^2 setting. In order to do so, we assume that $W_h \subset (\tilde{H}_p^1(\Omega))^2$. Now, define the following least-squares functional $J_h(v, \psi; f)$ on $V_h \times W_h$:

$$J_h(v, \psi; f)$$
$$= \sum_{j=0}^{2} \left(\|\nabla_\alpha \cdot \psi^j + v^j\|_{L^0(\Omega_j)}^2 + \|a_j \nabla_\alpha v^j - \psi^j\|_{L^2(\Omega_j)}^2 + h^{-1} \|[v]\|_{L^2(S_j)}^2 + \|[n \cdot \psi]\|_{L^2(S_j)}^2 \right)$$
$$+ \|a_1 T_1 v^1 - n_1 \cdot \psi^1 - f\|_{H^{-1/2}(\Gamma_1)}^2 + \|a_2 T^2 v^2 - n_2 \cdot \psi^2\|_{H^{-1/2}(\Gamma_2)}^2.$$

The approximate minimization problem is to find $u_h \in V_h$, $\phi_h \in W_h$ such that

$$J_h(u_h, \phi_h; f) = \min_{v_h \in V_h, \psi_h \in W_h} J_h(v_h, \psi_h; f). \tag{2.51}$$

It is obvious that (2.51) is equivalent to the following variational problem: find $u_h \in V_h$, $\psi_h \in W_h$ such that

$$A_h(u_h, \phi_h; v_h, \psi_h) = \langle f, a_1 T^1 v_h^1 - n_1 \cdot \psi_h^1 \rangle_{-\frac{1}{2}, \Gamma_1} \quad \text{for all } v_h \in V_h, \psi_h \in W_h, \tag{2.52}$$

where

$$A_h(u_h, \phi_h; v_h, \psi_h)$$
$$= \sum_{j=0}^{2} \left\{ (\nabla_\alpha \cdot \phi_h^j + u_h^j, \nabla_\alpha \cdot \psi_h^j + v_h^j)_{0, \Omega_j} + (a_j \nabla_\alpha u_h^j - \phi_h^j, a_j \nabla_\alpha v_h^j - \psi_h^j)_{0, \Omega_j} \right\}$$
$$+ \sum_{j=1}^{2} \left\{ h^{-1} \langle [u_h], [v_h] \rangle_{0, S_j} + \langle [n \cdot \phi_h], [n \cdot \psi_h] \rangle_{0, S_j} \right.$$
$$\left. + \langle a_j T_j u_h^j - n_j \cdot \phi_h^j, a_j T_j v_h^j - n_j \cdot \psi_h^j \rangle_{-\frac{1}{2}, \Gamma_j} \right\}.$$

Next, a convergence result for the TM case is presented [23].

THEOREM 2.5. *Let u_h, ϕ_h be the unique solution of (2.51) or (2.52). Assume that the solution (u, ϕ) of (2.10)–(2.16) is in $\tilde{H}^2(\Omega) \times (\tilde{H}^2(\Omega))^2$. Then*

$$\|u - u_h\|_{\tilde{H}^1(\Omega)} + \|\phi - \phi_h\|_{\tilde{H}(\text{div}, \Omega)} \le Ch^1 \left(\|u\|_{\tilde{H}^2(\Omega)} + \|\phi\|_{(\tilde{H}^2(\Omega))^2} \right).$$

The result indicates that as h goes to zero the computed solution converges to the exact solution at the rate of h. This result is stronger than Theorem 2.3.2, the optimal one for the standard finite element method. In fact, though the standard finite element method uniformly converges in the TM case, no convergence rate is available. It was shown in [23] that the condition number of the linear system resulting from (2.52) is $O(h^{-2})$. In the TE case, the noninterface negative one-norm least-squares finite element approach of Bramble, Lazarov, and Pasciak [28] may be adopted for solving the diffraction problem. Convergence and computational results have recently been obtained in [23, 15].

2.4 Inverse and Optimal Design Problems

Given the incident field and some information about the resulting outgoing fields, the inverse problem concerns the determination of the grating profile. The closely related optimal design problem is to determine a profile that optimizes some performance criteria. These types of problems in diffractive optics are an active area of current research. In keeping with the theme of this paper, our goal in this section will be to provide the reader with an overview of how variational methods can be applied. A sampling of other recent approaches is described at the end of this section.

All of the problems described here involve the "two-dimensional" geometries as outlined in section 2.2.3. Fully three-dimensional problems are more difficult both computationally and analytically.

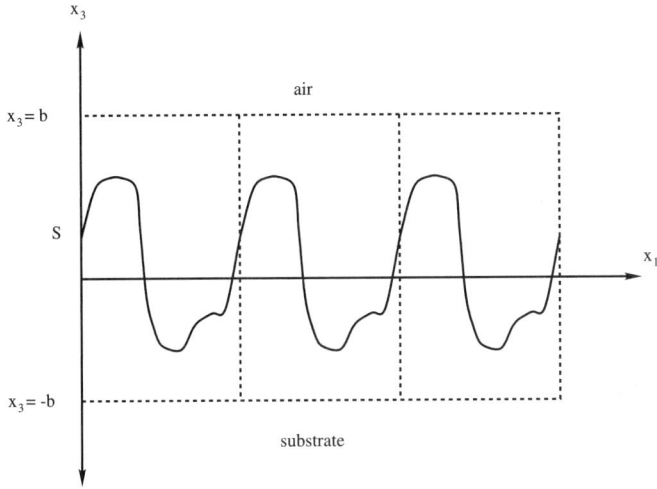

Figure 2.2: Geometrical configuration for inverse and optimal design problems.

2.4.1 An Inverse Shape Determination Problem

Consider a plane wave incident on a periodic structure from above as shown in Figure 2.2. The structure separates two regions. In one region, above the periodic structure, the dielectric coefficient ϵ is a fixed constant, as is the index of refraction k. The other region contains a perfectly reflecting material (or conductor). Given the incident field, an inverse diffraction problem is then to determine the periodic structure from the scattered field. Let the incident wave be of the form

$$u_I = e^{i\alpha x_1 - i\beta_1 x_3}, \tag{2.53}$$

where $\alpha = k\sin\theta$, $\beta_1 = k\cos\theta$, and $-\frac{\pi}{2} < \theta < \frac{\pi}{2}$ is the angle of incidence. As discussed in section 2.2.3, the direct model equation has a simple form,

$$(\Delta + k^2)u = 0, \tag{2.54}$$

$$u|_S = 0, \tag{2.55}$$

where S is the grating profile. We again seek quasiperiodic solutions to this problem, i.e., solutions u such that $ue^{-i\alpha x_1}$ are L-periodic for every x_3; here L is the period of the grating. Using the quasi-periodicity of the solution and the radiation condition that requires the boundedness of u as x_3 tends to infinity, we arrive at the boundary condition as in section 2.2.2:

$$\frac{\partial u}{\partial \nu}\Big|_{x_3=b} = B(u|_{x_3=b}) - 2i\beta e^{-i\beta b + i\alpha x_1}, \tag{2.56}$$

where

$$B(f) = \sum_{n \in Z} i\beta_j^n f^{(n)} e^{i(\alpha_n + \alpha)x_1}, \tag{2.57}$$

and α_n and β_n are as defined in section 2.2.2. The inverse problem is to determine S from the information $u|_{x_3=b}$.

A closely related problem is to determine the grating structure on some nonconductive optical material. In that case, one places optical detectors both above and below the material. Consequently, the measurements consist of information on the reflected wave and transmitted wave. Note that the boundary condition (2.55) should be replaced with a nonlocal boundary condition that is similar to (2.56). This inverse problem was proposed and studied in [20].

2.4.2 Uniqueness and Stability of the Interface

By counting the dimensions of the unknowns and data, it is easy to see that the inverse problem is underdetermined. Thus, in general, properties on uniqueness and stability are very hard, if not impossible, to establish. But because of the important impact of these properties on applications, characterizations of uniqueness and stability are required.

Here, we present a uniqueness result for the inverse problem. Let us assume that for the given incident field u_I, u_1, and u_2 solve the direct problem (2.54)–(2.56) with respect to $S_1 = \{x_3 = f_1(x_1)\}$ and $S_2 = \{x_3 = f_2(x_1)\}$, respectively. The functions f_1 and f_2 are assumed to be sufficiently smooth (say C^2) and L-periodic. Let $b > \max\{f_1(x_1), f_2(x_1)\}$ be a fixed constant. Denote $T = \max\{f_1(x_1), f_2(x_1)\} - \min\{f_1(x_1), f_2(x_1)\}$.

THEOREM 2.6. *Assume that $u_1(x_1, b) = u_2(x_1, b)$. Assume further that one of the following conditions is satisfied:*

(i) *k has a nonzero imaginary part;*
(ii) *k is real and T satisfies $k^2 < 2[T^{-2} + L^{-2}]$.*

Then $f_1(x_1) = f_2(x_1)$.

Thus in the case when k has nonzero imaginary part corresponding to a lossy medium, a global uniqueness result is available [9]. In the case with real k corresponding to a dielectric medium, one can only prove a local uniqueness result; i.e., any two surface profiles are identical if they generate the same diffraction patterns and the area in between the two profiles is sufficiently small. Moreover, the smallness of the area is characterized explicitly in terms of a condition that relates the index of refraction k, the period, and the maximum of the difference in height allowed for the two profiles; see [11] for details.

Uniqueness for the inverse diffraction problem in periodic structure was also studied in [59] for a dielectric medium, where a uniqueness theorem was proved by an approach for the general inverse scattering problem. The main idea was to prove the denseness of a set of special solutions by using many incident waves. For the optical applications we are interested in, one is only allowed to use single or a small number of incident plane waves. See [55] for some additional results along this direction.

In applications, it is impossible to make exact measurements. Stability is crucial in the practical reconstruction of profiles since it contains necessary information to determine to what extent the data can be trusted.

Chapter 2. Variational Methods for Diffractive Optics Modeling

Before stating the stability result, let us first introduce some notations. For any two domains D_1 and D_2 in $\mathbf{R}^2$, define $d(D_1, D_2)$, the Hausdorff distance between them, by

$$d(D_1, D_2) = \max\{\rho(D_1, D_2),\ \rho(D_2, D_1)\}, \qquad (2.58)$$

where

$$\rho(D_1, D_2) = \sup_{x \in D_1} \inf_{y \in D_2} |x - y|. \qquad (2.59)$$

Denote $D = \{x;\ f(x_1) < x_3 < b\}$ and a sequence of domains $D_h = \{x;\ f(x_1) + h\sigma_h(x_1)\mu(x_1) < x_3 < b\}$ for any $0 < h < h_0$, where $\mu(x_1)$ is the normal to $S = \{x_3 = f(x_1)\}$. Assume also that the boundary $S_h = \{x_3 = f(x_1) + h\sigma_h(x_1)\mu(x_1)\}$ is periodic of the same period L. Further, the function σ_h satisfies $|\sigma_h(x_1)| \le C$. Furthermore, for h_0 sufficiently small, the sequence of domains is assumed to satisfy

$$C_1 h \le d(D, D_h) \le C_2 h, \qquad (2.60)$$

where C_1 and C_2 are positive constants.

For the fixed incident plane wave u_I, assume that u and u_h solve the scattering problem with respect to periodic structures S and S_h, respectively. Then we have the following stability result.

THEOREM 2.7. *Under the above assumptions,*

$$d(D_h, D) \le C \| u_h |_{x_3=b} - u|_{x_3=b} \|_{H^{1/2}},$$

where the constant C may depend on the family $\{\sigma_h\}$.

The result indicates that for small h, if the measurements are $O(h)$ close to the true scattered fields in the $H^{1/2}$ norm, then D_h is $O(h)$ close to D in the Hausdorff distance. This result, as well as stability results for other models, was proved in [20].

More recent results on uniqueness and stability for the inverse diffraction problem in biperiodic structures may be found in [22, 7]. A complete account of the general theory and references for inverse scattering problems in general (nonperiodic) structures is available in Colton and Kress [36]. See Isakov [56] for a recent survey of other developments on general partial differential equation inverse problems.

2.4.3 Optimal Shape Design of the Interface

As opposed to the "inverse problem" of determining an unknown interface from a given diffracted field, the optimal design problem is that of determining an interface that gives the "best" approximation to some desired diffracted field. Thus, in the inverse problem, the interface that gave rise to the given diffracted field is implicitly assumed to exist. As outlined in the previous section, the main theoretical problems are then to determine whether it is uniquely and stably determined by the data. However, in the optimal design problem, an optimal interface may not be assumed a priori to exist. In fact, for a given desired diffracted field, there may be no interface in a given class of admissible designs that gives a best approximation. Hence the problem of existence becomes important. In this section, we will outline how one can handle questions of existence using the variational approach, for a simple shape design problem. The variational approach was introduced by Achdou and Pironneau [4, 5] for solving optimal photocell design problems. We note that until recently, most engineering approaches have used approximations to the underlying diffraction problem in order to simplify the optimal design problem [46, 51, 54].

We retain the same geometrical configuration described earlier, with a TE wave incident on a structure as shown in Figure 2.2. For convenience, from now on we assume the period L of the structure is 2π. The periodic curve $S \subset \Omega$ defines the grating profile. The material above S has refractive index k_1 and the material below S has index k_2. Both k_1 and k_2 are assumed to be real. Define

$$a_S(x) = \begin{cases} k_1^2 & \text{if } x \text{ is above } S, \\ k_2^2 & \text{if } x \text{ is below } S. \end{cases}$$

We then consider the problem

$$(\triangle_\alpha + \omega^2 a_S)u = 0 \quad \text{in } \Omega, \tag{2.61}$$

$$\left(T_1^\alpha - \frac{\partial}{\partial x_3}\right)u = 2i\beta_1 e^{-i\beta_1 b} \quad \text{on } \Gamma_1, \tag{2.62}$$

$$\left(T_2^\alpha - \frac{\partial}{\partial x_3}\right)u = 0 \quad \text{on } \Gamma_2. \tag{2.63}$$

With all parameters except the surface profile fixed, there is a fixed, finite number of propagating modes (each of which corresponds to an index n for which the propagation constant β_j^n is real valued). Define two sets of indices of propagating modes

$$\Lambda_j = \{n \in Z : \text{Im}\,(\beta_j^n) = 0\}, \quad j = 1, 2.$$

The set Λ_1 contains the indices of the reflected propagating modes; Λ_2 corresponds to the transmitted modes. The coefficients of each propagating reflected mode are determined by the Fourier components of the trace $u|_{\Gamma_1}$:

$$\begin{aligned} r_n &= u_n(b)e^{-i\beta_1 b} & \text{for } n \neq 0, \; n \in \Lambda_1, \\ r_0 &= u_0(b)e^{-i\beta_1 b} - \text{const.} & \text{for } n = 0, \end{aligned} \tag{2.64}$$

where $u_n(x_3) = \frac{1}{2\pi}\int_0^{2\pi} u(x_1,x_3)e^{-inx_1}dx_1$. Similarly, the coefficients of the propagating transmitted modes are

$$t_m = u_m(-b)e^{-i\beta_2 b} \quad \text{for } m \in \Lambda_2. \tag{2.65}$$

Writing the reflection and transmission coefficients as vectors,

$$r = (r_n)_{n \in \Lambda_1}, \quad t = (t_m)_{m \in \Lambda_2},$$

Chapter 2. Variational Methods for Diffractive Optics Modeling

we denote the pair $(r, t) = F$. The coefficients r_n and t_m, and hence F, are functions of the interface profile S. Denote this dependence by $F(a_S)$. Suppose that the desired "output vector" of diffraction coefficients is q; i.e., we wish to find S such that $F(a_S) = q$. One plausible way to formulate the optimal design problem is then

$$\min_{S \in \mathcal{A}} J(a_S) = \frac{1}{2}\|F(a_S) - q\|^2,$$

where $\|\cdot\|$ is the Euclidean norm. We ask the question, Does this problem have a solution S in the admissible class $\mathcal{A}$? The answer, of course, depends on the choice of $\mathcal{A}$. The first route one may consider is to begin with a very large class of admissible curves, so that potential optima are not excluded from the admissible set. Unfortunately, this may lead to minimizing sequences within the admissible class that do not converge. A solution is to "relax" the problem, enlarging the admissible set to include appropriate "mixtures" of materials. The relaxation method for general optimal design problems was first analyzed in [60]. In our case, relaxation can be accomplished as follows. Denote the set of all continuous simple curves contained in the domain Ω by $\mathcal{S}$. We allow any profile $S \in \mathcal{S}$ to be admissible. The set of admissible coefficients is then $\tilde{\mathcal{A}} = \{a_S : S \in \mathcal{S}\}$. Consider the set

$$\mathcal{A} = \{a = k_2^2 \gamma + k_1^2(1 - \gamma) : \gamma \in L^\infty(\Omega),\ 0 \leq \gamma \leq 1, \},$$

which could be described as the set of all mixtures of the two materials. The set $\mathcal{A}$ is the weak$*$ L^∞ closure of $\tilde{\mathcal{A}}$. In other words, given any $a \in \mathcal{A}$, there exists a sequence $\{a_{S_n}\} \subset \tilde{\mathcal{A}}$ such that

$$\int_\Omega a_{S_n} g \to \int_\Omega a g \quad \text{as } n \to \infty$$

for each fixed $g \in L^1(\Omega)$. Such a sequence can be constructed as follows. Given $n = 1, 2, \ldots$, one can divide Ω into a uniform grid of rectangles fine enough so that there exists a step function $a_n \in \mathcal{A}$, constant on each rectangle, and satisfying $|a_n - a| < 1/n$ except on a set of measure less than $1/n$. Now choose a curve $S_n \in \mathcal{S}$ which passes through each rectangle and is shaped in such a way that $|\int_G (a_{S_n} - a_n)| < 1/(nm)$ on each rectangle G, where m is the number of rectangles. It is now straightforward to show that with $g \in L^1(\Omega)$ fixed, we have $\int_\Omega (a_{S_n} - a) g \to 0$ as $n \to \infty$.

Notice that the variational formulation of the problem allows arbitrary bounded, measurable refractive index functions. Thus the map $F(\cdot)$ is still well defined for arguments $a \in \mathcal{A}$. We can then write down a "relaxed" formulation of the design problem

$$\min_{a \in \mathcal{A}} J(a) = \frac{1}{2}\|F(a) - q\|^2. \tag{2.66}$$

THEOREM 2.8. *For sufficiently low frequencies $\omega > 0$, the relaxed optimal design problem (2.66) admits at least one solution.*

Proof. The proof follows the direct method of the calculus of variations.

Choose a minimizing sequence $\{a_n\} \subset \mathcal{A}$. The set $\mathcal{A}$ is weak* compact, so there exists a subsequence (still denoted $\{a_n\}$) and an element $a \in \mathcal{A}$ such that $a_n \rightharpoonup^* a$. Denote by u_n the solution of problem (2.61)–(2.63) corresponding to a_n. One can easily see from the proof of Theorem 2.1 that for $\|a\|$ uniformly bounded and ω^2 sufficiently small, one can bound $\|u\|_{H^1(\Omega)}$ independently of the particular function a. Since $\|u_n\|_{H^1(\Omega)}$ is uniformly bounded, there exists a subsequence (still denoted u_n), and an element $u \in H^1(\Omega)$, with $u_n \rightharpoonup u$ weakly in $H^1(\Omega)$. Holding $v \in H^1(\Omega)$ fixed, we then have $u_n \bar{v} \to u\bar{v}$ strongly in $L^1(\Omega)$. Hence,

$$\int_\Omega a_n(u_n\bar{v} - u\bar{v}) \to 0 \quad \text{as } n \to \infty.$$

Also, since $a_n \rightharpoonup^* a$, we have $\int_\Omega a_n u\bar{v} \to \int_\Omega au\bar{v}$. Denote by $B_a(u,v)$ the sesquilinear form as defined in (2.30), with k^2 replaced by a. Since $B_{a_n}(u_n, v) = 0$ for all n and each fixed v, it then follows that $B_a(u,v) = 0$ for all $v \in H^1(\Omega)$. From Theorem 2.1, u is the unique solution to the variational problem.

Since $u_n \rightharpoonup u$ weakly in $H^1(\Omega)$, the traces $u_n|_{\Gamma_1}$ and $u_n|_{\Gamma_2}$ are weakly convergent in $H^{1/2}(\Gamma_j)$ to $u|_{\Gamma_1}$ and $u|_{\Gamma_2}$, respectively. By definition of the reflection and transmission coefficient vectors r (2.64) and t (2.65) and the fact that $F = (r,t)$, it follows immediately that $F(a_n) \to F(a)$. Thus a is a minimizer to problem (2.66). □

The primary disadvantage of the relaxed optimal design approach in diffractive optics applications is that optimal structures obtained in this way can be difficult or even impossible to fabricate. The "mixtures" of materials required may not be achievable on the microscale. Some numerical examples obtained using this approach can be found in [39, 40].

Another route to obtaining a well-posed minimization problem is to restrict the class of admissible curves in order to obtain a compact admissible set. One natural restriction would be to consider only interfaces S given by the graph of a bounded function $s(x_1)$. This suggests the problem

$$\min J(s) = \frac{1}{2}\|F(a_s) - q\|^2 \quad \text{subject to } \|s\|_{L^\infty} \leq b' < b, \tag{2.67}$$

where $F(a_s)$ is the diffraction pattern corresponding to the grating profile given by the graph of s. However, without further constraints on s, one would not expect that a solution to (2.67) would exist in general, due to the possibility of oscillatory minimizing sequences. Smoothness constraints on s are not appropriate, since manufacturing processes used in diffractive optics generally produce profiles with corners.

A particular quantity that measures oscillations but does not blow up for functions with "corners" is the total variation seminorm

$$TV(s) = \sup \sum_{k=1}^n |s(x_k) - s(x_{k-1})|,$$

Chapter 2. Variational Methods for Diffractive Optics Modeling   61

where the sup is taken over all subdivisions $0 = x_0 < x_1 < \cdots < x_n = 1$. We may then consider the problem

$$\min J(s) = \frac{1}{2}\|F(s) - q\|^2,$$
$$\text{subject to} \quad TV(s) \leq M, \quad (2.68)$$
$$\|s\|_{L^\infty} \leq b' < b.$$

THEOREM 2.9. *For sufficiently low-frequency incident waves, the constrained minimization problem* (2.68) *admits a solution s of bounded variation.*

Proof. We follow the proof in [41]. Consider the admissible set $\mathcal{S} = \{s : TV(s) \leq M, \|s\|_{L^\infty} \leq b'\}$. Let $\{s_n\} \in \mathcal{S}$ be a minimizing sequence for $J(s)$. By the compactness properties of the space BV of functions of bounded variation, there exists a subsequence (still denoted $\{s_n\}$) and an element $s \in \mathcal{S}$ such that $s_n \to s$ in L^1. One can establish by direct calculation that for $1 \leq p < \infty$,

$$\|a_{s_n} - a_s\|_{L^p(\Omega)} = C\|s_n - s\|_{L^1(0,1)}^{1/p}.$$

Thus the functions a_{s_n} converge weak* L^∞ to a_s. Now the proof follows exactly as in Theorem 2.8. □

A numerical approach to solving a slight modification of the minimization (2.68) was described in [41], along with several numerical experiments. The algorithm is based on a simple gradient descent method, using adjoint-state calculations to obtain the gradients. An example of an optimal design obtained from this method is shown in Figure 2.3. The design goal was to direct all energy from a normally incident plane wave into the +1 transmitted mode. The structure shown does so with 98% efficiency and is indicative of the nonintuitive designs that can sometimes be obtained with these methods.

Here we have considered only the optimal design problem in the case of TE polarization. The more difficult TM case was studied in [14], where existence of optimal solutions is proved for a properly formulated problem.

2.4.4 Other Approaches to Optimal Design

Elschner and Schmidt [43] also took a variational approach to the optimal design of gratings, but restricted the class of admissible designs to so-called "binary" structures (such structures can be represented by the graph of a step function taking on only two values). This simplification yields a finite-dimensional minimization problem, thus avoiding some of the technical issues encountered in the more general approach described above and producing practically realizable designs.

Prather [67] has applied simulated annealing and simulated quenching algorithms for the optimization of diffractive optical elements of finite extent and with nonperiodic geometry, using boundary element and hybrid finite element–boundary element methods to solve the underlying diffraction problem. One advantage of

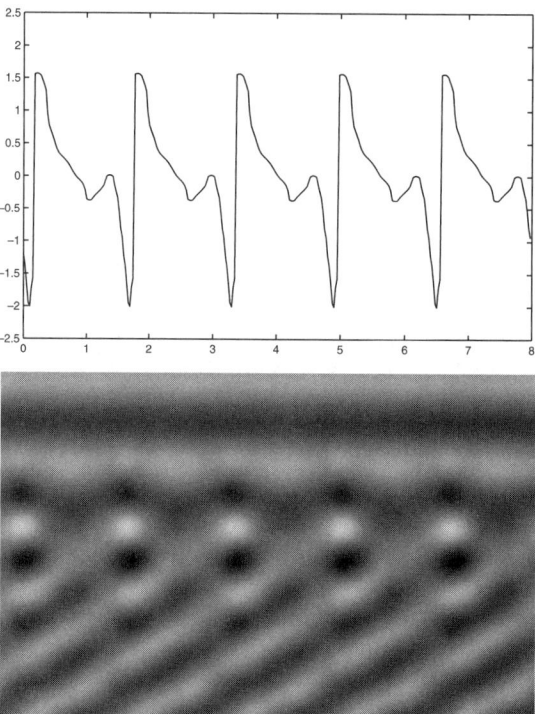

Figure 2.3: An optimal design obtained with the surface optimization formulation. Plot on the top is the curve separating materials $k_1 = 1$ above from $k_2 = 1.5$ below. Gray-scale plot on bottom shows the real part of the field u. Structure is of infinite extent—five periods are shown for clarity.

these approaches is that no derivative information on the cost functional is required. However, these algorithms are very computationally intensive.

Rudnaya, Santosa, and Chiareli [69] have recently studied the problem of designing an optical film that produces a desired intensity pattern from a plane wave source. The model is simplified by assuming the Kirchhoff approximation. The film is further assumed to alter the phase by an amount that depends only on wavelength, index of refraction, and thickness. By taking into account the manufacturing constraints, the optimal design problem is formulated as an integer programming problem with simple bounds and is solved by a variant of a generic algorithm.

2.5 Related Topics

We conclude the paper with a brief account of some related topics.

2.5.1 Finite Structures

Until now, we have only considered infinite periodic structures. In reality one can have only truncated or finite periodic structures. A natural question arises, What is the effect of the truncation on the wave fields? Direct numerical discretization of the truncated structure is possible, but is generally only computationally feasible for "small" structures with relatively few periods.

An interesting approach was proposed by Kriegsmann [61] and Kriegsmann and Scandrett [62] in the case of linear sound soft or hard acoustic media. The method was based on the analysis of different scales of the problem geometry. Suppose the structure length is L. In the near field, the scattering problem may be treated as one in an infinite periodic structure. In the far field, at a distance $\gg L$, the structure may be viewed as a bounded scatter. The idea was to derive the cross sections by matching the two expansions. The asymptotic matching techniques of Kriegsmann [61] can be applied with any method for solving the infinite periodic diffraction problem. Another important related problem is to study the optical effects when an infinite periodic structure is illuminated by an incident beam of finite width.

2.5.2 Second Harmonic Generation in Gratings

In addition to the linear structures described in sections 2.1–2.3, variational techniques can be productively applied to study certain nonlinear devices. In this section we describe a particular application to the phenomenon of "grating enhanced second harmonic generation." In physical experiments, either a nonlinear optical material is applied over a diffraction grating, or a grating structure is etched into the surface of a nonlinear optical crystal. Materials with relatively large second-order nonlinear susceptibility tensor (e.g., GaAs, GaSe, InSe) are used. When an intense "pump" beam is applied, a second harmonic field (at twice the pump frequency) is generated. Thus, for example, coherent blue light can be generated from a red pump laser. Additional physics and references on nonlinear optics may be found in [24, 70]. In most cases, the nonlinear susceptibility is very small, and the intensity of the second harmonic field is correspondingly weak. It has been found experimentally that the grating structure can significantly enhance the second harmonic conversion efficiency [68]. The enhancement occurs when the grating is operated near a resonance, for example, when a diffracted mode is directed parallel to the structure. Presumably the enhancement is due to increased pump field intensity within the nonlinear material.

Standard approaches to second harmonic generation problems rely on either the "undepleted pump approximation," which linearizes the problem and tends to lose accuracy as conversion efficiency increases [31], or the "slowly varying approximation," which tends to be inaccurate in the small structures we consider. Without making these approximations, in the simplest possible two-dimensional case the model equations can be written

$$\left[\Delta + (\omega k_1)^2\right] u = \chi_1 u^* v, \qquad (2.69)$$

$$\left[\Delta + (2\omega k_2)^2\right] v = \chi_2 u^2, \qquad (2.70)$$

where u is the "pump" field, v is the second harmonic field, and χ_1 and χ_2 are components of the second-order nonlinear susceptibility tensors. Using the same basic approach described in section 2.2.2 this system can be cast in variational form, with transparent boundary conditions similar to the linear case. Estimates involving Schauder's lemma combined with a contraction mapping argument yield the following result.

THEOREM 2.10. *Given a fixed "nonresonant" geometry and fixed incident wave, there exists a constant $\delta > 0$ such that if $\|\chi_1\|_{L^\infty}\|\chi_2\|_{L^\infty} \leq \delta$, then the problem (2.69)–(2.70), supplemented with appropriate radiation conditions, admits a unique solution $(u,v) \in H^1(\Omega)^2$.*

By "nonresonant" we mean that the solution operators correspoding to the decoupled linear problems are bounded. Thus in the physically realistic case of small nonlinear susceptibilities, the problem has a unique solution. This was established first for the case of nonlinear films in [17], then for the case of grating structures in [18]. In the same papers, a numerical approach was employed that combines finite elements with a fixed-point iteration.

The model described above is valid only for certain types of crystal structures in the nonlinear medium. If the crystal is such that the second harmonic field is generated in TM polarization, or is only excited by TM pump beams, then the model is more complicated and proving results like Theorem 2.10 is more difficult. Additional complications occur for other polarization modes and for parameter values near a resonant point. In fact, each of these complications represents cases of practical interest. Improved analytical and computational techniques are currently being developed with the hope of tackling these difficulties. See [16] for recent well-posedness results, [8] for results on nonlinear thin optical coatings, and [21] for optimal design of nonlinear films.

Acknowledgments

The research of the first author was partially supported by NSF grant DMS 98-03604, NSF University-Industry Cooperative Research Programs grants DMS 98-03809 and DMS 99-72292, and NSF Western Europe Programs grant INT 98-15798. The second author was supported by AFOSR. Effort sponsored by the Air Force Office of Scientific Research, Air Force Materiel Command, USAF, under grant F49620-95-1-0497. The U.S. government is authorized to reproduce and distribute reprints for governmental purposes notwithstanding any copyright notation herein. The views and conclusions contained herein are those of the authors and should not be interpreted as necessarily representing the official policies or endorsements, either expressed or implied, of the Air Force Office of Scientific Research or the U.S. government.

References

[1] T. Abboud, *Etude mathématique et numérique de quelques problèmes de diffraction d'ondes électromagnétiques*, Ph.D. Thesis, Ecole Polytechnique, Palaiseau, France, 1991.

[2] T. Abboud, *Electromagnetic waves in periodic media*, in Second International Conference on Mathematical and Numerical Aspects of Wave Propagation, R. Kleinman, T. Angell, D. Colton, F. Santosa, and I. Stakgold, eds., SIAM, Philadelphia, 1993, pp. 1–9.

[3] T. Abboud and J. C. Nédélec, *Electromagnetic waves in an inhomogeneous medium*, J. Math. Anal. Appl., 164 (1992), pp. 40–58.

[4] Y. Achdou, *Numerical optimization of a photocell*, Opt. Comput. Meth. Appl. Mech. Eng., 102 (1993), pp. 89–106.

[5] Y. Achdou and O. Pironneau, *Optimization of a photocell*, Optimal Control Appl. Meth., 12 (1991), pp. 221–246.

[6] R. Adams, *Sobolev Spaces*, Academic Press, New York, 1975.

[7] H. Ammari, *Uniqueness theorems for an inverse problem in a doubly periodic structure*, C. R. Acad. Sci. Paris Sér. I Math., 320 (1995), pp. 301–306.

[8] H. Ammari, G. Bao, and K. Hamdache, *The effect of thin coatings on second harmonic generation*, Electron. J. Differential Equations, 1999 (1999), pp. 1-13.

[9] G. Bao, *A uniqueness theorem for an inverse problem in periodic diffractive optics*, Inverse Problems, 10 (1994), pp. 335–340.

[10] G. Bao, *Finite elements approximation of time harmonic waves in periodic structures*, SIAM J. Numer. Anal., 32 (1995), pp. 1155–1169.

[11] G. Bao, *An inverse diffraction problem in periodic structures*, in Third International Conference on Mathematical and Numerical Aspects of Wave Propagation, G. Cohen, ed., SIAM, Philadelphia, 1995, pp. 694–704.

[12] G. Bao, *Numerical analysis of diffraction by periodic structures: TM polarization*, Numer. Math., 75 (1996), pp. 1–16.

[13] G. Bao, *Variational approximation of Maxwell's equations in biperiodic structures*, SIAM J. Appl. Math., 57 (1997), pp. 364–381.

[14] G. Bao and E. Bonnetier, *Optimal design of periodic diffractive structures in TM polarization*, Appl. Math. Optim., to appear.

[15] G. Bao, Y. Cao, and H. Yang, *Least-squares finite element computation of diffraction problems*, Math. Methods Appl. Sci., 23 (2000), pp. 1073–1092.

[16] G. Bao and Y. Chen, *A nonlinear grating problem in diffractive optics*, SIAM J. Math. Anal., 28 (1997), pp. 322–337.

[17] G. Bao and D. Dobson, *Second harmonic generation in nonlinear optical films*, J. Math. Phys., 35 (1994), pp. 1622–1633.

[18] G. Bao and D. Dobson, *Diffractive optics in nonlinear media with periodic structure*, European J. Appl. Math., 6 (1995), pp. 573–590.

[19] G. Bao and D. Dobson, *On the scattering by a biperiodic structure*, Proc. Amer. Math. Soc., 128 (2000), pp. 2715–2723.

[20] G. Bao and A. Friedman, *Inverse problems for scattering by periodic structures*, Arch. Rational Mech. Anal., 132 (1995), pp. 49–72.

[21] G. Bao and G. Li, *Optimal design in nonlinear optics*, in Encyclopedia of Optimization, P. M. Pardalos and C. A. Floudas, eds., Kluwer, Dordrecht, the Netherlands, to appear.

[22] G. Bao and Z. Zhou, *An inverse problem for scattering by a doubly periodic structure*, Trans. Amer. Math. Soc., 350 (1998), pp. 4089–4103.

[23] G. Bao and H. Yang, *A least-squares finite element analysis for diffraction problems*, SIAM J. Numer. Anal., 37 (2000), pp. 665–682.

[24] N. Bloembergen, *Nonlinear Optics*, Benjamin, New York, 1965.

[25] P. B. Bochev and M. D. Gunzburger, *Finite element methods of least-squares type*, SIAM Rev., 40 (1998), pp. 798–837.

[26] A. Bonnet-Ben Dhia and F. Starling, *Guided waves by electromagnetic gratings and non-uniqueness examples for the diffraction problem*, Math. Methods Appl. Sci., 17 (1994), pp. 305–338.

[27] M. Born and E. Wolf, *Principles of Optics*, 6th ed., Pergamon Press, Oxford, UK, 1980.

[28] J. H. Bramble, R. Lazarov, and J. E. Pasciak, *A least-squares approach based on a discrete minus one inner product for first order systems*, Math. Comp., 219 (1997), pp. 935–955.

[29] A. Brandt, *Multi-level adaptive solutions to boundary value problems*, Math. Comp., 31 (1977), pp. 333–390.

[30] S. C. Brenner and L. R. Scott, *The Mathematical Theory of Finite Element Methods*, Springer-Verlag, New York, 1994.

[31] E. Bringuier, A. Bourdon, N. Piccioli, and A. Chevy, *Optical second-harmonic generation in lossy media: Application to GaSe and InSe*, Phys. Rev. B, 49 (1994), pp. 16971–16982.

[32] O. Bruno and F. Reitich, *Numerical solution of diffraction problems: A method of variation of boundaries; II. Dielectric gratings, Pade approximants and singularities; III. Doubly-periodic gratings*, J. Opt. Soc. Amer. A., 10 (1993), pp. 1168–1175, 2307–2317, 2551–2562.

[33] O. Bruno and F. Reitich, *Accurate calculation of diffractive grating efficiencies*, in Mathematics in Smart Structures, H. T. Banks, ed., Proc. SPIE 1919, Bellingham, WA, 1993, pp. 236–247.

Chapter 2. Variational Methods for Diffractive Optics Modeling

[34] X. Chen and A. Friedman, *Maxwell's equations in a periodic structure*, Trans. Amer. Math. Soc., 323 (1991), pp. 465–507.

[35] P. G. Ciarlet, *The Finite Element Method for Elliptic Problems*, North-Holland, Amsterdam, 1978.

[36] D. Colton and R. Kress, *Inverse Acoustic and Electromagnetic Scattering Theory*, 2nd ed., Springer-Verlag, New York, 1998.

[37] J. A. Cox, D. C. Dobson, T. Ohnstein, and Z. D. Zook, *Optical performance of high-aspect LIGA gratings* II, Optical Engineering, 37 (1998), pp. 2878–2884.

[38] J. A. Cox, Z. D. Zook, T. Ohnstein, and D. C. Dobson, *Optical performance of high-aspect LIGA gratings*, Optical Engineering, 36 (1997), pp. 1367–1373.

[39] D. Dobson, *Optimal design of periodic antireflective structures for the Helmholtz equation*, European J. Appl. Math., 4 (1993), pp. 321–340.

[40] D. Dobson, *Exploiting ill-posedness in the design of diffractive optical structures*, in Mathematics in Smart Structures, H. T. Banks, ed., Proc. SPIE 1919, Bellingham, WA, 1993, pp. 248–257.

[41] D. Dobson, *Optimal shape design of blazed diffraction gratings*, Appl. Math. Optim., 40 (1999), pp. 61–78.

[42] D. Dobson and A. Friedman, *The time-harmonic Maxwell equations in a doubly periodic structure*, J. Math. Anal. Appl., 166 (1992), pp. 507–528.

[43] J. Elschner and G. Schmidt, *Analysis and Numerics for the Optimal Design of Binary Diffractive Gratings*, Preprint No. 323, Weierstrass Institute for Applied Analysis and Stochastics, Berlin, 1997.

[44] B. Engquist and A. Majda, *Absorbing boundary conditions for the numerical simulation of waves*, Math. Comp., 31 (1971), pp. 629–655.

[45] B. Engquist and A. Majda, *Radiation boundary conditions for acoustic and elastic wave calculations*, Comm. Pure Appl. Math., 32 (1979), pp. 313–375.

[46] M. W. Farn, *New iterative algorithm for the design of phase-only gratings*, in Computer and Optically Generated Holographic Optics, I. N. Cindrich and S. Lee, eds., Proc. SPIE 1555, Bellingham, WA, 1991, pp. 34–42.

[47] A. Friedman, *Mathematics in Industrial Problems*, IMA Vol. Math. Appl. 16, Springer-Verlag, New York, 1988.

[48] A. Friedman, *Mathematics in Industrial Problems, Part 3*, IMA Vol. Math. Appl. 38, Springer-Verlag, New York, 1991.

[49] A. Friedman, *Mathematics in Industrial Problems, Part 7*, IMA Vol. Math. Appl. 67, Springer-Verlag, New York, 1995.

[50] T. K. Gaylord and M. G. Moharam, *Analysis and applications of optical diffraction by gratings*, IEEE Proc., 73 (1985), pp. 894–937.

[51] R. W. Gerchberg and W. O. Saxton, *A practical algorithm for the determination of phase from image and diffraction plane pictures*, Optik, 35 (1972), pp. 237–246.

[52] D. Givoli, *Non-reflecting boundary conditions: a review*, J. Comp. Phys., 94 (1991), pp. 1–29.

[53] C. Goldstein, *A finite element method for solving Helmholtz type equations in waveguides and other unbounded domains*, Math. Comp., 39 (1982), pp. 309–324.

[54] Y. Han and C. A. Delisle, *Exact Surface Relief Profile of Kinoform Lenses from a Given Phase Function,* Proc. SPIE, 2152, Bellingham, WA, 1994.

[55] F. Hettlich and A. Kirsch, *Schiffer's theorem in inverse scattering theory for periodic structures*, Inverse Problems, 13 (1997), pp. 351–361.

[56] V. Isakov, *Inverse Problems for Partial Differential Equations*, Springer-Verlag, New York, 1998.

[57] T. Kato, *Perturbation Theory for Linear Operators*, corrected 2nd ed., Springer-Verlag, Berlin, 1980.

[58] A. Kirsch, *Diffraction by periodic structures*, in Proc. Lapland Conf. on Inverse Problems, L. Pävarinta and E. Somersalo, eds., Springer-Verlag, 1993, pp. 87–102.

[59] A. Kirsch, *Uniqueness theorems in inverse scattering theory for periodic structures*, Inverse Problems, 10 (1994), pp. 145–152.

[60] R. Kohn and G. Strang, *Optimal design and relaxation of variational problems I, II, III*, Comm. Pure Appl. Math., 39 (1986), pp. 113–137, 139–182, 353–377.

[61] G. A. Kriegsmann, *Scattering by acoustically large corrugated planar surfaces*, J. Acoust. Soc. Am., 88 (1990), pp. 492–495.

[62] G. A. Kriegsmann and C. L. Scandrett, *Large membrane array scattering*, J. Acoust. Soc. Am., 93 (1993), pp. 3043-3048.

[63] L. Li, *A model analysis of lamellar diffraction gratings in conical mountings*, J. Mod. Opt., 40 (1993), pp. 553–573.

[64] S. McCormick, *Multigrid Methods*, Frontiers in Applied Mathematics 3, SIAM, Philadelphia, 1987.

[65] J. C. Nédélec and F. Starling, *Integral equation methods in a quasi-periodic diffraction problem for the time-harmonic Maxwell's equations*, SIAM J. Math. Anal., 22 (1991), pp. 1679–1701.

[66] R. Petit, ed., *Electromagnetic Theory of Gratings,* Topics in Current Physics 22, Springer-Verlag, Heidelberg, 1980.

[67] D. W. Prather, *Analysis and Synthesis of Finite Aperiodic Diffractive Optical Elements Using Rigorous Electromagnetic Models*, Ph.D. Thesis, Department of Electrical Engineering, University of Maryland, College Park, 1997.

[68] R. Reinisch, M. Nevière, H. Akhouayri, J. Coutaz, D. Maystre, and E. Pic, *Grating enhanced second harmonic generation through electromagnetic resonances,* Opt. Eng., 27 (1988), pp. 961–971.

[69] S. Rudnaya, F. Santosa, and A. Chiareli, *Optimal design of a diffractive optical element,* in the Proceedings of Fourth International Conference on Mathematical and Numerical Aspects of Wave Propagation, J. DeSanto, ed., SIAM, Philadelphia, 1998, pp. 44–53.

[70] Y. R. Shen, *The Principles of Nonlinear Optics,* Wiley, New York, 1984.

[71] C. H. Wilcox, *Scattering Theory for Diffraction Gratings,* Appl. Math. Sci., 46, Springer-Verlag, New York, 1984.

Chapter 3

High-Order Boundary Perturbation Methods

Oscar P. Bruno and Fernando Reitich

3.1 Introduction

Perturbation theory is among the most useful and successful analytical tools in applied mathematics. Countless examples of enlightening perturbation analyses have been performed for a wide variety of models in areas ranging from fluid, solid, and quantum mechanics to chemical kinetics and physiology. The field of electromagnetic and acoustic wave propagation is certainly no exception. Many studies of these processes have been based on perturbative calculations where the role of the variation parameter has been played by the wavelength of radiation, material constants, or geometric characteristics. It is this latter instance of geometric perturbations in problems of wave propagation that we shall review in the present chapter.

Use of geometric perturbation theory is advantageous in the treatment of configurations which, however complex, can be viewed as deviations from simpler ones—those for which solutions are known or can be obtained easily. Many uses of such methods exist, including, among others, applications to optics, oceanic and terrain scattering, SAR imaging and remote sensing, and diffraction from ablated, eroded, or deformed objects; see, e.g., [47, 52, 56, 59, 62]. The analysis of the scattering processes involved in such applications poses challenging computational problems that require resolution of the interplay between highly oscillatory waves and interfaces. In the case of oceanic scattering, for instance, nonlinear water wave interactions and capillarity effects give rise to highly oscillatory modulated wave trains that are responsible for the most substantial portions of the scattering returns [35]. Similarly, diffraction gratings owe their remarkable optical properties to their submicron features, which are designed to resonate with the incident radiation [40].

The mathematical complexities associated with these problems have historically been tackled by a variety of numerical algorithms, including methods based on integral equations, differential formalisms, Rayleigh expansions, and finite differences or finite elements. These methods can consistently provide accuracies of the order of 1% of the incident energy [38]. Classical uses of perturbation theory in these contexts have been limited to low-order methods which provide similar accuracies for problems containing small deviations. The new uses of perturbation theory that we review in this chapter, on the other hand, rely on expansions of very high order in powers of a deviation parameter, denoted here by δ, and techniques of analytic continuation in the complex δ-plane. Specifically, Taylor series for the field quantities are obtained through differentiation of the Maxwell system with respect to δ. The possible (and common) divergence of the resulting series is handled through resummation techniques that exploit the analytic structure of the solution.

The resulting algorithms can resolve scattering returns with accuracies that are several orders of magnitude better than those given by classical methods. Such accuracies can play an important role in applications. For instance, the fine resolution provided by our algorithms has recently helped settle a longstanding controversy relating to polarized back-scattering returns from rough surfaces, which amount to very small fractions of the incident energy [55].

The advantages of the use of boundary perturbation methods for calculating scattering cross sections in problems of electromagnetic and acoustic wave propagation have been recognized for several decades, since the first-order calculation of Rayleigh [50]. Besides the simplicity of their implementations, perturbation approaches generally lead, quite efficiently, to very accurate results in their domain of applicability. Indeed, it was these characteristics that prompted a number of investigations in the last 30 years, mainly in the area of scattering by corrugated surfaces, and which resulted in a variety of low-order theories [1, 2, 25, 26, 32, 44, 62]. These, of course, are limited to fairly small departures from an exactly solvable geometry, and in particular, they cannot be applied to scatterers in the resonance regime, where the wavelength is comparable to a characteristic length of the scatterer [39, 57]. In an attempt to overcome this limitation, a high-order method was proposed by Lopez, Yndurain, and Garcia [36] in the context of atom scattering from crystal surfaces and later extended to the reflection of sonic and electromagnetic waves by a number of authors [22, 23, 24, 27, 37, 53, 64]. Interestingly, these advances led to an animated debate among researchers in the area regarding the validity of series expansions in a surface-roughness parameter. In fact, even though the higher order methods did apparently extend the domain of applicability of the perturbative approach in some cases, their convergence properties remained a mystery.

This was due, in part, to the lack of understanding of the changes in the fields upon boundary variations. For instance, Lopez, Yndurain, and Garcia assumed that the field scattered by a sinusoidal surface presented a singularity at a real (and negative) value of the surface height [36, p. 972]. Greffet et al. [22, 23, 24], on the other hand, have argued that the series expansions are restricted to the domain of validity of the well-known Rayleigh hypothesis [38] (see also section 3.2.2) and offered this as an explanation for the limitations in the amplitudes of the perturbations for which acceptable convergence was observed. Jackson, Winebrenner, and Ishimaru presented a different view when they suggested that such series may constitute "an asymptotic expansion, valid in the limit as the roughness goes to zero" [27, p. 968]. Actually, a related conjecture had been previously advanced by Uretsky [58] in the wake of a controversial paper by Meecham [42]. Indeed, although Meecham's approach was never implemented numerically, he was perhaps the first to suggest a series representation in powers of the surface roughness. Uretsky objected to Meecham's approach and, in connection with series expansions, wrote that "it is argued that there is no circle of convergence around the origin..." [58, p. 411]. Based on the fact that the kernels in every term of the Neumann series contain branch point singularities in the perturbation variable, he went on to claim that "these considerations suggest, in fact, that the solution for the sinusoidal surface does not continue analytically to the solution for a flat surface and that we must be wary of power-series expansions in any of the parameters of the problem."

It was only a few years ago that the controversy was finally resolved when we established [7] that electromagnetic and acoustic fields *do vary analytically* with respect to variations of a scattering surface and that, in fact, *no singularities* are present for real values of the perturbation parameter (see section 3.3). Our approach relied on a potential-theoretic formulation of the mathematical problem in a *holomorphic framework* which allowed us to show, in connection with Uretsky's objection, that in spite of the occurrence of Hankel functions and associated logarithmic branch points in the iterated integrals of the Neumann series, *the integrals themselves do not contain such singularities*. In the case of analytic surfaces, these results and their nontrivial extensions to three dimensions [6] guarantee the *joint* analyticity of the fields in the spatial and parameter variables. This is to be contrasted with the classical analyticity results of Calderón [14] and Coifman, MacIntosh, and Meyer [16], which relate only to the parametric dependence and which *do not* ensure the combined holomorphic regularity. Our search for such a result, on the other hand, was motivated by the need to rigorously justify the successive differentiation of the boundary conditions of the field with respect to the perturbation parameter at the varying interface. Indeed, such differentiations result in formulas that allow for the recursive calculation of the series expansion of the fields through the solution of a sequence of scattering problems on the unperturbed geometry and that could therefore be made into the basis of a numerical algorithm (see section 3.3.2).

Besides the independent interest of our theoretical results, they also provided an explanation for the (limited) performance that numerical algorithms based on boundary variations had presented until then [22, 23, 24, 27, 36, 37, 53, 64]. Indeed, we showed that in general these limitations were related to the smallness of the radius of convergence of the perturbation series, which prevented its use in many cases of practical importance, such as in the study of highly modulated diffractive gratings. More importantly, our theory also suggested possible means to enlarge the domain of applicability of such methods. In fact, it follows from our results that in the case of (two- or three-dimensional) periodic gratings, for instance, the scattered fields can be extended analytically to *a whole neighborhood of the real line in the complex plane of the roughness parameter*. This observation, in turn, implies that the restrictions on the convergence of the Taylor series are solely due to an unfavorable arrangement of the singularities of the fields as functions of such parameters. We therefore conjectured that their suitable rearrangement could allow for the analytic continuation of the series beyond its disk of convergence. And, in this regard, we subsequently established that such continuations may, in fact, be accurately and effectively performed through classical Padé approximation [3] or through our own summation mechanism based on conformal transformations [8, 11] (see section 3.4).

The purpose of this chapter is to present a detailed account of our high-order boundary perturbation theory and of the outcome of the resulting numerical codes. As we shall see, the capabilities of these algorithms extend well beyond those of classical perturbation procedures, and in fact they may provide, in many cases, results of an accuracy unparalleled by that given by other methods. Most of the results we present here have appeared in our papers [7, 8, 9, 10, 11, 12, 13] (an exception are those of section 3.5.3 on resonant mode calculations, which are the subject of ongoing investigations). We have attempted, however, to make this chapter self-contained, and in this spirit, our discussion begins in the following section with a number of preliminary remarks on Maxwell's equations, eigenfunction expansions (separation of variables), and far-field representations. Subsequently, in section 3.3, we present a precise description of our theoretical results on analyticity of scattered fields and of the derivation of the recursion associated with the calculation of their Taylor series representations. As we said, the resulting series may only converge in a limited regime but, as we now know, it can be continued analytically beyond its domain of convergence. Knowledge of the full Taylor series of course completely determines its analytic continuation; numerically stable and efficient methods for the calculation of analytic extensions of power series are presented in section 3.4. Finally, in section 3.5 we present numerical results for a variety of two- and three-dimensional scattering configurations, and we discuss some recent results extending our theory to evaluation of cavity eigenvalues.

3.2 Preliminaries

3.2.1 Maxwell's Equations

Consider a scattering configuration in which space is divided into two regions Ω^+ and Ω^- containing two different materials, such as air and a metal, of respective permittivities ϵ^+ and ϵ^-. The permeability of both materials is assumed to equal μ_0, the permeability of vacuum. In the cases we consider, the region Ω^+ is of infinite extent; the scatterer Ω^-, on the other hand, may be bounded or unbounded.

We wish to determine the pattern of diffraction that occurs when an electromagnetic plane wave

$$\vec{E}^i = \vec{A} e^{i(\alpha x_1 + \beta x_2 - \gamma x_3 - i\omega t)},$$
$$\vec{H}^i = \vec{B} e^{i(\alpha x_1 + \beta x_2 - \gamma x_3 - i\omega t)}$$

Chapter 3. High-Order Boundary Perturbation Methods

impinges upon Ω^-. Here, denoting by $\vec{k} = (\alpha, \beta, -\gamma)$ the wavevector, we have

$$\vec{A} \cdot \vec{k} = 0 \text{ and } \vec{B} = \frac{1}{\omega \mu_0} \vec{k} \times \vec{A}.$$

Dropping the factor $e^{-i\omega t}$, the time-harmonic Maxwell equations for the total fields read

$$\nabla \times \vec{E} = i\omega \mu_0 \vec{H}, \quad \nabla \cdot \vec{E} = 0,$$
$$\nabla \times \vec{H} = -i\omega \epsilon \vec{E}, \quad \nabla \cdot \vec{H} = 0. \tag{3.1}$$

In particular, the electromagnetic field

$$v = (\vec{E}, \vec{H})$$

satisfies the Helmholtz equations

$$\Delta v + (k^\pm)^2 v = 0 \text{ in } \Omega^\pm, \tag{3.2}$$

where $k^\pm = \omega \sqrt{\mu_0 \epsilon^\pm}$. The total electric and magnetic fields are given by

$$\vec{E} = \vec{E}^{out} = \vec{E}^i + \vec{E}^+, \quad \vec{H} = \vec{H}^{out} = \vec{H}^i + \vec{H}^+ \text{ in } \Omega^+ \text{ and}$$
$$\vec{E} = \vec{E}^{in} = \vec{E}^-, \quad \vec{H} = \vec{H}^{in} = \vec{H}^- \text{ in } \Omega^-,$$

where $(\vec{E}^+, \vec{H}^+)$ and $(\vec{E}^-, \vec{H}^-)$ are the reflected and refracted fields, respectively. At the interface

$$\Gamma = \partial \Omega^+ = \partial \Omega^-,$$

the field satisfies the transmission conditions

$$n \times (\vec{E}^{out} - \vec{E}^{in}) = 0, \quad n \times (\vec{H}^{out} - \vec{H}^{in}) = 0 \text{ on } \Gamma, \tag{3.3}$$

where n is normal to Γ. In case the region Ω^- is filled by a perfect conductor, the refracted fields vanish and the boundary conditions reduce to

$$n \times \vec{E}^{out} = n \times (\vec{E}^i + \vec{E}^+) = 0 \text{ on } \Gamma. \tag{3.4}$$

Finally, the field satisfies conditions of radiation at infinity, expressing the outgoing character of the scattered waves, which can be stated either in terms of the eigenfunction expansions of section 3.2.2, or, alternatively, in terms of the decay of the field at infinity; see, e.g., [4, 29, 48].

In the two-dimensional case in which Γ and the fields $\vec{E}, \vec{H}$ are independent of x_2 (and $\beta = 0$), the system of equations (3.1), (3.3) (or (3.1), (3.4)) can be reduced to a pair of decoupled equations for two scalar unknowns [38]. Indeed, the functions $u_1(x_1, x_3)$ and $u_2(x_1, x_3)$ equal to the transverse components E_{x_2} (field transverse electric, TE) and H_{x_2} (field transverse magnetic, TM), which satisfy (3.2), completely determine the electromagnetic field through (3.1). The boundary conditions (3.3), (3.4) can be translated into appropriate boundary conditions for the unknowns u_i. When Ω^- contains a perfect conductor, we have

$$u_1 = -e^{i\alpha x_1 - i\gamma x_3} \quad \text{and}$$

$$\frac{\partial u_2}{\partial n} = -\frac{\partial}{\partial n}\left(e^{i\alpha x_1 - i\gamma x_3}\right) \quad \text{on } \Gamma.$$

When Ω^- contains a finitely conducting metal or dielectric u_1 satisfies the transmission conditions

$$u_1^+ - u_1^- = -e^{i\alpha x_1 - i\gamma x_3} \quad \text{and}$$
$$\frac{\partial u_1^+}{\partial n} - \frac{\partial u_1^-}{\partial n} = -\frac{\partial}{\partial n}\left(e^{i\alpha x_1 - i\gamma x_3}\right) \quad \text{on } \Gamma,$$

while u_2 satisfies

$$u_2^+ - u_2^- = -e^{i\alpha x_1 - i\gamma x_3} \quad \text{and}$$
$$\frac{\partial u_2^+}{\partial n} - \left(\frac{k^+}{k^-}\right)^2 \frac{\partial u_2^-}{\partial n} = -\frac{\partial}{\partial n}\left(e^{i\alpha x_1 - i\gamma x_3}\right) \quad \text{on } \Gamma.$$

3.2.2 Eigenfunction Expansions

In addition to Taylor series, our analytic approach is based on the series expansions of the electromagnetic field which result from separation of variables. Such expansions are most frequently found in solutions associated with simple objects such as a circle, a sphere, or a semispace. This is in part due to the fact that, for such simple scatterers, the functions resulting from restriction of the separated solutions to the scattering boundaries form a complete orthonormal system, and thus boundary conditions can easily be accounted for by means of Fourier analysis.

It is interesting to note, however, that expansions in series of separated variables may be useful even when their restrictions to the boundary of the scatterer do not form an orthogonal system. The first occurrence of an approach of this type can be found in the work of Rayleigh [50]. After evaluating such expansions at the scattering boundaries, Rayleigh used appropriate approximations and found first-order corrections to the scattered field for geometries that result from small perturbations from an exactly solvable one. With the advent of computers attempts were made to extend Rayleigh's approach of evaluating series expansions at the boundary of the obstacles to general scattering solvers which do not assume small departures from an exact geometry. These attempts did not succeed since, indeed, the series may not converge at the obstacle boundaries; that is, *Rayleigh's hypothesis* may not be satisfied. This fact was first established by Petit and Cadilhac [49] by consideration of a sinusoidal corrugation on a plane.

Our method is not unrelated to Rayleigh's hypothesis. In fact, we established [7] the convergence of the eigenfunction expansions throughout the boundary for sufficiently small but otherwise arbitrary (analytic) perturbations of the exactly solvable

Chapter 3. High-Order Boundary Perturbation Methods

geometry. Whereas these "sufficiently small" perturbations for which Rayleigh's hypothesis can be used may be too small, they are certainly sufficient to allow for calculation of derivatives of any order with respect to boundary perturbations. Extension to the large perturbations that appear in practice can then be achieved by means of analytic continuation (cf. section 3.4).

The series expansions obtained from separation of variables are well known. For example, a solution to the two-dimensional Helmholtz equation outside a circular cylinder is given, in polar coordinates, by an expansion of the form

$$u(\rho, \theta) = \sum_{r=-\infty}^{\infty} B_r (-i)^r H_r^{(1)}(k\rho) e^{ir\theta}, \qquad (3.5)$$

where $H_r^{(1)}$ denotes the first Hankel function of order r. The principle of conservation of energy can be given a simple form in terms of the amplitudes B_r in this expansion. Indeed, any solution u to a scattering problem from a perfectly conducting obstacle *of any shape* admits a representation (3.5) outside a cylinder containing the scatterer, and we have

$$\sum_r |B_r|^2 + \text{Re}\left(\sum_r B_r\right) = 0. \qquad (3.6)$$

Relation (3.6), which holds independently of whether or not the series (3.5) converges at the boundary of the obstacle, can be made into a useful estimator for errors in the numerical calculation of the fields; see section 3.5.

For a solution in three-dimensional space and outside a sphere we have

$$\vec{E}^+(R, \theta, \phi) = \sum_{r=0}^{\infty} \sum_{s=-r}^{r} \vec{B}_{rs} h_r^{(1)}(kR) P_r^s(\cos(\theta)) e^{is\phi},$$

where (R, θ, ϕ) are spherical coordinates, P_r^s are the Legendre functions of the first kind, and $h_r^{(1)}$ are the first spherical Hankel functions [29].

Finally, let us consider scatterers that are given by a biperiodic corrugation of a plane

$$\Omega^- = \{x_3 < f(x_1, x_2)\}, \qquad (3.7)$$

where f is a biperiodic function of periods d_1 and d_2 in the variables x_1 and x_2, respectively. These configurations arise naturally in optics applications in the form of diffraction gratings designed, for instance, to alter (reflect, absorb) incident light at specific wavelengths. In this case, the periodicity of the structure implies that the fields must be (α, β) quasi-periodic; i.e., they must verify equations of the form

$$v(x_1 + d_1, x_2, x_3) = e^{i\alpha d_1} v(x_1, x_2, x_3) \text{ and } v(x_1, x_2 + d_2, x_3) = e^{i\beta d_2} v(x_1, x_2, x_3).$$

Then, separation of variables leads to expansions of the form

$$\vec{E}^\pm = \sum_{r=-\infty}^{\infty} \sum_{s=-\infty}^{\infty} \vec{B}_{r,s}^\pm e^{i\alpha_r x_1 + i\beta_s x_2 \pm i\gamma_{r,s}^\pm x_3}.$$

The expansion for $\vec{E}^+$ (respectively, $\vec{E}^-$) converges to the field in the region $\{x_3 > \max(f(x_1, x_2))\}$ (respectively, $\{x_3 < \min(f(x_1, x_2))\}$). Here we have put

$$\alpha_r = \alpha + rK_1, \quad \beta_s = \beta + sK_2, \quad \alpha_r^2 + \beta_s^2 + (\gamma_{r,s}^\pm)^2 = (k^\pm)^2, \tag{3.8}$$

where $\gamma_{r,s}^\pm$ is determined by $\mathrm{Im}(\gamma_{r,s}^\pm) > 0$ or $\gamma_{r,s}^\pm \geq 0$,

$$(k^\pm)^2 = \omega^2 \epsilon^\pm \mu_0,$$

and

$$K_1 = \frac{2\pi}{d_1}, \quad K_2 = \frac{2\pi}{d_2}.$$

It is clear from (3.8) that only a finite number of modes propagate away from the grating, since the remaining modes decay exponentially. The main quantities of interest here are the *grating efficiencies*

$$e_{r,s}^\pm = \frac{|B_{r,s}^\pm|^2 \gamma_{r,s}^\pm}{\gamma_{0,0}^+}$$

for the finitely many propagating modes, i.e., the modes (r,s) such that $\gamma_{r,s}^\pm$ is real. The principle of conservation of energy can be stated as follows: if we let $U^\pm$ denote the set of indices corresponding to the nonevanescent modes in $\Omega^\pm$, then

$$\sum_{(r,s)\in U^+} e_{r,s}^+ + \sum_{(r,s)\in U^-} e_{r,s}^- = 1, \tag{3.9}$$

provided the dielectric constants ϵ^+ and ϵ^- are real.

In the two-dimensional case we shall consider gratings of the form

$$\Omega^- = \{x_3 < f(x_1)\},$$

for which the expansion above reduces to

$$u^\pm = \sum_{r=-\infty}^{\infty} B_r^\pm e^{i\alpha_r x_1 \pm \gamma_r^\pm x_3}. \tag{3.10}$$

The principle of conservation of energy now reads

$$\sum_{r\in U^+} e_r^+ + \sum_{r\in U^-} e_r^- = 1, \tag{3.11}$$

where the efficiencies are now given by $e_r^\pm = \gamma_r^\pm |B_r^\pm|^2 / \gamma_0^+$.

3.3 Analyticity and Taylor Coefficients

3.3.1 Overview

As we said, our algorithms are based on a theorem of analyticity of the electromagnetic field with respect to boundary variations [7] (see also [6]). To describe our

Chapter 3. High-Order Boundary Perturbation Methods

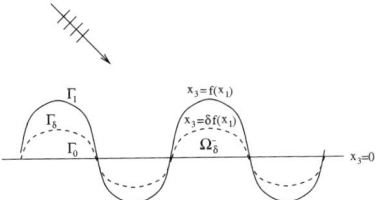

Figure 3.1: The geometry.

results assume Ω_δ^- is a family of scatterers, one for each value of the real parameter δ; see, e.g., Figure 3.1 and equation (3.14) below. Further, assume that the boundaries Γ_δ of these obstacles admit a parameterization

$$\vec{r} = H(s, \delta), \tag{3.12}$$

where the function H is jointly analytic in the spatial parameter s ($s = (s_1, s_2)$ for two-dimensional scattering interfaces) and the perturbation parameter δ. Our theorems state that both the values $v = v(\vec{r}, \delta)$ of the electromagnetic field at a fixed point in space as well as the values at a point *on the varying boundary* depend analytically on the boundary variations. More precisely, if $\vec{r}$ is a point in space away from Γ_δ and $\vec{r}_\delta \in \Gamma_\delta$ is a point on the interface which varies analytically with δ, then $v(\vec{r}, \delta)$ is jointly analytic in $(\vec{r}, \delta)$, and $v(\vec{r}_\delta, \delta)$ is an analytic function of δ for all real values of δ for which the surface (3.12) does not self-intersect.

It follows from these theorems that the field can be expanded in a series in powers of δ,

$$v^\pm(\vec{r}, \delta) = \sum_{n=0}^\infty v_n^\pm(\vec{r}) \delta^n, \tag{3.13}$$

which converges for δ small enough. The vector field $v_n^\pm$ satisfies Maxwell's equations (3.1) as well as conditions of radiation at infinity. They also satisfy boundary conditions on $\Gamma_0 = \Gamma_\delta|_{\delta=0}$, which can be obtained by differentiation, as we show below. Such differentiations and use of the chain rule are permissible, as it follows from the analyticity theorems mentioned above and related extension theorems [6, 7]. The solution of the boundary value problems for the v_n's then easily leads to a numerical algorithm for the calculation of the scattered field.

3.3.2 Recursive Formulas

In this section we derive the recursive formulas that allow for the successive calculation of the coefficient functions $v_n^\pm$ in (3.13). The general validity of such formulas for arbitrary scattering configurations follows from the results quoted above. On the other hand, the most convenient form of the recursion for numerical implementation does, of course, depend on the geometric arrangements of interest. For instance, while Cartesian coordinates are best suited for calculations on diffraction gratings when these are viewed as modulations of a plane, polar or spherical coordinates constitute the clear choice when dealing with bounded scatterers as perturbations of circular cylinders and spheres, respectively. In every case these formulas, together with appropriate mechanisms of analytic continuation (see section 3.4), will form the basis of our numerical approach.

Let us first consider the simplest case of a family of two-dimensional perfectly conducting and d-periodic diffraction gratings

$$\Omega_\delta^- = \{x_3 < \delta f(x_1)\}, \tag{3.14}$$

for which the far-field eigenfunction expansions take the form (cf. (3.10))

$$u(x_1, x_3, \delta) = u^+(x_1, x_3, \delta) = \sum_{r=-\infty}^{\infty} B_r(\delta) e^{i\alpha_r x_1 + \gamma_r x_3} \tag{3.15}$$

with

$$\alpha_r = \alpha + \frac{2\pi}{d} r \quad \text{and} \quad \alpha_r^2 + \gamma_r^2 = k^2.$$

From section 3.3.1 we have an expansion

$$u(x_1, x_3, \delta) = \sum_{n=0}^{\infty} u_n(x_1, x_3) \delta^n,$$

which converges for sufficiently small values of δ. The functions

$$u_n(x_1, x_3) = \frac{1}{n!} \frac{\partial^n u}{\partial \delta^n}(x_1, x_3, 0)$$

satisfy the Helmholtz equation

$$\Delta u_n + k^2 u_n = 0 \text{ in } \{(x_1, x_3) : x_3 > 0\} \tag{3.16}$$

and conditions of radiation at infinity. Thus, an expansion analogous to (3.15) holds for u_n itself,

$$u_n(x_1, x_3) = \sum_{r=-\infty}^{\infty} d_{n,r} e^{i\alpha_r x_1 + i\gamma_r x_3}. \tag{3.17}$$

We clearly have

$$d_{n,r} = \frac{1}{n!} \frac{d^n B_r}{d\delta^n}\bigg|_{\delta=0},$$

so that the amplitudes of the various modes are given by the Taylor series

$$B_r(\delta) = \sum_{n=0}^{\infty} d_{n,r} \delta^n. \tag{3.18}$$

Chapter 3. High-Order Boundary Perturbation Methods   81

Our approach is based on evaluation of the Taylor series (3.18), whose coefficients $d_{n,r}$ can be obtained recursively. Such recursive formulas follow from the *explicit* successive solution of (3.16) subject to appropriate boundary conditions on the plane $x_3 = 0$. These conditions can be easily derived from the values of the field at the scattering surfaces which, in this case, may be written in the form

$$u(x_1, \delta f(x_1), \delta) = -e^{i\alpha x_1 - i\gamma \delta f(x_1)}.$$

Indeed, differentiation of this relation at $\delta = 0$ simply gives

$$u_n(x_1, 0) = -\frac{(-i\gamma f(x_1))^n}{n!} e^{i\alpha x_1} - \sum_{l=0}^{n-1} \frac{f(x_1)^{n-l}}{(n-l)!} \frac{\partial^{n-l} u_l}{\partial x_3^{n-l}}(x_1, 0). \qquad (3.19)$$

Now, let the Fourier series of the function $f(x_1)$ be given by

$$f(x_1) = \sum_{r=-F}^{F} C_{1,r} e^{i2\pi r x_1/d}$$

with either finite or infinite F. Then, substitution in (3.19) of u_l ($0 \le l \le n$) and their x_3-derivatives as calculated from (3.17) permits us to find all the coefficients $d_{n,r}$ in terms of the coefficients $d_{k,r}$ with $k < n$ and the Fourier series coefficients $C_{l,r}$ of the function $f(x_1)^l/l!$:

$$\frac{f(x_1)^l}{l!} = \sum_{r=-lF}^{lF} C_{l,r} e^{i2\pi r x_1/d}.$$

In fact, from (3.19) we have

$$\sum_{r=-\infty}^{\infty} d_{n,r} e^{i\alpha_r x_1} = -(-i\gamma)^n \left(\sum_{r=-nF}^{nF} C_{n,r} e^{i\alpha_r x_1} \right)$$
$$- \sum_{l=0}^{n-1} \left(\sum_{s=-(n-l)F}^{(n-l)F} C_{n-l,s} e^{i2\pi s x_1/d} \right) \left(\sum_{q=-\infty}^{\infty} d_{l,q} (i\gamma_q)^{n-l} e^{i\alpha_q x_1} \right),$$

or equivalently,

$$\sum_{r=-\infty}^{\infty} d_{n,r} e^{i\alpha_r x_1} = -\sum_{r=-nF}^{nF} (-i\gamma)^n C_{n,r} e^{i\alpha_r x_1}$$
$$- \sum_{l=0}^{n-1} \sum_{s=-(n-l)F}^{(n-l)F} \sum_{q=-\infty}^{\infty} C_{n-l,s} d_{l,q} (i\gamma_q)^{n-l} e^{i\alpha_{s+q} x_1}$$
$$= -\sum_{r=-nF}^{nF} (-i\gamma)^n C_{n,r} e^{i\alpha_r x_1}$$
$$- \sum_{l=0}^{n-1} \sum_{q=-\infty}^{\infty} \sum_{r=q-(n-l)F}^{q+(n-l)F} C_{n-l,r-q} d_{l,q} (i\gamma_q)^{n-l} e^{i\alpha_r x_1}.$$

Thus, we have

$$d_{n,r} = -(-i\gamma)^n C_{n,r} - \sum_{l=0}^{n-1} \sum_{q=r-(n-l)F}^{r+(n-l)F} C_{n-l,r-q}(i\gamma_q)^{n-l}d_{l,q}, \qquad (3.20)$$

where, of course,

$$d_{0,0} = -1,$$

as it follows from the law of reflection onto a planar surface.

Analogous formulas can be derived for two-dimensional bounded obstacles when viewed, for instance, as perturbations of a circular cylinder or any other exactly solvable geometry. More precisely, for obstacles whose boundaries are given, in polar coordinates, by

$$\Gamma_\delta = \{(\rho,\theta) : \rho = a + \delta f(\theta)\}, \qquad (3.21)$$

the recursion reads

$$d_{n,q} = -k^n \sum_{p=q-nF}^{q+nF} C_{n,q-p}(-i)^{p-q}\frac{d^n J_p}{dz^n}(ka)/H_q^{(1)}(ka)$$

$$- \sum_{l=0}^{n-1} k^{n-l} \sum_{p=q-(n-l)F}^{q+(n-l)F} d_{l,p} C_{n-l,q-p}(-i)^{p-q}\frac{d^{n-l}H_p^{(1)}}{dz^{n-l}}(ka)/H_q^{(1)}(ka), \qquad (3.22)$$

where again $d_{n,r}$ denotes the Taylor coefficients of the amplitudes B_r (cf. (3.18)), which are now defined through (3.5) and

$$\frac{f(\theta)^l}{l!} = \sum_{r=-lF}^{lF} C_{l,r} e^{ir\theta}.$$

In this case, the calculation is initialized by means of the relations

$$d_{0,q} = -J_q(ka)/H_q^{(1)}(ka),$$

which follow from the exact expressions for the field scattered by a circular cylinder of radius a (see [4]). For the general formulas corresponding to the vector scattering problem in three dimensions we refer the reader to [10] for details on the treatment of biperiodic gratings and to [13] for those on three-dimensional bounded scatterers.

3.3.3 Calculation of Taylor Coefficients

Recursive formulae such as (3.20) and (3.22) allow us to compute Taylor coefficients of a scattering problem to arbitrarily high orders. To demonstrate the calculation, let us take a simple case of a two-dimensional grating (3.14) with

$$f(x_1) = 2\cos(Kx_1) = e^{iKx_1} + e^{-iKx_1}, \qquad (3.23)$$

Chapter 3. High-Order Boundary Perturbation Methods

for which $F = 1$, $C_{n,k} = 0$ if $n - k$ is odd, and $C_{n,k} = \frac{1}{n!}\binom{n}{\frac{n-k}{2}}$ if $n - k$ is even. Then, it is easy to see from (3.20) that the nonzero coefficients $d_{n,r}$ are

$$
\begin{array}{ccccccccc}
& & & & d_{0,0} = -1 & & & & \\
& & & d_{1,-1} & & d_{1,1} & & & \\
& & d_{2,-2} & & d_{2,0} & & d_{2,2} & & \\
& d_{3,-3} & & d_{3,-1} & & d_{3,1} & & d_{3,3} & \\
d_{4,-4} & & d_{4,-2} & & d_{4,0} & & d_{4,2} & & d_{4,4} \\
& & & & \text{etc.} & & & &
\end{array}
$$

If an Nth-order approximation to the Taylor series in (3.18) is to be computed, it is clearly unnecessary to produce *all* of the coefficients $d_{n,r}$ for $n \leq N$. In fact, to compute B_0 with $N = 4$ we need only generate

$$
\begin{array}{ccccc}
& & d_{0,0} = -1 & & \\
& d_{1,-1} & & d_{1,1} & \\
d_{2,-2} & & d_{2,0} & & d_{2,2} \\
& d_{3,-1} & & d_{3,1} & \\
& & d_{4,0} & &
\end{array}
\tag{3.24}
$$

Of course, in general we wish to obtain not only B_0 but a number w of amplitudes that will in turn be used to calculate the fields with a given accuracy; the arrays of derivatives must be augmented accordingly. Similar considerations apply to functions f given by an arbitrary finite Fourier series.

In practice, and in order to reduce the number of operations, one can choose to truncate the inner sum in equations such as (3.20) by setting $d_{k,q} = 0$ for $|q| > q_0$. The parameter q_0 has to be chosen judiciously. Our experiments indicate that in the case of the sinusoidal profile (3.23), no truncation is permissible (e.g., one cannot take $q_0 < 2$ in (3.24)). On the other hand, for the grating associated with the function

$$f(x_1) = 2\cos(Kx_1) + \frac{1}{5}\cos(3Kx_1), \tag{3.25}$$

the effect of the higher order harmonics generated by the second summand can be truncated. In other words, even though the actual formula (3.20) for this profile involves frequencies roughly as high as $\frac{3}{2}N + w$, one can take $q_0 = \frac{N}{2} + w$ with errors comparable to roundoff. This is related to the fact that the height-to-period ratio h/d of the first term in (3.25) is larger than the one for the second term. Thus, in the general case of a general Fourier series, q_0 should be chosen so that no truncation would occur if all but the principal terms in it (i.e., the ones with the largest h/d's) were neglected. Naturally, as the height-to-period ratio of a secondary term approaches those for the principal terms, the value of q_0 should be increased. The ultimate choice of q_0 must be made by consideration of the actual errors as measured by the defect in the energy balance, convergence, reciprocity, or other criteria.

In the case of the sinusoidal grating considered above, closed form expressions for the coefficients $C_{n,r}$ of (3.20) were found. This is, of course, not possible if the boundary of the scatterer is given by an arbitrary function f. By appropriate truncation we may always assume that f is given by a *finite* Fourier series containing modes with orders between $-F$ and F, say; simple iterated multiplication of the series of f then yields a very stable algorithm for the calculation of $C_{n,r}$. Convergence as F is increased is then the criterion for an appropriate choice of this parameter.

Now the issue arises that calculation of the complete powers of the Fourier series of f may be prohibitively expensive in the three-dimensional cases—even when F is as small as, say, $F = 10$. It is therefore fortunate that, again, appropriate truncations can be used with errors comparable with roundoff. The procedure is very simple indeed: if the Fourier series of f^n has been computed, then all modes of order r with $|r| > q_1$ are set to zero, and the resulting series is multiplied by the series of f. The result is then taken as an approximation to f^{n+1}. Of course, the choice of the parameter q_1 depends on the particular scatterer, on F, on q_0, and on the order n of the Taylor approximation. We can assume that appropriate values of these parameters have been found when further increases in them do not lead to improvements in the numerical error.

To complete our algorithms we must now consider the question of summation of series such as (3.18). As we have said, the functions $B_r(\delta)$ are analytic in a common neighborhood of the real axis, and therefore the series in (3.18) certainly has a positive radius of convergence. It turns out that this series diverges (or converges slowly) for many cases of interest, and we thus need appropriate numerical schemes for analytic continuation in the complex δ plane. This is the subject of the following section.

3.4 Approximation of Analytic Functions

Our understanding of the problem of calculation of the electromagnetic field by means of analytic continuation can be presented at two different levels of detail. On one hand we may accurately state that Padé approximants have produced better accuracy than other approximators in all the applications of our method we have encountered. We therefore view Padé approximation, which is briefly described below, as an integral element of our algorithms. An interesting insight into the approximation problem, on the other hand, can be gained by consideration of the spectrum of singularities of the field *as a function of the perturbation parameter δ*. Indeed, singularities play a major role in the approximation problem. They determine the radius of convergence of the Taylor series and they are closely related to the speed of convergence of Padé approximants [3]. Further, the spectrum of singularities determines the conditioning in the values of the Padé approximants [11], and even partial information on such singularities may be used in some cases to produce a rather dramatic improvement in the Padé problem; a simple example of this phenomenon is presented at the end of this section. Our current knowledge of the analytic structure of the field in the δ plane is discussed in what follows.

Chapter 3. High-Order Boundary Perturbation Methods 85

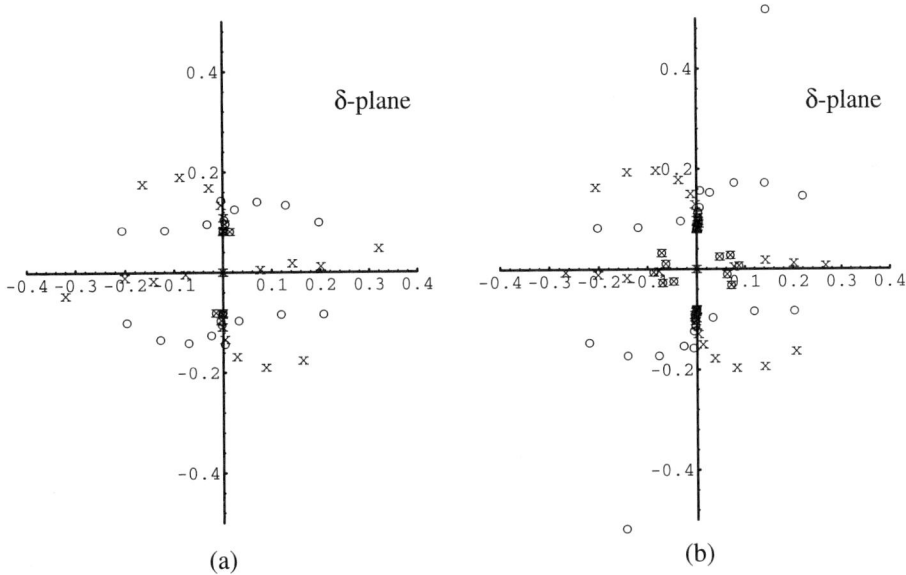

Figure 3.2: Poles (o) and zeros (x) of the Padé approximants of $B_1(\delta)$: (a) [28/28]-approximant; (b) [48/48]-approximant.

The $[L/M]$ Padé approximant of a function

$$B(\delta) = \sum_{n=0}^{\infty} d_n \delta^n \qquad (3.26)$$

is defined (see [3]) as a rational function

$$[L/M] = \frac{a_0 + a_1\delta + \cdots + a_L\delta^L}{1 + b_1\delta + \cdots + b_M\delta^M}$$

whose Taylor series agrees with that of B up to order $L + M + 1$. A particular $[L/M]$ approximant may fail to exist but, generically, $[L/M]$ Padé approximants exist and are uniquely determined by L, M, and the first $L + M + 1$ coefficients of the Taylor series of B. Padé approximants have some remarkable properties of approximation of analytic functions from their Taylor series (3.26) for points far outside their radii of convergence; see, e.g., [3]. They can be calculated by first solving a set of linear equations for the denominator coefficients b_i, and then using simple formulae to compute the numerator coefficients a_i. For convergence studies and numerical calculation of Padé approximants, see [3, 5, 21].

In Figure 3.2 we show the location of the zeros of the numerators and denominators of the [28/28] and [48/48] Padé approximants to the coefficient $B_1(\delta)$ for the perfectly conducting grating with profile

$$f_\delta(x_1) = \delta(e^{i2\pi x_1} + e^{-i2\pi x_1}) = 2\delta \cos(2\pi x_1)$$

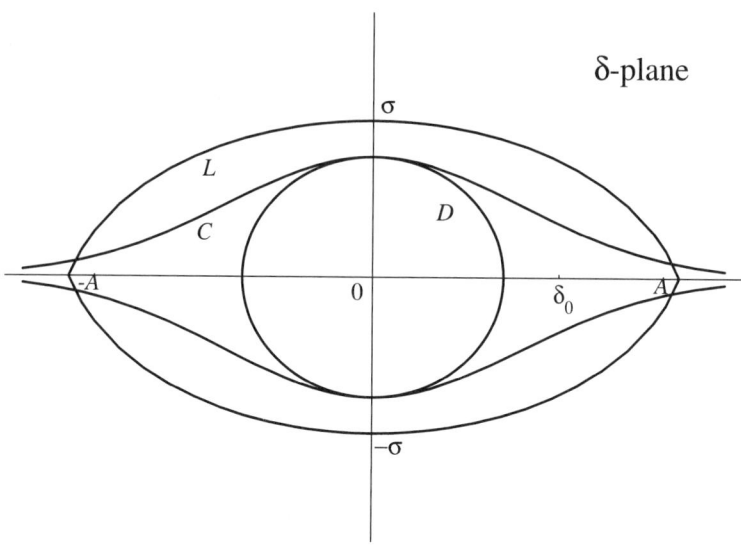

Figure 3.3: The region C of analyticity of the Rayleigh coefficients $B_r^{\pm}(\delta)$ and a lens-shaped region L that is conformally transformed onto the right half-plane via $g(\delta) = (\frac{A-\delta}{A+\delta})^\alpha$.

under normal incidence and with light of wavelength $\lambda = 0.4368$. In this figure, a circle (o) represents a zero of the denominator, which is a singularity of the approximant provided it is not crossed out by a corresponding zero (x) in the numerator. Very similar pictures are obtained for other amplitudes B_r and for approximants of other orders. Now, it is well known that, rather generally, the singularities of the Padé approximants approach the singularities of the function they approximate [3]. Thus, Figure 3.2 provides us with an approximation to the domain of analyticity of the diffracted fields. Note that no singularities occur on the real axis, as expected from the theoretical discussion of section 3.3.1.

A domain of analyticity C which resembles the one suggested by Figure 3.2 was proposed in [8]; see Figure 3.3. This picture led us to devise a summation mechanism based on conformal transformations that we called *enhanced convergence* [8, 11]; see also [54]. Given a function $B(\delta)$ and a complex number δ_0, the method of enhanced convergence uses conformal mappings to produce an appropriate arrangement of the singularities of B and the point δ_0, so that a truncated Taylor series can be used to calculate $B(\delta_0)$.

Suppose we wish to compute the function $B(\delta)$ at a point δ_0 which lies outside the circle of convergence D of the Taylor series of B around $\delta = 0$ (see Figure 3.3). The series is divergent at δ_0. If we consider, however, the composition of B with a conformal transformation,

$$\xi = g(\delta),$$

Table 3.1: Comparison between direct and enhanced convergence. First-order efficiency for a sinusoidal grating in TE polarization, $\delta = 0.075$.

	Direct		Enhanced	
n	e_1	ϵ	e_1	ϵ
10	0.5034545E–02	1.5E–01	0.1735547E–01	8.3E–02
15	0.1073396E–01	5.3E–02	0.1204894E–01	1.3E–05
20	0.1045474E–01	4.0E–03	0.1157684E–01	5.1E–04
25	0.1140532E–01	3.2E–03	0.1163189E–01	3.1E–05
30	0.1161950E–01	2.2E–04	0.1163870E–01	–8.4E–06
35	0.1165000E–01	1.1E–05	0.1163798E–01	–7.3E–07
40	0.1162548E–01	–3.2E–05	0.1163793E–01	–4.4E–08
45	0.1163059E–01	5.1E–05	0.1163793E–01	2.2E–08
50	0.1164204E–01	1.2E–05	0.1163793E–01	2.0E–09
55	0.1164090E–01	–1.7E–05	0.1163793E–01	–3.7E–11
60	0.1163616E–01	–4.0E–06	0.1163793E–01	–4.5E–11

Table 3.2: Comparison between direct and enhanced convergence. First-order efficiency for a sinusoidal grating in TE polarization, $\delta = 0.1$.

	Direct		Enhanced	
n	e_1	ϵ	e_1	ϵ
10	0.3897061E+00	–7.2E–01	0.4018525E+00	–4.5E–01
15	0.4713435E+01	–1.4E+01	0.1649737E+00	–7.3E–02
20	0.3635994E+01	–1.5E+01	0.1568326E+00	3.1E–02
25	0.3848452E+01	–1.2E+01	0.1687898E+00	4.8E–03
30	0.3357765E+00	–5.5E+00	0.1698172E+00	3.0E–04
35	0.1276039E+01	–2.4E+01	0.1701259E+00	–4.0E–04
40	0.4143993E+02	–2.2E+02	0.1700261E+00	–1.3E–04
45	0.1665033E+03	–1.6E+03	0.1699795E+00	1.3E–05
50	0.1406855E+04	–8.2E+03	0.1699760E+00	1.1E–05
55	0.7075471E+04	–6.9E+04	0.1699781E+00	2.5E–06
60	0.7847034E+05	–4.1E+05	0.1699792E+00	–4.8E–07

the singularities and the point $\xi_0 = g(\delta_0)$ at which the function is sought will show a different arrangement *in the ξ plane*, and ξ_0 may lie inside the circle of convergence of the composite function $B(g^{-1}(\xi))$. If so, a truncated Taylor series of the composite function can be summed to yield the value $B(\delta_0)$. Even if δ_0 lies inside the circle of convergence D, this procedure may result in improved convergence rates [8, 11]. In Tables 3.1 and 3.2, for example, we compare the convergence of the power series for $e_1(\delta) = |B_1(\delta)|^2 \beta_1/\beta_0$ about $\delta = 0$ ("Direct") with that of $e_1(\xi) = |B_1(\xi)|^2 \beta_1/\beta_0$ ("Enhanced") obtained by means of an appropriate conformal change of variables [8]. Again we consider the sinusoidal grating scattering problem mentioned above. We observe that even in the case $\delta = 0.075$, in which the direct series converges, the convergence is substantially enhanced by the conformal mapping. In case $\delta = 0.1$ the direct series does not converge.

The performance of this summation method depends strongly on the parameters A and σ of Figure 3.3. The optimality of a choice of these parameters can be checked through the defect ϵ in the energy relation (3.11)

$$\epsilon = 1 - \sum_{n \in U^+} e_n^+.$$

($e_n^- = 0$ here, since we are dealing with a perfectly conducting grating.) Alternatively, the optimal values of these parameters can be calculated [9] by consideration of the poles shown in Figure 3.2. The results of these calculations are in close agreement. Further, note the position of singularities closest to the origin in Figure 3.2, which implies a radius of convergence consistent with the convergence results of the direct series in Tables 3.1 and 3.2. The agreements found in these calculations constitute an important consistency check in our theory. In addition, they add substantial credibility to our view that the analytic structure in the δ plane is well approximated by representations such as that in Figure 3.2.

As we said, Padé approximation does exhibit better numerical performance than enhanced convergence; in the examples of Tables 3.1 and 3.2, for instance, Padé approximants permit us to obtain two additional significant digits [9]. Interestingly, enhanced convergence can be used to improve the performance of Padé approximation. Indeed, the theory in [11] shows that the relative arrangement of the singularities of an analytic function is closely related to the numerical conditioning of the corresponding Padé problem. Further, a conformal change of variables on a function $B(\delta)$ can lead to a dramatic improvement in this conditioning. (The Padé approximants of the functions in the transformed variables will be referred to as enhanced Padé approximants.) Since the main numerical weakness of Padé approximation is its ill conditioning, it is reasonable to expect that enhanced approximants could lead to improvement in our solutions of scattering problems. Unfortunately, we have not yet succeeded in devising a numerical implementation of these ideas in the context of scattering calculations that will meet a basic requirement of the approach, namely, the need for *accurate* values of the Taylor coefficients of the composite functions. Composition of the corresponding series, which certainly suffices in applications such as those of Tables 3.1 and 3.2, is not appropriate for the enhanced Padé application. Indeed, composition of power series leads to a loss of significant digits in the Taylor coefficients. This accuracy loss interacts with the conditioning problem of Padé approximation in such a way that no substantial improvements in the calculated values are obtained. If accurate values of the coefficients of the composite functions can be found, on the other hand, then very substantial improvements can be obtained, as shown in an example below. Thus, further improvement in the performance of our algorithms could result if an accurate method for computation of the enhanced coefficients were found.

As we said, enhanced convergence may help obtain a remarkable improvement in the performance of Padé approximation. Let us consider, for example, Table 3.3, which shows the values of the $[\frac{N}{2}/\frac{N}{2}]$ Padé and enhanced Padé approximants for the function $f(z) = \log(1 + z)$; see [11] for details. This table shows that enhanced Padé approximants produce up to 13 correct digits of $\log(21)$ while ordinary Padé

Chapter 3. High-Order Boundary Perturbation Methods

Table 3.3: $[\frac{N}{2}/\frac{N}{2}]$ approximants for $\log(1+z)$.

N	z	$\log(1+z)$	Padé	Enh. Padé
20	20	3.044522437723	3.043989111079	3.043988784141
40			3.044612164211	3.044522360574
60			3.044477040660	3.044522437596
80			3.044175772366	3.044522437727
100			3.044463021924	3.044522437722
120			3.044489520809	3.044522437724
140			3.044496462919	3.044522437723
160			3.044619592662	3.044522437723
180			3.044362344599	3.044522437723
20	200	5.30330	5.03582	5.03577
40			5.32614	5.28588
60			5.17831	5.30093
80			5.08690	5.30276
100			5.16939	5.30305
120			5.18899	5.30324
140			5.19792	5.30328
160			5.70660	5.30329
180			5.13885	5.30330

fractions do not produce more than the first 4 digits. It is very remarkable, in any case, that the Padé approximation is so stable and that it produces these four digits for N up to at least $N = 180$, in spite of the tremendous ill-conditioning of the denominator problem. Table 3.3 also shows the values of both approximants at $z = 200$; again, an improvement is observed.

3.5 Applications

The difficulty associated with the numerical solution of a scattering problem depends roughly on two elements: the magnitude of the ratio P/λ of the "size" (characteristic length) of the scatterer to the wavelength on one hand, and the oscillations and/or lack of smoothness exhibited by its boundary, on the other. In short, numerical complexity arises from the need to account accurately for oscillatory behavior of fields and interfaces. In what follows we present applications of our method which test its performance in problems of various degrees of difficulty, and we compare the results of these algorithms with those given by classical methods.

As we have noted, our approach is applicable to configurations that may be viewed as perturbations from an exactly solvable geometry. Such perturbations need not be small, as may be seen from the examples that follow, and our analytic method may be considered as a rather general one. In many cases of practical importance our approach yields results that are several orders of magnitude more accurate than those given by classical methods, such as the method of moments and other algorithms based on the solution of integral equations.

3.5.1 Two-Dimensional Problems

We have tested our analytic method in a variety of problems of diffraction by gratings [9], and indeed, our method was originally intended as a grating solver. We therefore discuss applications to grating problems first. To substantiate the general applicability of our approach, we also include here an example of a calculation of the radar cross section (RCS) generated by a bounded cylindrical obstacle, and a corresponding example in three dimensions is given in section 3.5.2. In this and the following sections the error estimator ϵ is the defect in the energy balance criterion. For example, in the grating configurations ϵ is defined as the defect in the relations (3.9) and (3.11), and for two-dimensional dielectric gratings, ϵ is given by

$$\epsilon = 1 - \sum_{r \in U^+} e_r^+ - \sum_{r \in U^-} e_r^-$$

as calculated by the numerical solver. For a perfectly conducting bounded obstacle in two dimensions ϵ is defined as the relative error in the calculated value of the left-hand side of equation (3.6):

$$\epsilon = \frac{\left|\sum_r |B_r|^2 + \operatorname{Re}\left(\sum_r B_r\right)\right|}{\sum_r |B_r|^2}. \tag{3.27}$$

These defects provide an accurate measure of the relative error in the quantities of interest. Indeed, unlike other approaches, our method does not verify energy balance exactly, so that its defect is in fact a good accuracy estimator. We have verified this through convergence tests in a large number of examples; e.g., compare efficiency errors and energy defects in the results of Tables 3.1 and 3.2. Of course, the energy balance criterion is only valid in the absence of absorption. When dealing with lossy scatterers we generally turn to estimating the errors directly by means of convergence tests, as in the first application in section 3.5.2.

In our first example we consider a problem of diffraction by a sinusoidal dielectric grating (3.14) with

$$f(x_1) = 2\cos(2\pi x_1/d)$$

and period $d = 1\mu m$. (Note that the height h of the corrugations, that is, the vertical distance from the highest peak to the lowest valley, is given by $h = 4\delta$.) The grating has a refractive index $\nu_0 = 2$ and is illuminated under normal incidence with light of wavelength $\lambda = 0.83\mu m$. Table 3.4 contains results given by our algorithms for the reflected and transmitted energy R and T, which results from a unit input energy. This case was treated in [18] by means of integral equations and the method of moments; there, the authors report the following values of R and T for $h = 0.2\mu m$:

$$R = 0.117274,$$

$$T = 0.882759,$$

Table 3.4: Reflected and transmitted energies for a sinusoidal grating with index of refraction $\nu_0 = 2$, under normal incidence with a wavelength-to-period ratio $\lambda/d = 0.83$: [20/20] Padé approximants. Left: TE polarization; right: TM polarization.

h/d	R	T	ϵ	h/d	R	T	ϵ
0.00	0.111111	0.888889	$-2.2\text{E}-16$	0.00	0.111111	0.888889	$-2.2\text{E}-16$
0.10	0.114926	0.885074	$0.0\text{E}+00$	0.10	0.104046	0.895954	$-2.2\text{E}-16$
0.20	0.117282	0.882718	$0.0\text{E}+00$	0.20	0.086355	0.913645	$0.0\text{E}+00$
0.30	0.104871	0.895129	$1.6\text{E}-14$	0.30	0.062807	0.937193	$-1.6\text{E}-12$
0.40	0.080184	0.919816	$9.8\text{E}-11$	0.40	0.039117	0.960883	$-1.1\text{E}-08$
0.50	0.055902	0.944098	$1.0\text{E}-07$	0.50	0.025636	0.974363	$-6.5\text{E}-07$
0.60	0.038983	0.961015	$-1.3\text{E}-06$	0.60	0.023655	0.976517	$1.7\text{E}-04$
0.70	0.029619	0.969848	$-5.3\text{E}-04$	0.70	0.021333	0.982000	$3.3\text{E}-03$
0.80	0.024083	0.972233	$-3.7\text{E}-03$	0.80	0.016013	0.999951	$1.6\text{E}-02$

and

$$\epsilon = 1 - (T + R) = 3.3 \times 10^{-5};$$

compare the left side of Table 3.4.

As is the case here, the analytic method yields results of very good definition in most grating problems of interest. Other results for grating problems obtained by discretization of boundary integral equations include, for example, those of Van den Berg [60] and Pavageau and Bousquet [46]. These authors considered a perfectly conducting sinusoidal grating for values of h/d ranging from 0.3 to 0.56 and illuminated with light of wavelength $\lambda = 0.4368$. They report errors of the order of 10^{-5} for a ratio of 0.3 and of order 10^{-3} for ratios of 0.4 and 0.56. The corresponding errors in the analytic method are of order 10^{-12}, 10^{-8}, and 10^{-6}; similar improvements in performance over other methods have been found in calculations containing lossy gratings and nonsmooth (e.g., triangular) profiles [9]. It must be pointed out that the largest ratio $h/d = 0.56$ considered here is larger than those corresponding to gratings actually used in applications [40]. For even deeper gratings, say, ratios of 0.7 and beyond (and for this wavelength), our method in its present form rapidly breaks down due to numerical ill-conditioning, while the integral method deteriorates more slowly, and it gives results with errors of a few percent for gratings with heights as large as $h/d = 1$. As we pointed out in section 3.4, remedies for conditioning problems in our algorithms may result from a more detailed consideration of the analytic properties and singularities of the electromagnetic field; see, e.g., Table 3.3.

In our second example we study scattering by the perfectly conducting obstacle of Figure 3.4(a) (a two-dimensional bounded scatterer without symmetries) to demonstrate the wider applicability of our methods. The boundary of this obstacle is given by (3.21) with $f(\theta) = 0.125\sin(4\theta) - 0.15\sin(3\theta)$ and $\delta = 1$. Figure 3.4(b) gives the amplitude of the bistatic far field coefficient Φ for the TE polarized configuration indicated in Figure 3.4(a) with perimeter-to-wavelength ratio $P/\lambda = 20$.

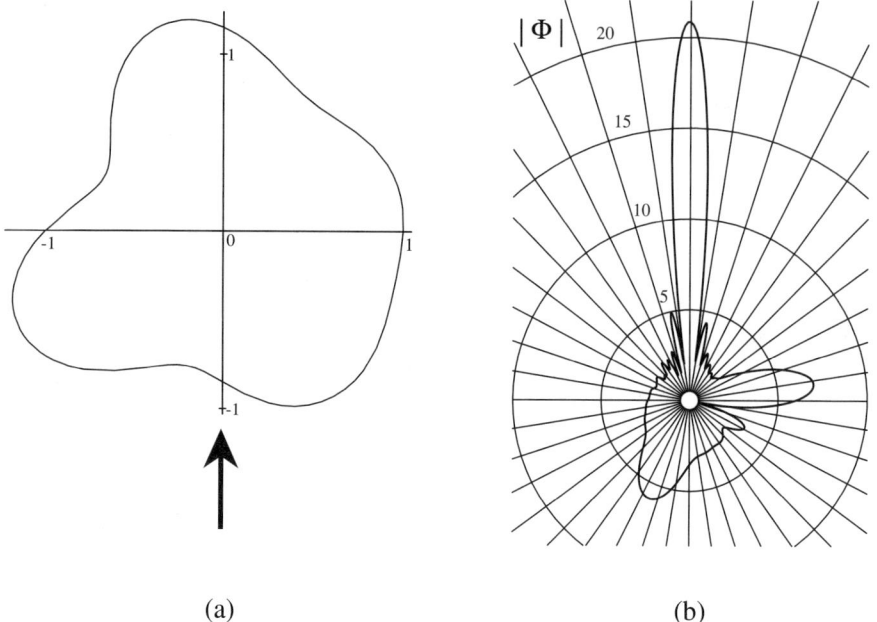

(a)   (b)

Figure 3.4: (a) A two-dimensional bounded scattering configuration without symmetries, (b) the amplitude of its bistatic far-field coefficient Φ. TE polarization, $P/\lambda = 20$.

3.5.2 Three-Dimensional Problems

In this section, we present results of applications of our approach to truly three-dimensional geometries. We begin again with the case of optical diffraction gratings, where we consider the full problem of electromagnetic scattering (Maxwell's equations) off biperiodic surfaces. After a discussion of smooth (sinusoidal) metallic gratings, we turn to a challenging example of a configuration that presents edges and corners. Finally, we again demonstrate the generality of the approach by briefly recounting the outcome of our numerical methods in the case of (acoustic) scattering by three-dimensional bounded bodies.

Bisinusoidal gratings in gold have been used in the experimental and numerical studies on total absorption; see [41]. These are gratings of the form (3.7) with

$$f(x_1, x_2) = \left[\cos\left(\frac{2\pi x_1}{d_1}\right) + \cos\left(\frac{2\pi x_2}{d_2}\right)\right] \quad \text{(see Figure 3.5)};$$

note that the groove depth is again given by $h = 4\delta$. In [41] the authors treated this problem by means of the integral method of [17]; they considered gratings with depths of $h = 0.040$, $h = 0.055$, and $h = 0.070$ and with periods $d_1 = d_2$ varying from 0.60 to 0.64. In Figure 3.6 we show the results given by our algorithm for these problems. Qualitative agreement with the results in [17, 41] is observed, but some discrepancies occur. For example, in contrast with Figure 7.17 of [41], our

Chapter 3. High-Order Boundary Perturbation Methods

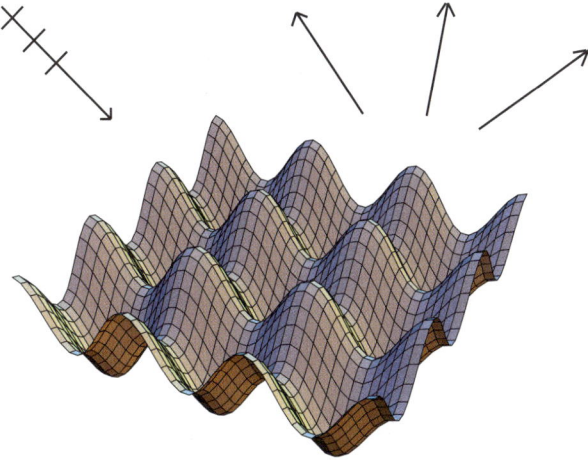

Figure 3.5: Section of a three-dimensional bisinusoidal diffraction grating.

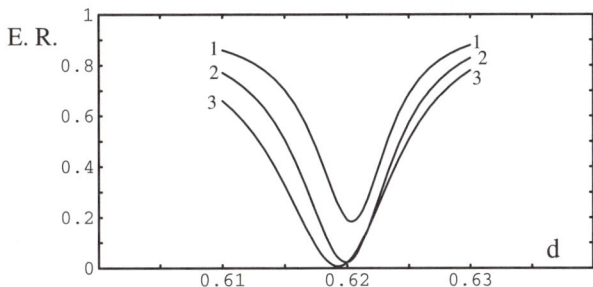

Figure 3.6: Energy reflected by the bisinusoidal grating of Figure 3.5 (in gold), with normally incident light of wavelength 0.65μm. Curve 1: $h = 0.040\mu$m; curve 2: $h = 0.055\mu$m; curve 3: $h = 0.070\mu$m: [6/6] Padé approximants.

curves 2 and 3 coincide at $d = 0.62\mu$m. This prompted us to analyze the accuracy of our results. We found that, for this range of parameters, our method yields extremely accurate results, with errors in the reflected energy ("E.R.") that are better than 10^{-14}. This can be seen in Table 3.5, which contains a convergence study for the values of the reflected energy at $d = 0.62\mu$m for the curves labeled 2 and 3 in Figure 3.6. Note that an accuracy better than eight digits is obtained by an approximation of order 13. The accuracy of the integral method in this problem ($h = 0.055$ and $h = 0.070$) has been estimated to be of the order of two digits [17]. To demonstrate the range of parameters in which our method can be applied, we include a third column in Table 3.5 showing the values of E.R. for a much deeper grating profile of height $h = 0.500\mu$m, for which $h/d = 0.806$. We see that even in this case, the results are quite accurate: the errors are of the order of 10^{-4} for a [6/6] approximant ($n = 13$) and of 10^{-6} for a [14/14] approximant ($n = 29$). (Padé approximants with $n = 15, 19, 23, 27$, and 31 are singular for this problem.)

Table 3.5: Convergence study of the reflected energy for the example in Figure 3.5 (gold). The period is fixed at 0.62μm and the wavelength at 0.65μm. $[\frac{n-1}{2}/\frac{n-1}{2}]$ Padé approximants.

n	$h = 0.055\mu$m	$h = 0.070\mu$m	$h = 0.500\mu$m
13	0.0227882361359963	0.0226057361431067	0.84146746
17	0.0227882361334883	0.0226057359874209	0.84202623
21	0.0227882361334891	0.0226057359874838	0.84219841
25	0.0227882361334900	0.0226057359874644	0.84260919
29	0.0227882361334896	0.0226057359874220	0.84197301
33	0.0227882361334896	0.0226057359874253	0.84197398

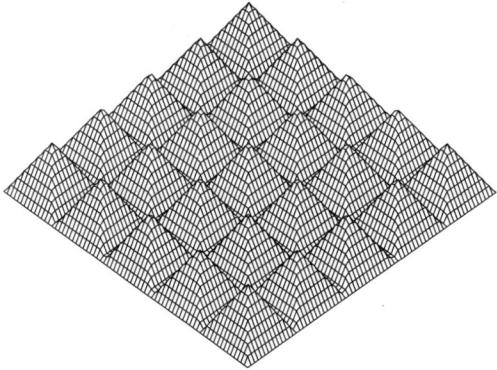

Figure 3.7: A three-dimensional array of square pyramids.

Next, let us consider the problem proposed in [17], that is, a crossed grating of rectangular pyramids with periods $d_1 = 1.50$ and $d_2 = 1$, of height $h = 0.25$, under incident light of wavelength $\lambda = 1.533$ and with incidence angles given as follows: cylindrical angle $\phi = 45°$; azimuthal angle $\theta = 30°$, and polarization with the electric field in the vertical plane $\phi = 45°$. This is an interesting configuration, which contains a three-dimensional obstacle with corners and edges. A schematic representation of the grating is given in Figure 3.7. As we have said, our algorithm requires the boundary of the scatterer to be approximated by a finite Fourier series, thus effectively rounding its edges. For comparison purposes we show, in Figure 3.8, one element of the grating of Figure 3.7 together with its Fourier series approximation with $F = 10$. We can conclude that we have obtained the exact solution for the actual pyramid grating within a given error tolerance when convergence within that tolerance is observed as the number of Fourier modes in the approximation is increased.

As we explained in section 3.3.3, it is necessary here to choose appropriately the truncation parameters F, q_0, and q_1 as well as the order n of the approximation. In the cases below we found that convergence to the actual numerical solution within the error estimates indicated in Table 3.6 is achieved with $F = 5$, $q_0 = q_1 = 20$, and

Chapter 3. High-Order Boundary Perturbation Methods

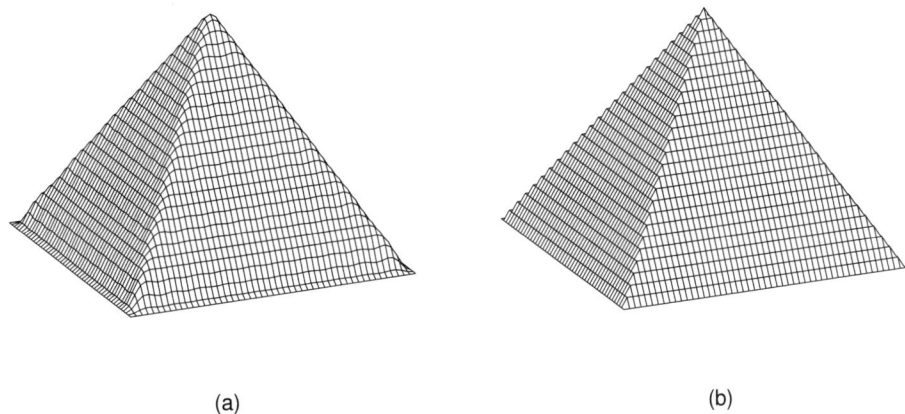

(a)   (b)

Figure 3.8: An element of the grating of Figure 3.7: (a) Fourier approximation with $F = 10$; (b) exact.

Table 3.6: Efficiencies for the grating of pyramids described in the text: [4/4] Padé approximants.

h	$e_{r,0,0}$	$e_{r,-1,0}$	$e_{t,0,0}$	$e_{t,1,0}$	$e_{t,-1,0}$	$e_{t,0,-1}$	$e_{t,-1,-1}$	ϵ
0.00	0.02525	0.00000	0.97475	0.00000	0.00000	0.00000	0.00000	$-1\mathrm{E}{-}16$
0.05	0.02500	0.00010	0.97432	0.00012	0.00013	0.00029	0.00004	$-2\mathrm{E}{-}12$
0.10	0.02425	0.00041	0.97306	0.00049	0.00050	0.00115	0.00015	$-8\mathrm{E}{-}10$
0.15	0.02304	0.00092	0.97099	0.00107	0.00110	0.00256	0.00032	$-2\mathrm{E}{-}08$
0.20	0.02143	0.00161	0.96817	0.00185	0.00192	0.00445	0.00056	$-2\mathrm{E}{-}07$
0.25	0.01951	0.00246	0.96465	0.00280	0.00294	0.00679	0.00086	$-1\mathrm{E}{-}06$
0.30	0.01737	0.00341	0.96051	0.00388	0.00415	0.00948	0.00120	$-6\mathrm{E}{-}06$
0.35	0.01511	0.00441	0.95582	0.00506	0.00553	0.01247	0.00157	$-2\mathrm{E}{-}05$
0.40	0.01284	0.00540	0.95068	0.00631	0.00709	0.01567	0.00196	$-5\mathrm{E}{-}05$
0.45	0.01064	0.00633	0.94515	0.00758	0.00882	0.01901	0.00237	$-1\mathrm{E}{-}04$
0.50	0.00858	0.00713	0.93933	0.00885	0.01070	0.02241	0.00278	$-2\mathrm{E}{-}04$

$n = 9$. Indeed, additional calculations with $F = 10$ and with $q_0 = q_1 = 22$, 24, and 30 result in no changes in the values shown in Table 3.6—exception being made for small changes in the error estimator. This suggests that the solution obtained with $F = 5$ is, within the accuracy shown in Table 3.6, the exact solution to the sharp-edge problem under consideration.

In [17] the authors treated this problem by means of their integral algorithm, and they reported the results of Table 3.7. In addition to results given by the integral method, which is the one the authors recommend, they also presented calculations performed by means of two other solvers based on solution of ordinary differential equations. These differential methods are known to be unstable and generally less competitive than those based on solutions of integral equations [61]. Interestingly, in this particular case one of the differential algorithms produced good results with defect ϵ of order $2 \cdot 10^{-5}$; the other produced $2 \cdot 10^{-3}$. Compare our results in Table 3.6.

Table 3.7: Efficiencies for the grating of pyramids described in the text; values given in reference [17].

h	$e_{r,0,0}$	$e_{r,-1,0}$	$e_{t,0,0}$	$e_{t,1,0}$	$e_{t,-1,0}$	$e_{t,0,-1}$	$e_{t,-1,-1}$	ϵ
0.25	0.01984	0.00254	0.96219	0.00299	0.00303	0.00704	0.00092	1E−03

Table 3.8: Energy and error values for scatterers $\rho = a + \delta \left[\frac{3}{8} \cos(\phi) \sin(\theta) \left(4 - 5\sin^2(\theta)\right)\right]$ (cf. Fig. 3.9).

Padé	δ	Energy	ϵ (Taylor)	ϵ (Padé)
[5/5]	0.5	1.7853305	2.5E−03	5.3E−07
[5/5]	1.0	2.0208911	9.0E−01	1.6E−04

Finally, we present some of the results we have obtained on problems of scattering by acoustically soft bounded bodies [13]. For these applications the scatterers were realized as (large) perturbations of a sphere of radius a: in spherical coordinates (ρ, θ, ϕ) (θ = azimuthal angle and ϕ = polar angle) their boundaries are given by

$$\Gamma_\delta = \{(\rho, \theta, \phi) : \rho = a + \delta f(\theta, \phi)\},$$

where f is an arbitrary function of (θ, ϕ) and δ is the perturbation parameter.

In Table 3.8 we present calculations and error estimates for the field reflected by the heart-shaped scatterer of Figure 3.9 and related variations of it. The parameter ϵ in the table is, as in the previous applications, a convenient error estimator given by the defect in the energy balance. The values of the fields were obtained by means of direct summation of the Taylor series of the fields (Taylor) and by means of Padé approximation (Padé). As is apparent from an examination of the error estimates in Table 3.8, very accurate results have been obtained in these cases through Padé approximation, providing a substantial improvement over direct summation of the Taylor series (which, in fact, *diverges* for $\delta = 1.0$).

3.5.3 Eigenvalue Calculations

The scattering algorithms described above are based on our results on the analytic dependence of the fields and their boundary values upon boundary variations. As we have explained (section 3.3), these results, in turn, rely heavily on the uniqueness of solutions to the scattering problem, which allows for the inversion of the differential operators within suitable spaces of holomorphic functions. A number of important problems in electromagnetics and acoustics, on the other hand, give rise to models for which this uniqueness property does not hold. This is the case, for instance, in the study of cavities and waveguides (see, e.g., [20, 29] and the references cited there), where *eigenvalue problems* naturally arise. For example, the determination of these eigenvalues and eigenfunctions plays a central role in the assessment of the

Chapter 3. High-Order Boundary Perturbation Methods

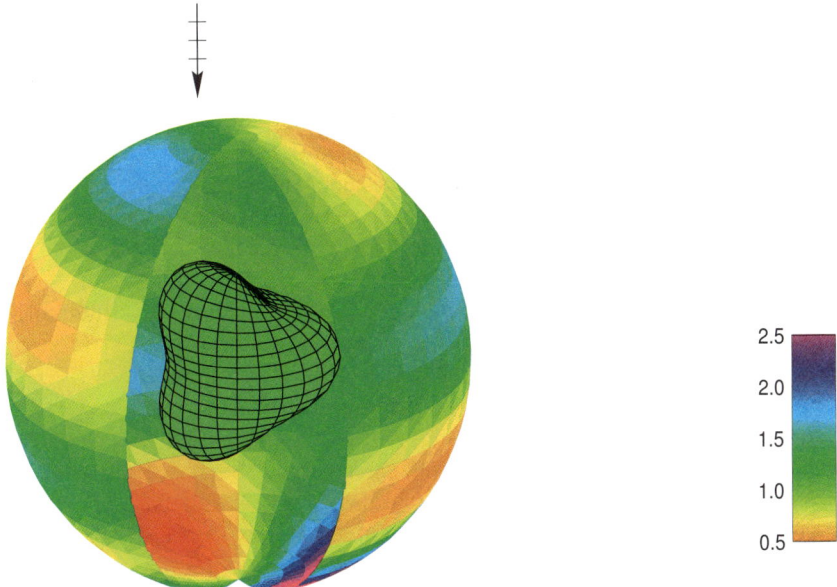

Figure 3.9: An acoustically soft scatterer with boundary $\rho = a + \delta \left[\frac{3}{8}\cos(\phi)\sin(\theta) \times \left(4 - 5\sin^2(\theta)\right)\right]$: far field for $ka = 1$ and $\delta = 0.5$, *normalized* to that produced by the sphere $\rho = a$.

quality ("Q-factor") of resonators for use in lasers and other optical systems [15, 34]. Thus, substantial literature has been devoted to the shape dependence of Q-factors, that includes the proposition of a number of numerical approaches for their estimation (see, e.g., [63] for an FDTD calculation and [43, 45] for results of raytracing).

The general observation that small to moderate shape changes can have dramatic effects on the properties of conservative and leaky cavities (see, e.g., [19] for a recent experiment) has generated considerable interest in the development of perturbation methods that might shed light on this process [15, 28, 31, 33]. All the work to date, however, has resulted only in low (first- and second-) order theories, and thus it is restricted to very small perturbations. In fact, these results have prompted the suggestion [43] that perturbative methods may be applicable only within this limited range. We know, however, that appropriate uses of analytic function theory can substantially enlarge the domain of applicability of these methods, at least in scattering calculations. And, motivated by these results, we have recently embarked on a project to develop a high-order boundary perturbation theory that will be applicable to the calculation of normal and quasi-normal modes of arbitrarily shaped resonators.

As we had anticipated, the development of such a theory entailed the unraveling of significant new theoretical and algorithmic challenges that were not present in our scattering applications and that certainly do not arise in connection with low-order approximations. First, on the theoretical side, the issue of analyticity in spatial and perturbation variables (cf. section 3.3) remained to be settled. Indeed, although the analyticity of eigenvalues could be derived from classical results [30, 51], the possibility of choosing corresponding eigenfunctions that are *jointly* analytic in space and parameter variables—so that formal boundary differentiations are permissible—needed to be established. The classical results, however, did prove useful in our effort to demonstrate this latter property of eigenfunctions, as they allowed us to translate the problem to that of uncovering analytic extensions of suitable volume potentials (generated by the appropriate Green's functions).

These theoretical considerations justified, once again, all boundary differentiations and the consequent derivation of appropriate recursive formulas (cf. section 3.3.2). Still, this procedure presented us with yet another unprecedented complication. Indeed, new difficulties arose as we attempted to derive formulas for the continuation of *multiple* eigenvalues of the unperturbed configuration which evolve, upon shape deformation, as separate, *distinct* eigenvalues. Such is the case, for instance, for the simplest case of the Laplace operator on a (two-dimensional) disk. Our theoretical results, on the other hand, guaranteed the existence of an appropriate choice of eigenfunctions so that the problem reduced to finding an efficient process for their identification.

To describe this process, let us consider eigenvalue problems for Laplace's equation in domains with boundaries (3.21). That is, we seek solutions $(u(\vec{x};\delta), k(\delta))$ of

$$\begin{cases} \Delta_x u(\vec{x};\delta) + k(\delta)^2 u(\vec{x};\delta) = 0 & \text{for } \vec{x} \in \Omega_\delta \equiv \{r \leq a + \delta f(\theta)\}, \\ \text{and} \quad u(\vec{x};\delta) = 0 & \text{for } \vec{x} \in \partial\Omega_\delta. \end{cases}$$

Then, writing

$$u(\vec{x};\delta) = \sum_{k \geq 0} u_k(\vec{x})\delta^k \quad \text{and} \quad k(\delta)^2 = \sum_{k \geq 0} q_k \delta^k,$$

we find the recursive relations

$$\begin{cases} \Delta u_n(r,\theta) + q_0 u_n(r,\theta) = -\sum_{p=0}^{n-1} q_{n-p} u_p(r,\theta) & \text{for } r \leq a, \\ \text{and} \quad u_n(a,\theta) = -\sum_{m=0}^{n-1} \frac{f(\theta)^{n-m}}{(n-m)!} \frac{\partial^{n-m} u_m}{\partial r^{n-m}}(a,\theta). \end{cases} \quad (3.28)$$

The function u_0 corresponds, of course, to an eigenfunction of the Laplacian in the disk Ω_0 and is therefore given by

$$u_0(r,\theta) = \alpha_0 J_M(q_0^{1/2} r) e^{iM\theta} + \beta_0 J_M(q_0^{1/2} r) e^{-iM\theta}, \quad \alpha_0, \beta_0 \in \mathbb{R}, \quad (3.29)$$

where J_M denotes the Bessel function of order $M \geq 0$, and the unperturbed eigenvalue q_0 satisfies

$$J_M(q_0^{1/2} a) = 0.$$

Chapter 3. High-Order Boundary Perturbation Methods

Note that the constants α_0, β_0 in (3.29) are arbitrary (any linear combination provides an eigenfunction in Ω_0) since q_0 has multiplicity 2. However, upon boundary deformations this double eigenvalue will, in general, "split" into two simple ones, each having only a one-dimensional family of associated eigenfunctions. The requirement of analyticity (or even continuity) in δ of these eigenfunctions will thus force a very particular choice of the constants α_0, β_0. As we shall see below, however, this choice may not be apparent until several Taylor coefficients in the expansion for $(u(\cdot, \delta), k(\delta))$ have been derived, as the aforementioned splitting may occur at any order in δ, depending on the perturbation function $f(\theta)$. For this reason, our algorithm to find the coefficients $(u_n(\cdot), q_n)$ in (3.28) proceeds in several steps, which we now discuss briefly.

Step 1. Assume

$$u_0(r,\theta) = J_M(q_0^{1/2} r) e^{iM\theta}. \tag{3.30}$$

Note that, as already stated, this assumption will generally be inconsistent with the desired analytic dependence of eigenfunctions on the perturbation parameter and will therefore have to be reconsidered at a latter stage of the algorithm (see Step 5). Moreover, even if we assume u_0 to have this form, the question arises as to how to expand the successive derivatives u_n. A possible choice for basis functions is the actual eigenfunctions of the unperturbed geometry. However, such a choice would result in infinite series representations, and we therefore choose instead to define functions

$$\psi_{k,l}(r) = \frac{r^k}{(4q_0)^{k/2} k!} J_{l+k}(q_0^{1/2} r) \quad \text{and} \quad \phi_{k,l}(r,\theta) = \psi_{k,|l|}(r) e^{il\theta}.$$

These functions are characterized by

$$P_l(\psi_{k,l}) = \psi_{k-1,l} \quad \text{and} \quad \mathcal{L}(\phi_{k,l}) = \phi_{k-1,l}, \tag{3.31}$$

where

$$P_l(\cdot) = \partial_r^2 + \frac{1}{r}\partial_r + \left(q_0 - \frac{l^2}{r^2}\right) \quad \text{and} \quad \mathcal{L}(\cdot) = \Delta + q_0.$$

And, if $f(\theta)$ has a Fourier series expansion

$$f(\theta) = \sum_{l=-F}^{F} C_{1,l} e^{il\theta},$$

we will seek u_n of the form

$$u_n(r,\theta) = \sum_{\substack{0 \le k \le n \\ M-(n-k)F \le l \le M-(n-k)F}} d_{k,l}^n \phi_{k,l}(r,\theta). \tag{3.32}$$

Step 2. Find q_n. Multiplying the first equation in (3.28) by $\overline{u_0}$ and integrating on $\{r \leq a\}$ we find that

$$q_n = \frac{1}{A^0_{1,|M|}} \left[-\sum_{\substack{1 \leq p \leq n-1 \\ 0 \leq k \leq p}} q_{n-p} d^p_{k,M} A^0_{k+1,|M|} + \sum_{\substack{0 \leq m \leq n-1 \\ 0 \leq k \leq m \\ M-\min(m-k,n-m)F \leq s \leq M+\min(m-k,n-m)F}} C_{n-m,M-s} d^m_{k,s} A^{n-m}_{k,|s|} \right],$$

where

$$A^m_{n,l} = \partial^m_r \psi_{n,l} \bigg|_{r=a} \quad \text{and} \quad \frac{f(\theta)^n}{n!} = \sum_{l=-nF}^{nF} C_{n,l} e^{il\theta}.$$

Step 3. Find $d^n_{k,l}$ for $1 \leq k \leq n$ (all except $k = 0$). For this, we use the differential equation for u_n (cf. (3.28)) and the properties (3.31), which imply that

$$\mathcal{L}(u_n) = \sum_{k,l} d^n_{k,l} \phi_{k-1,l} = -\sum_p q_{n-p} \sum_k d^p_{k,l} \phi_{k,l}$$

and therefore

$$d^n_{k,l} = -\sum_{p=k-1}^{n-1} q_{n-p}\, d^p_{k-1,l}.$$

Step 4. Find $d^n_{0,l}$ and *check for eigenvalue "splitting."* For this, let $v_n = \sum_l d^n_{0,l} \phi_{0,l}$ and recall that $\phi_{0,l}(r,\theta) = J_l(q_0^{1/2} r) e^{il\theta}$ solves $(\Delta + q_0)\phi_{0,l} = 0$. Then, from (3.28), (3.32),

$$\begin{cases} \Delta v_n(r,\theta) + q_0 v_n(r,\theta) = 0 & \text{for } r \leq a, \\ v_n(a,\theta) = -\sum_{m=0}^{n-1} \frac{f(\theta)^{n-m}}{(n-m)!} \partial^{n-m}_r u_m(a,\theta) - \sum_{\substack{1 \leq k \leq n \\ l}} d^n_{k,l} \phi_{k,l}(a,\theta). \end{cases}$$

Thus, if B^n_l denotes the lth Fourier coefficient of the boundary values $v_n(a,\theta)$,

$$v_n(a,\theta) = \sum_l B^n_l e^{il\theta}, \tag{3.33}$$

then the solution takes the form

$$v_n(r,\theta) = \sum_l \frac{B^n_l}{J_{|l|}(q_0^{1/2} a)} J_l(q_0^{1/2} r) e^{il\theta}. \tag{3.34}$$

Now, the choice of q_n (cf. Step 2) guarantees that the coefficient B_M^n of the (resonant) mode $l = M$ in (3.33) vanishes (recall $J_M(q_0^{1/2}a) = 0$). However, the coefficient B_{-M}^n may or may not vanish. And, in fact, it can be shown that

$$\text{the eigenvalue ``splits''} \iff B_{-M}^n \neq 0.$$

Therefore, the procedure follows different paths depending on the value B_{-M}^n. If $B_{-M}^n = 0$, then we may indeed define v_n as in (3.34) and continue: replace $n \to n+1$ and go back to Step 2. Otherwise, if $B_{-M}^n \neq 0$ we proceed to the next step.

Step 5. "Recalculate" (the correct value of) q_n and choose appropriate eigenfunctions to order 0. As we said, once the splitting has been identified it needs to be accounted for by an appropriate choice of the constants α_0 and β_0 in (3.29) and the determination of coefficients q_n^+ and q_n^- corresponding to each of the two distinct eigenvalues, which have been found (cf. Step 4) to split at order n (i.e., $q_k^+ = q_k^- = q_k$ for $1 \leq k \leq n-1$). To this end, we first note that if we begin the procedure, in Step 1, with u_0 in (3.30) replaced by

$$u_0^1 = \alpha_0 u_0 + \beta_0 \overline{u_0},$$

we get that the corresponding higher-order coefficients u_k will be given by

$$u_k^1 = \alpha_0 u_k + \beta_0 \overline{u_k} \qquad (1 \leq k \leq n-1),$$

which satisfy

$$\begin{cases} \Delta u_k^1(r,\theta) + q_0 u_k^1(r,\theta) = -\sum_{p=0}^{k-1} q_{k-p} u_k^1(r,\theta) & \text{for } r \leq a \\ \text{and} \quad u_k^1(a,\theta) = -\sum_{m=0}^{k-1} \frac{f(\theta)^{k-m}}{(k-m)!} \frac{\partial^{k-m} u_m^1}{\partial r^{k-m}}(a,\theta). \end{cases}$$

To find q_n, α_0, and β_0, we multiply the first equation above for $k = n$ by $(\frac{u_0}{u_0})$ and integrate in $\{r \leq a\}$. After some straightforward calculations we find that

$$q_n^\pm = q_n \pm \frac{|B_{-M}^n|}{\psi_{1,|M|}(a)}, \qquad \alpha_0^\pm = \mp |B_{-M}^n| \quad \text{and} \quad \beta_0^\pm = B_{-M}^n, \qquad (3.35)$$

where the superscripts $\pm$ differentiate the two (simple) eigenvalues. Thus, we define

$$u_k^{1,\pm} = \alpha_0^\pm u_k + \beta_0^\pm \overline{u_k} \quad \text{for } 0 \leq k \leq n-1,$$

and now $u_n^{1,\pm}$ can be computed as in Steps 3 and 4, since the choices in (3.35) guarantee that the new $B_{\pm M}^n$ vanish.

Step 6. Calculate $q_{n+1+\nu}$ and choose appropriate eigenfunctions to order $1+\nu$ for each $\nu \geq 0$ (iterate in ν). For this, assume that for a fixed integer $\nu \geq 0$ we have calculated

$$\{u_k^{\nu+1,\pm}\}_{k=0}^{n+\nu} \quad \text{and} \quad \{q_k^\pm\}_{k=0}^{n+\nu} \qquad (q_k^\pm = q_k \text{ for } k = 0, \ldots, n-1).$$

We want to define $q_{n+\nu+1}^\pm$ and new corrections $\{u_k^{\nu+2,\pm}\}_{k=1}^{n+\nu+1}$, where

$$u_k^{\nu+2,\pm} = u_k^{\nu+1,\pm} \quad \text{for } k = 0, \ldots, \nu.$$

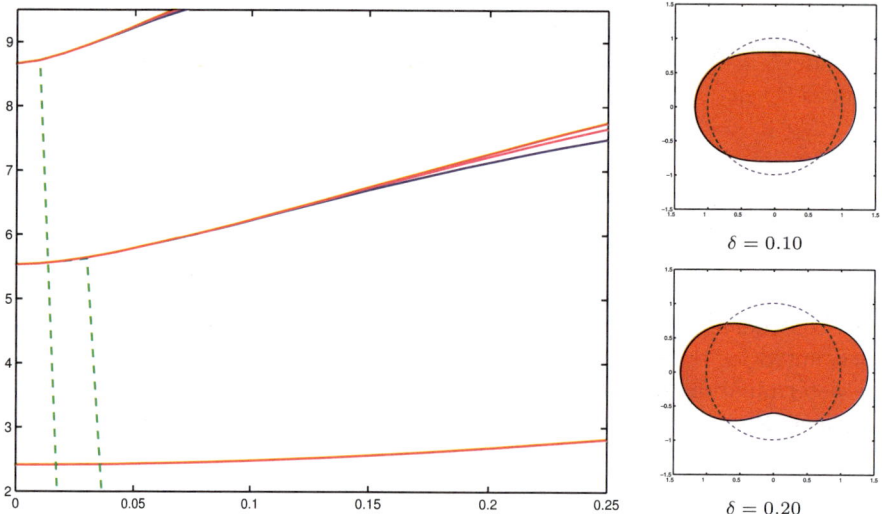

Figure 3.10: Continuation of three zeros of J_0 as eigenfrequencies for perturbations with $f(\theta) = 2\cos(2\theta)$. *No splitting*. Dashed green: Taylor series (order 28); solid blue: Padé [10/10]; solid magenta: Padé [12/12]; solid red: Padé [14/14]. Inset: domains $r = 1 + \delta f(\theta)$ for $\delta = 0.10$ and $\delta = 0.20$.

For instance, if $\nu = 0$, we want to define $q_{n+1}^{\pm}$ and corrections $\{u_k^{2,\pm}\}_{k=0}^{n+1}$; however, $u_0^{2,\pm} = u_0^{1,\pm}$ has already been appropriately chosen in the previous step. We shall then look for solutions $u_k^{\nu+2,\pm}$ of the form

$$u_k^{\nu+2,\pm} = u_k^{\nu+1,\pm} \qquad \text{for } k = 0, \ldots, \nu,$$
$$u_k^{\nu+2,\pm} = u_k^{\nu+1,\pm} + \alpha_{\nu+1}^{\pm} u_{k-\nu-1} + \beta_{\nu+1}^{\pm} \overline{u}_{k-\nu-1} \qquad \text{for } k = \nu+1, \ldots, n+\nu,$$

where the u_m, $0 \le m \le n-1$, are those that were originally found (Steps 1–4). As before (Step 5), from the orthogonality conditions it is possible to find the values of $q_{n+1+\nu}^{\pm}$, $\alpha_{\nu+1}^{\pm}$, and $\beta_{\nu+1}^{\pm}$. And, finally, $u_{n+\nu+1}^{\nu+2,\pm}$ can be determined as in Steps 3 and 4.

We have implemented the above algorithm and have conducted experiments for a variety of perturbations $f(\theta)$. Figures 3.10–3.13 depict the results of our codes in cases were no splitting occurs and were the eigenvalues separate at orders $n = 1$, 2, and 4, respectively. We conclude with Table 3.9 where we record the outcome of convergence studies for the case of first-order splitting (Figure 3.11). These results show that very accurate approximations can be achieved for substantial deformations which may be well beyond the disk of convergence of the Taylor series.

Chapter 3. High-Order Boundary Perturbation Methods   103

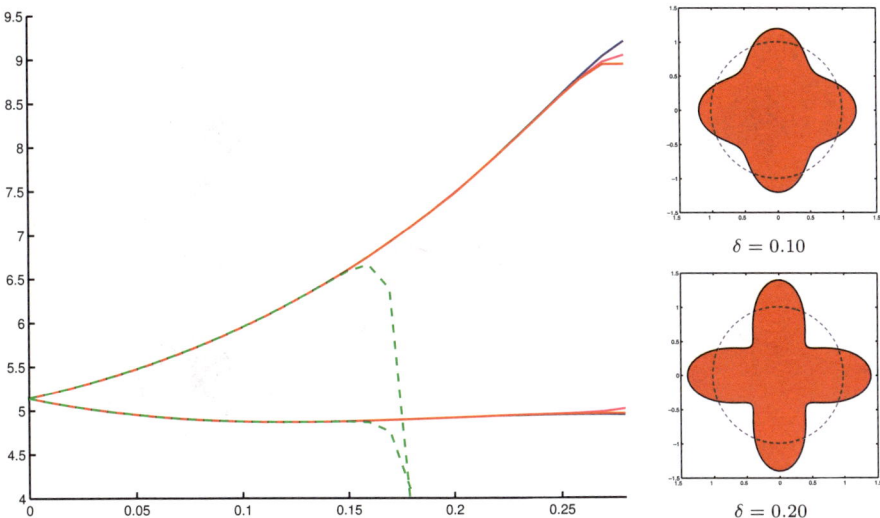

Figure 3.11: Continuation of a zero of J_2 as eigenfrequencies for perturbations with $f(\theta) = 2\cos(4\theta)$. *Splitting at order* 1. Dashed green: Taylor series (order 28); solid blue: Padé [10/10]; solid magenta: Padé [12/12]; solid red: Padé [14/14]. Inset: domains $r = 1 + \delta f(\theta)$ for $\delta = 0.10$ and $\delta = 0.20$.

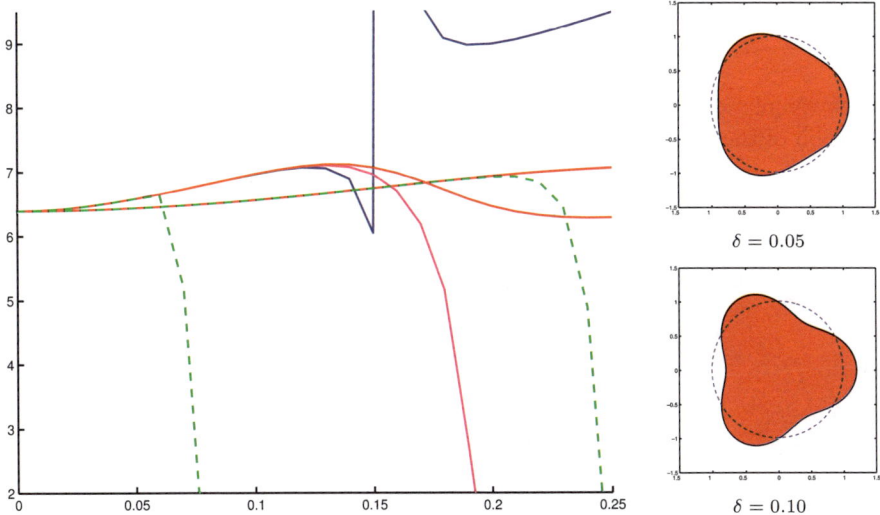

Figure 3.12: Continuation of a zero of J_3 as eigenfrequencies for perturbations with $f(\theta) = 2\cos(3\theta)$. *Splitting at order* 2. Dashed green: Taylor series (order 28); solid blue: Padé [10/10]; solid magenta: Padé [12/12]; solid red: Padé [14/14]. Inset: domains $r = 1 + \delta f(\theta)$ for $\delta = 0.05$ and $\delta = 0.10$.

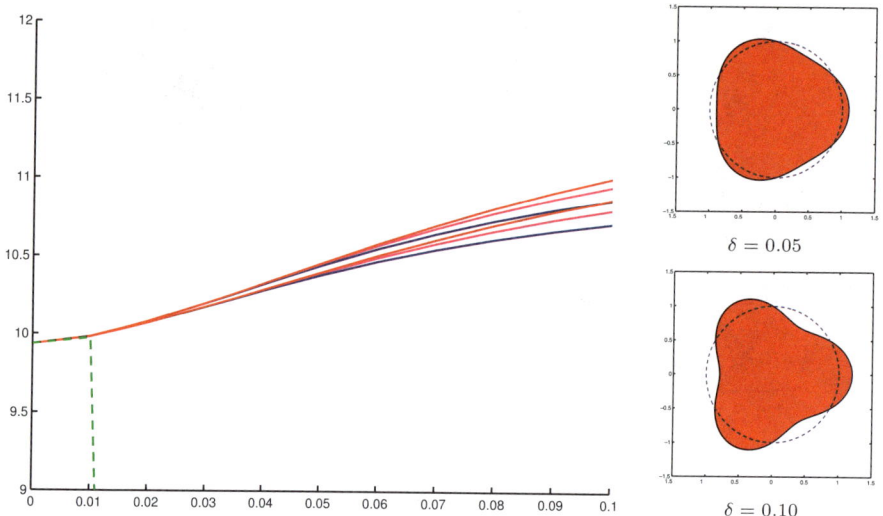

Figure 3.13: Continuation of a zero of J_6 as eigenfrequencies for perturbations with $f(\theta) = 2\cos(3\theta)$. *Splitting at order* 4. Dashed green: Taylor series (order 28); solid blue: Padé [10/10]; solid magenta: Padé [12/12]; solid red: Padé [14/14]. Inset: domains $r = 1 + \delta f(\theta)$ for $\delta = 0.05$ and $\delta = 0.10$.

Table 3.9: Numerical values for the example of Figure 3.11: $f(\theta) = 2\cos(4\theta)$. Continuation of the zero $z = 5.1356223$ of $J_2(z)$.

N	δ	Padé ($[N/2, N/2]$)	Taylor (N)
4	0.10	5.932459814	5.933746721
8	0.10	5.939888161	5.939685615
12	0.10	5.940103147	5.940002737
16	0.10	5.940063852	5.940052067
20	0.10	5.940063913	5.940061333
24	0.10	5.940063912	5.940063302
28	0.10	5.940063912	5.940063758

N	δ	Padé ($[N/2, N/2]$)	Taylor (N)
4	0.20	7.250946252	7.290785243
8	0.20	7.430618958	7.338717672
12	0.20	7.476836436	7.147414978
16	0.20	7.459583127	6.463497406
20	0.20	7.462505263	2.837427216
24	0.20	7.462303523	Diverges

References

[1] G. V. Anand and M. K. George, *Normal mode sound propagation in an ocean with sinusoidal surface waves*, J. Acoust. Soc. Amer., 80 (1986), pp. 238–243.

[2] G. V. Anand and M. K. George, *Normal mode sound propagation in an ocean with random narrow–band surface waves*, J. Acoust. Soc. Amer., 94 (1993), pp. 279–292.

[3] G. A. Baker and P. Graves-Morris, *Padé Approximants,* 2nd ed., Encyclopedia Math. Appl. 59, Cambridge University Press, Cambridge, UK 1996.

[4] J. J. Bowman, T. B. A. Senior, and P. L. E. Uslenghi, *Electromagnetic and acoustic scattering by simple shapes*, Revised Printing, Hemisphere Publishing, New York, 1987.

[5] C. Brezinski, *Procedures for estimating the error in Padé approximation*, Math. Comp., 53 (1965), pp. 639–648.

[6] O. P. Bruno and P. Laurence, *On the MHD equations in a three-dimensional toroidal geometry*, Comm. Pure Appl. Math., 49 (1996), pp. 717–764.

[7] O. P. Bruno and F. Reitich, *Solution of a boundary value problem for Helmholtz equation via variation of the boundary into the complex domain*, Proc. Roy. Soc. Edinburgh, 122A (1992), pp. 317–340.

[8] O. P. Bruno and F. Reitich, *Numerical solution of diffraction problems: A method of variation of boundaries*, J. Opt. Soc. Amer. A, 10 (1993), pp. 1168–1175.

[9] O. P. Bruno and F. Reitich, *Numerical solution of diffraction problems: A method of variation of boundaries II. Dielectric gratings, Padé approximants and singularities*, J. Opt. Soc. Amer. A, 10 (1993), pp. 2307–2316.

[10] O. P. Bruno and F. Reitich, *Numerical solution of diffraction problems: A method of variation of boundaries III. Doubly periodic gratings*, J. Opt. Soc. Amer. A, 10 (1993), pp. 2551–2562.

[11] O. P. Bruno and F. Reitich, *Approximation of analytic functions: A method of enhanced convergence*, Math. Comp., 63 (1994), pp. 195–213.

[12] O. P. Bruno and F. Reitich, *Calculation of electromagnetic scattering via boundary variations and analytic continuation*, ACES J., 11 (1996), pp. 17–31.

[13] O. P. Bruno and F. Reitich, *Boundary-variation solutions for bounded-obstacle scattering in three dimensions*, J. Acoust. Soc. Amer., 104 (1998), pp. 2579–2583.

[14] A. P. Calderón, *Cauchy integrals on Lipschitz curves and related operators*, Proc. Nat. Acad. Sci. USA, 75 (1977), pp. 1324–1327.

[15] R. K. Chang and A. J. Campillo, *Optical Processes in Microcavities*, World Scientific, Singapore, 1996.

[16] R. R. Coifman, A. MacIntosh, and Y. Meyer, *L'integrale de Cauchy définit un opérateur borné sur L^2 pour les courbes Lipschitziennes*, Ann. of Math., 116 (1982), pp. 361–387.

[17] G. H. Derrick, R. C. McPhedran, D. Maystre, and M. Nevière, *Crossed gratings: A theory and its applications*, in Electromagnetic Theory of Gratings, R. Petit, ed., Springer-Verlag, Berlin, 1980, pp. 227–275.

[18] D. C. Dobson and J. A. Cox, *An integral equation method for biperiodic diffraction structures*, in Proc. SPIE 1545, Bellingham, WA, 1991, pp. 106–113.

[19] C. Gmachl, F. Carpasso, E. E. Narimanov, J. U. Nöckel, A. D. Stone, J. Faist, D. L. Sivco, and A. Y. Cho, *High-power directional emission from microlasers with chaotic resonators*, Science, 280 (1998), pp. 1556–1564.

[20] G. Goubau, *Electromagnetic Waveguides and Cavities*, Pergamon Press, London, 1961.

[21] P. Graves-Morris, *The numerical calculation of Padé approximants*, in Lecture Notes in Math. 765, L. Wuytack, ed., Springer-Verlag, Berlin, 1979, pp. 231–245.

[22] J. J. Greffet, *Scattering of electromagnetic waves by rough dielectric surfaces*, Phys. Rev. B, 37 (1988), pp. 6436–6441.

[23] J. J. Greffet, C. Baylard and P. Versaevel, *Diffraction of electromagnetic waves by crossed gratings: A series solution*, Opt. Lett., 17 (1992), pp. 1740–1742.

[24] J. J. Greffet and Z. Maassarani, *Scattering of electromagnetic waves by a grating: A numerical evaluation of the iterative-series solution*, J. Opt. Soc. Amer. A, 7 (1990), pp. 1483–1493.

[25] E. Y. Harper and F. M. Labianca, *Perturbation theory for scattering of sound from a point source by a moving rough surface in the presence of refraction*, J. Acoust. Soc. Amer., 57 (1975), pp. 1044–1051.

[26] E. Y. Harper and F. M. Labianca, *Scattering of sound from a point source by a rough surface progressing over an isovelocity ocean*, J. Acoust. Soc. Amer., 58 (1975), pp. 349–364.

[27] D. R. Jackson, D. P. Winebrenner, and A. Ishimaru, *Comparison of perturbation theories for rough-surface scattering*, J. Acoust. Soc. Amer., 83 (1988), pp. 961–969.

[28] M. S. Janaki and B. Dasgupta, *Eigenmodes in a toroidal cavity of elliptic cross section*, IEEE Trans. Microwave Theory Tech., 44 (1996), pp. 1147–1150.

[29] D. S. Jones, *The theory of electromagnetism*, MacMillan, New York, 1964.

[30] T. Kato, *Perturbation theory for linear operators*, 2nd ed., Springer-Verlag, Berlin, 1980.

[31] G. C. Kokkorakis and J. A. Roumeliotis, *Electromagnetic eigenfrequencies in a spheroidal cavity*, J. Electromagnetic Waves Appl., 11 (1997), pp. 279–292.

[32] W. A. Kuperman and F. Ingenito, *Attenuation of the coherent component of sound propagating in shallow water with rough boundaries*, J. Acoust. Soc. Amer., 61 (1977), pp. 1178–1187.

[33] H. M. Lai, P. T. Leung, K. Young, P. W. Barber, and S. C. Hill, *Time-independent perturbation for leaking electromagnetic modes in open systems with application to resonances in microdroplets*, Phys. Rev. A, 41 (1990), pp. 5187–5198.

[34] L. Levi, *Applied Optics*, Wiley, New York, 1968.

[35] Y. Liu, S. J. Frasier, and R. E. McIntosh, *Measurement and classification of low-grazing-angle radar sea spikes*, IEEE Trans. Antennas and Propagation, 46 (1998), pp. 27–40.

[36] C. Lopez, F. J. Yndurain, and N. Garcia, *Iterative series for calculating the scattering of waves from hard corrugated surfaces*, Phys. Rev. B, 18 (1978), pp. 970–972.

[37] A. A. Maradudin, *Iterative solutions for electromagnetic scattering by gratings*, J. Opt. Soc. Amer., 73 (1983), pp. 759–764.

[38] D. Maystre, *Rigorous vector theories of diffraction gratings*, in Progress in Optics, E. Wolf, ed., North Holland, Amsterdam, 1984, pp. 3–67.

[39] D. Maystre and M. Nevière, *Electromagnetic theory of crossed gratings*, J. Optics, 9 (1978), pp. 301–306.

[40] D. Maystre, M. Nevière, and R. Petit, *Experimental verifications and applications of the theory*, in Electromagnetic Theory of Gratings, R. Petit, ed., Springer-Verlag, Berlin, 1980, pp. 159–223.

[41] R. C. McPhedran, G. H. Derrick, and L. C. Botten, *Theory of crossed gratings*, in Electromagnetic Theory of Gratings, R. Petit, ed., Springer-Verlag, Berlin, 1980, pp. 227–275.

[42] W. C. Meecham, *On the use of the Kirchoff approximation for the solution of reflection problems*, J. Rational Mech. Anal., 5 (1956), pp. 323–334.

[43] A. Mekis, J. U. Nöckel, G. Chen, A. D. Stone, and R. K. Chang, *Ray chaos and Q spoiling of lasing droplets*, Phys. Rev. Lett., 75 (1995), pp. 2682–2685.

[44] A. H. Nayfeh and O. R. Asfar, *Parallel-plate waveguide with sinusoidally perturbed boundaries*, J. Appl. Phys., 45 (1974), pp. 4797–4800.

[45] J. U. Nökel, A. D. Stone, and R. K. Chang, *Q spoiling and directionality in deformed ring cavities*, Optics Letters, 19 (1994), pp. 1693–1695.

[46] J. Pavageau and J. Bousquet, *Diffraction par un réseau conducteur nouvelle méthode de résolution*, Optica Acta, 17 (1970), pp. 469–478.

[47] W. H. Peake, *Theory of radar return from terrain*, IRE Nat'l Conv., 7 (1959), pp. 27–41.

[48] R. Petit, *A tutorial introduction*, in Electromagnetic Theory of Gratings, R. Petit, ed., Springer-Verlag, Berlin, 1980, pp. 1–52.

[49] R. Petit and M. Cadilhac, *Sur la diffraction d'une onde plane par un réseau infinitement conducteur*, C. R. Acad. Sci. Paris Sér. B, 262 (1966), pp. 468–471.

[50] Lord Rayleigh, *The Theory of Sound*, Vol. 2, Dover, New York, 1945.

[51] F. Rellich, *Perturbation Theory of Eigenvalue Problems*, Gordon and Breach Science Publishers, New York, 1969.

[52] S. O. Rice, *Reflection of electromagnetic waves from slightly rough surfaces*, Comm. Pure Appl. Math., 4 (1951), pp. 351–378.

[53] J. Roginsky, *Derivation of closed-form expressions for the T matrices of Rayleigh–Rice and extinction-theorem perturbation theories*, J. Acoust. Soc. Amer., 90 (1991), pp. 1130–1137.

[54] R. E. Scraton, *A note on the summation of divergent power series*, Proc. Cambridge Philos. Soc., 66 (1969), pp. 109–114.

[55] A. Sei, O. Bruno, and M. Caponi, *Study of polarization dependent scattering anomalies with applications to oceanic scattering*, Radio Sci., 34 (1999), pp. 385–411.

[56] N. C. Skaropoulos and D. P. Chrissoulidis, *General perturbative solution to wave scattering from a soft random cylindrical surface*, J. Acoust. Soc. Amer., 106 (1999), pp. 596–604.

[57] D. Talbot, J. B. Titchener, and J. R. Willis, *The reflection of electromagnetic waves from very rough interfaces*, Wave Motion, 12 (1990), pp. 245–260.

[58] J. L. Uretsky, *The scattering of plane waves from periodic surfaces*, Ann. Phys., 33 (1965), pp. 400–427.

[59] G. R. Valenzuela, *Theories for the interaction of electromagnetic and oceanic waves—A review*, Boundary-Layer Meteor., 13 (1978), pp. 61–85.

[60] P. M. Van den Berg, *Diffraction theory of a reflection grating*, Appl. Sci. Res., 24 (1971), pp. 261–293.

[61] P. Vincent, *Differential Methods*, in Electromagnetic Theory of Gratings, R. Petit, ed., Springer-Verlag, Berlin, 1980, pp. 101–121.

[62] J. R. Wait, *Perturbation analysis for reflection from two-dimensional periodic sea waves*, Radio Sci., 6 (1971), pp. 387–391.

[63] C. Wang, B.-Q. Gao, and C.-P. Deng, *Accurate study of Q-factor of resonator by a finite-difference time-domain method*, IEEE Trans. Microwave Theory and Techniques, 43 (1995), pp. 1524–1529.

[64] A. Wirgin, *Scattering from hard and soft corrugated surfaces: Iterative corrections to the Kirchhoff approximation through the extinction theorem*, J. Acoust. Soc. Amer., 85 (1989), pp. 670–679.

Chapter 4

Mathematical Reflections on the Fourier Modal Method in Grating Theory

Lifeng Li

4.1 Introduction

The modal method is one of the most effective methods for modeling diffraction of electromagnetic waves by periodic gratings. Its basic idea is quite simple: The electromagnetic fields are first solved as eigenfunctions of Maxwell's equations in the interior of the grating region where the periodic permittivity variation occurs. The total fields are then expressed as superpositions of the eigenfunctions. The final solution of the grating problem is obtained by matching the electromagnetic boundary conditions at the boundaries of the grating region. The most crucial step in the modal method is the solution of the eigenfunctions. In physical and engineering terms the eigenfunctions of a mechanical or electromagnetic system are called modes, hence the name *modal method*.

Among many existing approaches to seeking the eigenfunctions, one approach is most favored in the applied optics community. This approach expands both the fields and the permittivity function into Fourier series, thus transforming the electromagnetic boundary value problem in real space into a matrix eigenvalue problem in the denumerably infinite Fourier space. Many names have been given by various authors to this popular method. A probably incomplete list includes the BKK (Burckhardt, Kaspar, and Knop; see below) method, Fourier expansion method, Fourier modal method, Moharam–Gaylord method, rigorous coupled-wave approach, and waveguide method. To this author, the name Fourier modal method (FMM) is the most appropriate because it accurately describes the character of the method.

As far as the analysis of gratings is concerned, Tamir, Wang, and Oliver [1, 2, 3] seem to have been the first, in the mid-1960s, to use the FMM. However, because they only studied volume gratings of sinusoidal permittivity variation in the transverse electric (TE) polarization, their treatment did not involve true matrix operations. The similar problem in transverse magnetic (TM) polarization was studied at almost the same time by Yeh, Casey, and Kaprielian [4]. Their work was on the propagation characteristics of the waves in a sinusoidally stratified medium. Therefore, it did not require the complete solution of a matrix eigenvalue problem. Burckhardt [5] was the first, in 1966, to present a full matrix formulation of the FMM. Although he also confined himself to sinusoidally stratified media, his formulation was general enough to allow other forms of permittivity stratification, as realized by Kaspar [6] in 1973. Not knowing the work of Burckhardt, in 1975 Peng, Tamir, and Bertoni [7] redeveloped and, more important, made an attempt to justify the validity of, the FMM. However, their conclusion appears to be overly conservative. Apparently unaware of the work of Peng, Tamir, and Bertoni, Knop [8] later applied the method to rectangular surface-relief gratings anyway. Between 1981 and 1986, Moharam and Gaylord applied the FMM to analyze slanted volume gratings [9], surface-relief gratings of arbitrary profiles in TE and TM polarizations [10, 11] and in conical mountings [12], and to metallic gratings [13]. The problem of one-dimensionally periodic anisotropic gratings was first treated by Rokushima and Yamakita [14] in 1983. The problem of two-dimensionally periodic (crossed) isotropic gratings was analyzed by Han et al. [15] and by Noponen and Turunen [16].

Nowadays the FMM has become one of the most versatile methods for modeling diffraction gratings. However, the problem of slow convergence, which occurred with metallic gratings whenever the electric field vector had a component perpendicular to the permittivity discontinuities in the grating region [17], plagued the method in most of its history. In the case of one-dimensional isotropic gratings, the problem was recently cured, more or less empirically or accidentally, by Lalanne and Morris [18] and by Granet and Guizal [19]. Their work has led me to establish three theorems on Fourier factorization that govern the expression for the Fourier coefficients of a product of discontinuous periodic functions [20]. The mathematical result in turn has yielded greatly improved convergence of the FMM when it is applied to crossed gratings [21], anisotropic gratings [22], and several other cases.

The popularity of the FMM can probably be ascribed more to its simplicity than to its versatility. However, this does not mean that the method is well understood from a mathematical standpoint. To the contrary, until recently there was not even a mathematical proof that the method converges, let alone an estimate of its convergence rate. Section 4.2 introduces the FMM for the simplest grating problem: the diffraction of a plane wave by a rectangular grating. For formulations of the FMM for other types of grating problems, the reader is referred to the references cited above. This chapter is devoted to two mathematical aspects of the FMM, which are both necessitated by the inescapable matrix truncation in numerical analysis. In section 4.3, the validity of using the solutions of a truncated matrix eigenvalue problem to approximate the exact solutions is established for

Chapter 4. The Fourier Modal Method in Grating Theory

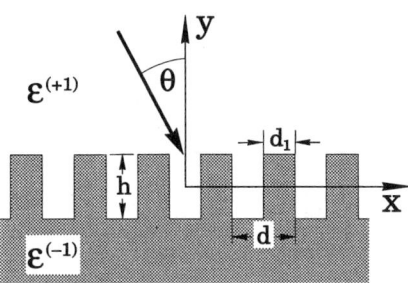

Figure 4.1: A one-dimensionally periodic rectangular grating.

the grating problem introduced in section 4.2. In section 4.4, the three theorems on Fourier factorization are presented, followed by a comprehensive discussion of their graphical interpretation, their relationship to physical problems, and their numerical implications. At the end of section 4.4 three examples are provided to illustrate the procedures of applying these theorems to solve grating problems and to demonstrate the improvement in convergence as a result of using the factorization theorems.

4.2 Mathematical Formulation of FMM for Rectangular Gratings

Figure 4.1 depicts a one-dimensionally periodic grating, which consists of a rectangularly corrugated surface separating two isotropic, nonmagnetic media with permittivities $\epsilon^{(+1)}$ and $\epsilon^{(-1)}$. A monochromatic plane wave with its wave vector contained in the xOy plane is incident on the grating. The two-dimensional space is divided into three regions. In the upper ($y > h/2$) and lower ($y < -h/2$) regions, the expressions of the electromagnetic waves are readily given by plane wave expansions, i.e., the so-called Rayleigh expansions [23]. Such expansions are employed in almost all rigorous mathematical treatments of diffraction gratings, including the FMM.

Because the grating structure is invariant along the z-direction, the diffraction problem can be solved in two independent polarizations, TE/TM (the electric/magnetic vector parallel to the grooves). From Maxwell's equations, we have

$$\partial_y E_z = ik_0\mu H_x, \qquad (4.1a)$$
$$\partial_x E_z = -ik_0\mu H_y, \qquad (4.1b)$$
$$\partial_x H_y - \partial_y H_x = -ik_0\varepsilon E_z \qquad (4.1c)$$

for TE polarization and

$$\partial_y H_z = -ik_0\varepsilon E_x, \qquad (4.2a)$$
$$\partial_x H_z = ik_0\varepsilon E_y, \qquad (4.2b)$$
$$\partial_x E_y - \partial_y E_x = ik_0\mu H_z \qquad (4.2c)$$

for TM polarization. In the above, E_w and H_w with $w = x, y$, and z are the vector components of the electric and magnetic fields, k_0 is the vacuum wavenumber, and $\mu = 1$. The weight of grating analysis lies mostly in solving the above equations in the grating region ($|y| < h/2$), where $\epsilon(x+d) = \epsilon(x)$. The simplicity of the rectangular grating allows the total fields inside the grating region to be written as superpositions of modes whose y-dependence is exponential: $\exp(i\lambda y)$, where λ is the eigenvalue. From (4.1) and (4.2) the eigenvalue problems follow easily:

$$[\partial_x^2 + k_0^2 \mu \varepsilon - \rho]E_z = 0, \tag{4.3}$$

$$E_z(x+d) = e^{i\alpha_0 d} E_z(x)$$

for TE polarization and

$$\varepsilon \left[\partial_x \left(\frac{1}{\varepsilon}\partial_x\right) + k_0^2 \mu\right] H_z = \rho H_z, \tag{4.4}$$

$$H_z(x+d) = e^{i\alpha_0 d} H_z(x)$$

for TM polarization, where $\alpha_0 = k_0[\epsilon^{(+1)}\mu]^{1/2}\sin\theta$ and $\rho = \lambda^2$. The solution of (4.3) or (4.4), together with

$$H_x = \frac{\lambda}{k_0 \mu} E_z \tag{4.5}$$

or

$$E_x = -\frac{\lambda}{k_0 \varepsilon} H_z, \tag{4.6}$$

will provide the necessary field components for matching the boundary conditions with the Rayleigh expansions along $y = \pm h/2$. The coefficients of the Rayleigh expansions give the efficiencies and polarization parameters of the diffraction orders directly. Hence the whole grating problem is solved.

The FMM solves the TE eigenvalue problem (4.3) by expanding $\epsilon(x)$ and $E_z(x)$ into Fourier series and Floquet–Fourier series

$$\varepsilon(x) = \sum_n \varepsilon_n e^{inKx}, \tag{4.7}$$

$$E_z(x) = \sum_n E_{zn} e^{i\alpha_n x}, \tag{4.8}$$

respectively, where $K = 2\pi/d$ and $\alpha_n = \alpha_0 + nK$. In this chapter summation signs without explicit limit indications extend from $-\infty$ to $+\infty$. Substituting (4.7) and (4.8) into (4.3) yields

$$(\rho + \alpha_n^2)E_{zn} = k_0^2 \mu \sum_m \varepsilon_{n-m} E_{zm}. \tag{4.9}$$

Thus, the periodic boundary value problem (4.3) in real space is turned into an algebraic eigenvalue problem in Fourier space. The implementation of FMM in TE polarization is straightforward and historically uneventful, whereas the treatment of TM polarization, as mentioned in the previous section, has gone through some interesting changes. We will discuss TM polarization in section 4.4.

4.3 Applicability of Reduction Method

Since the order of the matrix eigenvalue problem in the FMM is infinite and the exact solution is in general not available, numerical solutions are sought. The common approach is to solve a problem of finite order obtained by judiciously truncating the infinite matrix. Mathematically this approximation method is called a reduction method. It is applicable provided that the approximate solutions tend to the exact ones as the order of the matrix increases. Although the FMM has been widely used for a long time, few authors have seriously considered the legitimacy of employing the reduction method. For a brief survey of how the earlier researchers handled this issue, the reader is referred to my 1999 article [24]. In short, most researchers have put their good faith in the physics of the problem or in the satisfactory performance of their numerical codes. However, there are at least three good reasons to investigate the applicability of the reduction method.

(1) To develop the FMM into a scientific method (not just an engineering tool), it is necessary to place it on a solid mathematical ground. Physical understanding cannot supplant mathematical rigor. Numerical experiments cannot constitute mathematical proofs. After all, there are examples of a sequence of finite matrices whose eigenvalues exist but do not converge to the eigenvalues of the corresponding infinite matrix [25].

(2) The applicability study will pave the road to a more refined analysis of the convergence characteristics of the method, which may lead to improvements to the present implementation or discovery of new approximation methods.

(3) Such a study potentially could also suggest a priori criteria for determining optimum truncation.

For simplicity, let us consider only the eigenvalue problem (4.9). The aim of this section is to prove the existence of the solutions of (4.9) and to establish the applicability of the reduction method in seeking the approximate solutions.

In deriving (4.9) we have assumed the legitimacy of two operations on the infinite series (4.7) and (4.8). The first is the term-by-term differentiation of (4.8), and the second is the expansion of the product of the two infinite series by using Laurent's multiplication rule. Based on physical considerations, E_z should be bounded and have a continuous first-order derivative and a piecewise continuous second-order derivative; therefore, the first operation is permissible [26, p. 169]. In the domain of infinite mathematics, according to Zygmund [27], the multiplication of two Fourier series can always be carried out using Laurent's rule if the functions that the series represent are square integrable. The function E_z described above and the permittivity function $\epsilon(x)$ of interest in physics both satisfy this condition. As a matter of fact, in practical situations $\epsilon(x)$ is either continuous (volume grating) or piecewise continuous (surface-relief grating). In the domain of finite mathematics imposed by matrix truncation, as will be shown in section 4.4, Laurent's rule is not always the best rule to use even when the two functions being multiplied together are both square integrable. Nonetheless, the analysis in section 4.4 confirms that in the present circumstance the use of Laurent's rule is valid. Equation (4.9) can be rewritten as

$$\sum_m \left(\delta_{nm} + \frac{\varepsilon'_{n-m}}{k_0^2 \mu \varepsilon_0 - \alpha_n^2 - \rho} \right) E_{zm} = 0, \qquad (4.10)$$

where δ_{mn} is the Kronecker symbol, $\epsilon'_{n-m} = k_0^2 \mu \epsilon_{n-m}$ if $n \neq m$ and $\epsilon'_0 = 0$, and we have assumed that $\rho \neq k_0^2 \mu \epsilon_0 - \alpha_n^2$ for all n.

In order to prove the convergence of the solutions to equation (4.10), we use the theory of determinant of infinite order. Although the theory of eigenvalue problems can be formulated independent of determinant theory and the results of this section can probably be proven by using more modern mathematical theory, the use of determinant theory here makes the analysis elementary. For the history of determinants of infinite order, the reader may consult Riesz [28] and Bernkopf [29].

Before I present the convergence theorems on infinite determinants, let us first go through some definitions. We denote by $A = \{A_{ik}; i, k = \ldots, -2, -1, 0, +1, +2, \ldots\}$ a matrix of infinite order. The matrix elements A_{ik} will also be written as $A_{ik} = \delta_{ik} + a_{ik}$. The matrix of finite order obtained by truncating off those elements A_{ik} for which $|i|, |k| > M$ is denoted by $A^{(M)}$, and its determinant by $D^{(M)}$. In what follows the superscript (M) is used invariably to indicate something that is associated with the truncated matrix. The symmetrical truncation about $i = k = 0$ is for convenience rather than necessity. The determinant of A, denoted D, is defined as the limit of $D^{(M)}$ as $M \to \infty$, if it exists. D is said to be absolutely convergent if $D^{(M)}$ is absolutely convergent as $M \to \infty$ even if A_{ii} are replaced by $1 + |a_{ii}|$. The minor of D obtained by setting, on row i, the kth element to unity and all other elements to zero is denoted by $\binom{i}{k}$.

The first convergence theorem on infinite determinants is due to Poincaré [30]. The following is a version of the theorem extended by von Koch [31].

THEOREM 4.1 (Poincaré–von Koch). *For the determinant of A and all of its minors to be absolutely convergent, it is sufficient that*

$$\sum_{i,k} |a_{ik}| < \infty. \qquad (4.11)$$

If condition (4.11) is satisfied, the determinant remains absolutely convergent when a row or a column of A is replaced by a sequence of bounded numbers.

A determinant that satisfies condition (4.11) is said to be in *normal* form.

For volume gratings, Theorem 4.1 is sufficient for proving that the determinant of the coefficient matrix in (4.10) is absolutely convergent. Indeed, the matrix is already in the required form, and for volume gratings $\epsilon'_n \leq O(1/n^2)$. However, for surface-relief gratings Theorem 4.1 is insufficient because condition (4.11) is not satisfied with $\epsilon'_n = O(1/n)$. It is a pity that this theorem is all that one can easily find in advanced textbooks and reference books such as [26] and [32]. In these books, in order to have a normal determinant, the authors assume that the coefficient function in the differential equations like (4.3) is differentiable everywhere, a luxury that one cannot afford in many practical situations.

Thanks to von Koch, many alternative convergence tests have been in existence for a century. Let us suppose that the matrix elements A_{ik} are functions of a parameter τ in a domain T in the complex plane. The following are the results given by von Koch in 1892.

THEOREM 4.2 (von Koch [31]). *For the determinant of $A(\tau)$ and all of its minors to be absolutely and uniformly convergent in a domain T, it is sufficient that there exists a sequence of nonzero numbers, $\{x_i\}$, such that $\Sigma_{i,k}|a_{i,k}(\tau)x_i/x_k|$ is uniformly convergent in T. If the above condition is satisfied, the determinant remains absolutely and uniformly convergent when a row or a column of $A(\tau)$ is replaced by a sequence of bounded numbers.*

Von Koch further showed that for a system of linear equations with an absolutely convergent determinant, many results of classical linear algebra are unchanged. For example, Laplace expansion and Cramer's rule are still valid, and the necessary and sufficient condition for a homogeneous system to have nonzero solutions is that the determinant be zero.

With Theorem 4.2 we can prove that the coefficient matrix in (4.10) is absolutely convergent, even in the case of surface-relief gratings. To this aim, let us define $\Omega(R,r)$ to be a domain that consists of a disk of radius R centered at the origin in the complex plane of τ but excludes all of the small disks of radius r centered at $k_0^2\mu\epsilon_0 - \alpha_n^2, n = 0, \pm 1, \ldots$. For the sequence $\{x_i\}$ in the theorem we can simply take $x_n = n$ if $n \neq 0$, and $x_0 = 1$. It is an elementary exercise to show that in a given $\Omega(R,r)$, no matter how large R is and how small r is, the double series $\Sigma_{i,k}|a_{i,k}(\tau)x_i/x_k|$ is uniformly convergent in Ω, thus proving the opening statement of this paragraph. The determinant $D(\tau)$ defines an analytical function of τ in Ω. The solution of $D(\tau) = 0$ gives the eigenvalues. For each eigenvalue ρ, the elements of the corresponding eigenvector, $E_{zn}(\rho)$, can be expressed in terms of one of themselves, for example, E_{z0}. Suppose that $\binom{0}{0} \neq 0$. Then by Cramer's rule

$$E_{zn}(\rho) = \frac{\binom{0}{n}(\rho)}{\binom{0}{0}(\rho)} E_{z0}, \qquad (4.12)$$

where we have indicated the dependence of the minors on ρ.

Thus far we have proven the existence of the solutions of the infinite eigenvalue problem (4.10). Our final objective is to prove that the exact solutions can be approximated by using a reduction method; i.e., $\rho^{(M)}$ and $\{E_{zn}^{(M)}[\rho^{(M)}]\}$ converge to ρ and $\{E_{zn}(\rho)\}$ as $M \to \infty$. The desired results are assured by the uniform convergence of the determinant and its minors. Based on a theorem on uniform convergence [33], $|D(\tau_M) - D^{(M)}(\tau_M)| \to 0$ as $M \to \infty$ for any sequence of numbers $\{\tau_M\}$ in Ω. In particular, we can choose $\tau_M = \rho^{(M)}$, a solution of $D^{(M)}(\tau) = 0$. Then, we have $D[\tau^{(M)}] \to 0$; that is, in $\Omega(R,r)$ the solutions of $D^{(M)}(\tau) = 0$ tend to the solutions of $D(\tau) = 0$ as $M \to \infty$. Note that we can make the radius r in $\Omega(R,r)$ sufficiently small so that all zeros of $D(\tau)$ and $D^{(M)}(\tau)$ inside the disk of radius R fall outside the small disks. Next, let $\rho_0^{(M)}$ and ρ_0 be the zeros of $D^{(M)}(\tau)$ and $D(\tau)$, respectively, such that $\rho_0^{(M)} \to \rho_0$. The eigenvector $\{E_{zn}^{(M)}[\rho_0^{(M)}]\}$ of the truncated eigenproblem has an expression like (4.12). To prove that $\{E_{zn}^{(M)}[\rho_0^{(M)}]\} \to \{E_{zn}(\rho_0)\}$ as $M \to \infty$, it is sufficient to show that $\binom{0}{n}^{(M)}[\rho_0^{(M)}] \to \binom{0}{n}(\rho_0)$. To this aim, we write

$$\left| \binom{0}{n}^{(M)}[\rho_0^{(M)}] - \binom{0}{n}(\rho_0) \right| \leq \left| \binom{0}{n}^{(M)}[\rho_0^{(M)}] - \binom{0}{n}[\rho_0^{(M)}] \right|$$
$$+ \left| \binom{0}{n}[\rho_0^{(M)}] - \binom{0}{n}(\rho_0) \right|. \qquad (4.13)$$

Then the desired result follows because the first term on the right-hand side tends to zero for $\binom{0}{n}^{(M)}(\rho)$ converging uniformly, and the second term tends to zero for $\binom{0}{n}(\rho)$ continuous. This completes our proof of the applicability of the reduction method to the FMM.

It is important to point out that the convergence of the eigensolutions in the FMM is nonuniform over the whole complex plane. To clarify the above statement, let us suppose that the zeros of $D^{(M)}(\tau)$ and $D(\tau)$ are ordered such that $|\rho_1^{(M)}| \leq |\rho_2^{(M)}| \leq \cdots \leq |\rho_{2M+1}^{(M)}|$ and $|\rho_1| \leq |\rho_2| \leq \cdots$, respectively. Then for any fixed integer $n > 0$, the first n eigenvalues of $A^{(M)}$ converge to the first n eigenvalues of A, as $M \to \infty$. However, because asymptotically $|\rho_n| \sim (nK)^2$ for large n, the higher order eigenvalues of $A^{(M)}$ (e.g., $\lambda_{2M+1}^{(M)}$) do not converge. The nonuniform convergence of the eigensolutions has been observed in numerical experiments [17] and is believed to play an important role in determining the overall convergence characteristics of the FMM.

For completeness, we are obligated to verify that the eigenfunctions of (4.10) indeed meet the requirements that allowed us to derive the eigenproblem in the first place. Substituting the expression of E_{zn} given by (4.12) into the right-hand side of (4.9), and using the fact that ϵ_{n-m} is uniformly bounded and that the sum of the minors $\binom{0}{n}$ is absolutely convergent, we see that $E_{zn} \leq O(1/n^2)$. This implies that the function $E_z(x)$ that has E_{zn} as its Fourier coefficients is continuous. Thus, the Laurent multiplication in (4.9) is justified. Furthermore, the right-hand side of (4.9) now is the Fourier coefficient of a piecewise continuous function. Therefore, the Fourier series having $\alpha_n^2 E_{zn}$ as its coefficients converges everywhere. In other words, the term-by-term differentiation is justified. Substituting $E_{zn} \leq O(1/n^2)$ into the right-hand side gives $E_{zn} = O(1/n^3)$ if $\epsilon_n' = O(1/n)$, as expected.

The results given in this section are preliminary. For more details the reader may consult [24]. From a theoretical point of view much work needs to be done in this area. For example, instead of being content with merely knowing that the FMM converges, it is more profitable, but also more difficult, to investigate the precise convergence rate and its parametric dependence on various grating variables. In addition, after the eigensolutions are determined, an inhomogeneous infinite system of linear equations that resulted from matching the boundary conditions along the boundaries of the grating region has to be solved in order to calculate the diffraction

Chapter 4. The Fourier Modal Method in Grating Theory

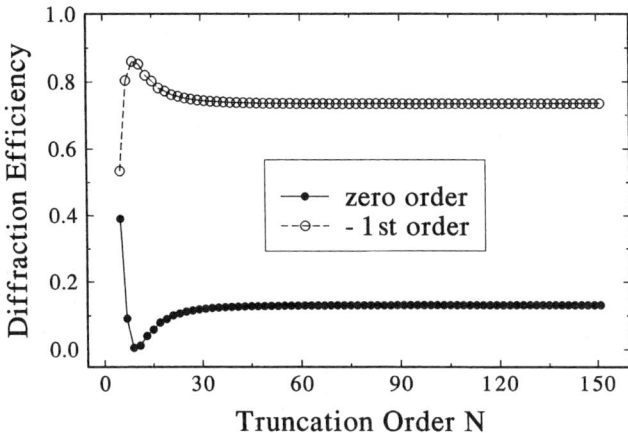

Figure 4.2: Convergence of the diffraction efficiency of a metallic rectangular grating in TE polarization. $\theta = 30°, \epsilon^{(+1)} = 1.0, \epsilon^{(-1)} = (0.22 + i6.71)^2, d = h = 2d_1 = \lambda_0$.

amplitudes. Here, too, it is desirable to prove the legitimacy of truncating the matrices.

4.4 Fourier Factorization

This section is devoted to the theory of Fourier factorization and its applications to grating analysis. Fourier factorization is a term that I coined to mean the expression of function $h(x)$ or its Fourier coefficients in terms of the Fourier coefficients of some functions of $f(x)$ and $g(x)$, where $h(x) = f(x)g(x)$. This subject is relevant here because the multiplication of the Floquet–Fourier series of the unknown electromagnetic field components and the Fourier series of the periodic permittivity function must be carried out in the FMM.

In section 4.4.1, we first consider a numerical example that illustrates the importance of making correct Fourier factorization in numerical analysis. The precise statement of the factorization problem is given in section 4.4.2. Section 4.4.3 presents the mathematical theorems on Fourier factorization. Section 4.4.4 contains a comprehensive discussion of the significance of the mathematical results. In section 4.4.5 the mathematical results are illustrated by numerical examples of solving three different types of grating problems.

4.4.1 Numerical Examples

Consider a rectangular grating as shown in Figure 4.1 with the following parameters: $\epsilon^{(+1)} = 1.0, \epsilon^{(-1)} = (0.22 + i6.71)^2, h = d = 2d_1 = \lambda_0$, and $\theta = 30°$, where λ_0 is the vacuum wavelength. Figure 4.2 shows the satisfactory convergence of the two diffraction orders as the matrix order (truncation order N; see section 4.4.5 for its definition) increases, when the incident plane wave is TE polarized. The eigenvalue

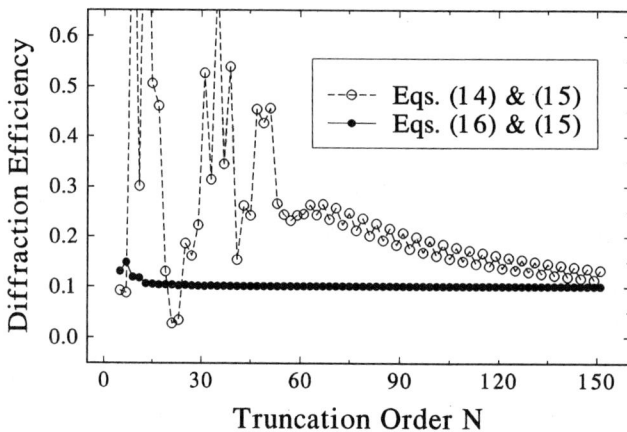

Figure 4.3: Convergence of the first-order diffraction efficiency of the grating in Figure 4.2 in TM polarization.

problem used is given by (4.9). Applying Laurent's rule to (4.4) and (4.6), we obtain

$$\sum_m [\![\varepsilon]\!]_{nm} \sum_p \left(k_0^2 \mu \delta_{mp} - \alpha_m \left[\!\!\left[\frac{1}{\varepsilon} \right]\!\!\right]_{mp} \alpha_p \right) H_{zp} = \rho H_{zn}, \qquad (4.14)$$

$$E_{xn} = -\frac{\lambda}{k_0} \sum_m \left[\!\!\left[\frac{1}{\varepsilon} \right]\!\!\right]_{nm} H_{zm}, \qquad (4.15)$$

where $[\![f]\!]$ is the Toeplitz matrix generated by the Fourier coefficients of function f such that $[\![f]\!]_{mn} = (f)_{m-n}$. The convergence of the first-order diffraction efficiency, for the same grating but in TM polarization, produced by using the combination of (4.14) and (4.15) is shown by the data points represented by the empty circles in Figure 4.3. Evidently, the convergence is not at all satisfactory. The problem of slow convergence for metallic gratings in TM polarization continued for many years, until the publication of [18] and [19]. In place of (4.14), these authors used

$$\sum_m \left[\!\!\left[\frac{1}{\varepsilon} \right]\!\!\right]^{-1}_{nm} \sum_p (k_0^2 \mu \delta_{mp} - \alpha_m [\![\varepsilon]\!]^{-1}_{mp} \alpha_p) H_{zp} = \rho H_{zn}, \qquad (4.16)$$

where $[\![f]\!]^{-1}$ is the inverse matrix of $[\![f]\!]$, and they achieved fast convergence as shown by the solid dots in Figure 4.3. This numerical example demonstrates the importance of Fourier factorization. From a numerical point of view Laurent's rule is not always applicable.

4.4.2 Statement of the Problem

Let P be the set of square-integrable, piecewise continuous, piecewise smooth, periodic functions of x with period 2π. The periodic functions in many physical

Chapter 4. The Fourier Modal Method in Grating Theory   121

problems, certainly the permittivity function and the modal functions in grating problems, are contained in $\boldsymbol{P}$. Since we allow the functions to be discontinuous, we will speak of convergence of their Fourier series only in the restrictive sense as defined by Hardy [34]. Furthermore, a statement such as "$f(x)$ is represented by its Fourier series" or "a trigonometric series converges to $f(x)$" will be understood in the pointwise sense. The value of a function at a point of discontinuity is taken to be the arithmetic mean of the left and right limiting values toward that point.

For every pair of functions $f(x) \in \boldsymbol{P}$ and $g(x) \in \boldsymbol{P}$,

$$h(x) = f(x)g(x) \tag{4.17}$$

is obviously also in $\boldsymbol{P}$. It is a well-established result that the Fourier coefficients of $h(x)$ can be obtained from the Fourier coefficients of $f(x)$ and $g(x)$ by Laurent's rule [27]:

$$h_n = \sum_{m=-\infty}^{+\infty} f_{n-m} g_m. \tag{4.18}$$

The Fourier factorization of $h(x)$ is given by

$$h(x) = \sum_{n=-\infty}^{+\infty} h_n e^{inx} = \sum_{n=-\infty}^{+\infty} \sum_{m=-\infty}^{+\infty} f_{n-m} g_m e^{inx}. \tag{4.19}$$

To be more precise, (4.19) should be understood in the following sense:

$$h(x) = \lim_{N \to \infty} \sum_{n=-N}^{N} \left(\lim_{M \to \infty} \sum_{m=-M}^{M} f_{n-m} g_m \right) e^{inx}. \tag{4.20}$$

The way in which the above equation is written emphasizes that the two limits are independent of each other and that the inner limit is to be taken first.

To solve a practical problem on a computer, truncation of infinite series is inevitable. Similar to its usage in section 4.3, superscript (M) here is used to denote the symmetrically truncated partial sums. Corresponding to (4.18) and (4.19), we have

$$h_n^{(M)} = \sum_{m=-M}^{M} f_{n-m} g_m, \tag{4.21}$$

$$h^{(M)}(x) = \sum_{n=-M}^{M} h_n^{(M)} e^{inx}. \tag{4.22}$$

Equation (4.21) defines the finite version of Laurent's multiplication rule. Equation (4.22) is a partial sum that uses the approximate Fourier coefficients. Similarly, we denote by $h_M(x)$ the partial sum of $h(x)$ that uses the exact Fourier coefficients,

$$h_M(x) = \sum_{n=-M}^{M} h_n e^{inx}. \tag{4.23}$$

Note that in (4.22) the same positive integer M is used both for the summation bounds and for the superscript of the coefficients. This is the most commonly adopted truncation arrangement in numerical analysis. This condition is of fundamental importance to the validity of the theorems to be given in the next section.

If $h^{(\infty)}(x) = h(x)$, i.e., if $h^{(M)}(x)$ converges to $h(x)$ pointwise as $M \to \infty$, we say that $f(x)g(x)$ can be Fourier factorized; otherwise we say that it cannot be Fourier factorized. The theorems in the next section answer the questions when and how a product of two functions in $\boldsymbol{P}$ can be Fourier factorized. Although the mathematical theory on multiplication of Fourier series is well documented [34], to the best of my knowledge, the special and practically important problem that is posed by the simultaneous truncation has not been addressed in the mathematical literature.

4.4.3 Theorems on Fourier Factorization

Suppose that $f(x) \in \boldsymbol{P}$ and $g(x) \in \boldsymbol{P}$. We define

$$U_f = \{x_j \mid f(x_j + 0) \neq f(x_j - 0), \quad x_j \in [0, 2\pi), \quad j = 1, 2, \ldots\} \tag{4.24}$$

to be the set of the abscissae of the discontinuities of $f(x)$, and U_g to be the same for $g(x)$. Then

$$U_{f,g} = U_f \cap U_g \tag{4.25}$$

is the set of the abscissae of the *concurrent* discontinuities of $f(x)$ and $g(x)$. By assumption, both U_f and U_g have a finite number of elements. If $h(x)$ is such that

$$h(x_p - 0) = h(x_p + 0) \quad (x_p \in U_{f,g}), \tag{4.26}$$

then $f(x)$ and $g(x)$ are said to have a pair of *complementary* jumps at x_p. The amount of discontinuity of f at x_j is denoted $\hat{f}_j$,

$$\hat{f}_j = f(x_j + 0) - f(x_j - 0). \tag{4.27}$$

Note that at a concurrent discontinuity point x_p,

$$\lim_{M \to \infty} f_M(x_p) \lim_{M \to \infty} g_M(x_p) \neq \lim_{M \to \infty} h_M(x_p) \quad (x_p \in U_{f,g}). \tag{4.28}$$

The above expression is incompatible with (4.17), which is already a sign for possible problems in the Fourier factorization of products with concurrent discontinuities.

THEOREM 4.3. *If $f(x) \in \boldsymbol{P}$ and $g(x) \in \boldsymbol{P}$ do not have any concurrent jump discontinuities and $h_n^{(M)}$ is given by the finite Laurent's rule, then $h^{(M)}(x) \to h_M(x)$ uniformly on the whole real axis as $M \to \infty$.*

THEOREM 4.4. *If $f(x) \in \boldsymbol{P}$ and $g(x) \in \boldsymbol{P}$ have concurrent jump discontinuities and $h_n^{(M)}$ is given by the finite Laurent's rule, then*

$$h^{(M)}(x) = h_M(x) - \sum_{x_p \in U_{f,g}} \frac{\hat{f}_p \hat{g}_p}{2\pi^2} \Phi_M(x - x_p) + o(1), \tag{4.29}$$

Chapter 4. The Fourier Modal Method in Grating Theory   123

where the term o(1) uniformly tends to zero and

$$\Phi_M(x) = \sum_{n=1}^{M} \frac{\cos nx}{n} \sum_{|m|>M} \frac{1}{m-n}. \tag{4.30}$$

Furthermore,

$$\lim_{M\to\infty} \Phi_M(x) = \begin{cases} 0 & (x \neq 0), \\ \dfrac{\pi^2}{4} & (x = 0). \end{cases} \tag{4.31}$$

THEOREM 4.5. *Let S be a subinterval or a collection of subintervals of $[0, 2\pi)$ and $\overline{S}$ be its complement (S or $\overline{S}$ may be empty). If all discontinuities of $f(x)$ and $g(x)$ are complementary, $f(x) \neq 0$, and $f(x)$ satisfies either one of the two following conditions:*
(a) $\mathrm{Re}[1/f]$ does not change sign in $[0, 2\pi)$, $\mathrm{Re}[1/f] \neq 0$ in S, and $\mathrm{Im}[1/f]$ does not change sign in $\overline{S}$;
(b) $\mathrm{Im}[1/f]$ does not change sign in $[0, 2\pi)$, $\mathrm{Im}[1/f] \neq 0$ in S, and $\mathrm{Re}[1/f]$ does not change sign in $\overline{S}$; then $h^{(M)}(x) \to h_M(x)$ uniformly on the whole real axis as $M \to \infty$, provided that the coefficients $h_n^{(M)}$ in (4.22) are calculated by using

$$h_n^{(M)} = \sum_{m=-M}^{M} \left[\!\!\left[\frac{1}{f}\right]\!\!\right]_{nm}^{(M)-1} g_m. \tag{4.32}$$

The rule expressed in (4.32) will be called the inverse rule since it involves the algebraic inverse and then the matrix inverse of function f.

The proofs of the theorems are given in Appendix A at the end of this chapter.

4.4.4 Discussion

Theorem 4.4 says that if f and g have concurrent jumps, complementary or otherwise, and Laurent's rule is used, then $h^{(\infty)}(x)$ does not represent the original function $h(x)$ in the pointwise sense. In the neighborhood of a point $x_p \in U_{f,g}$ the difference between $h^{(M)}(x)$ and $h_M(x)$ is proportional to $\Phi_M(x - x_p)$. The function $\Phi_M(x)$ has some interesting properties. Its limit $\Phi_\infty(x)$ is a periodic function that is $\pi^2/4$ at $x = 0$ and zero everywhere else in $[0, 2\pi)$. Because the sum of a uniformly convergent infinite series of continuous terms should be continuous, but $\Phi_\infty(x)$ is discontinuous at $x = 0$, the convergence of $\Phi_M(x)$ cannot be uniform in the neighborhood of $x = 0$. $\Phi_M(x)$ is unique in the sense that if there is another function, $\Phi'_M(x)$, that satisfies (4.29), then the difference between $\Phi_M(x)$ and $\Phi'_M(x)$ must converge uniformly to zero everywhere.

A few graphs of $\Phi_M(x)$ may help the reader to see its general behavior. Figure 4.4 shows three plots of $\Phi_M(x)$ in the neighborhood of $x = 0$ for $M = 10, 100$, and 1000. Note that although the same vertical axis is used for all three plots, the

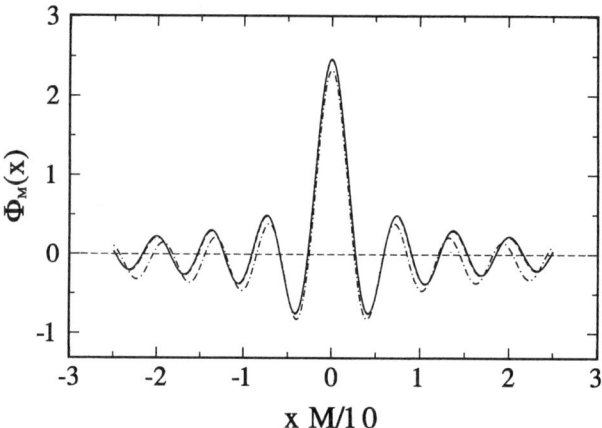

Figure 4.4: Behavior of function $\Phi_M(x)$ in the neighborhood of $x = 0$. Three curves are plotted, with $M = 10$ (dotted-and-dashed line), $M = 100$ (dashed line), and $M = 1000$ (solid line). The latter two are almost indistinguishable. The value of the main peak tends to $\pi^2/4$ as $M \to \infty$.

horizontal axis is scaled differently for each plot. Although there are noticeable differences between the cases of $M = 10$ and $M = 100$, no discernable differences can be easily detected between the cases of $M = 100$ and $M = 1000$. In other words, in the neighborhood of $x = 0$, the graph of $\Phi_{nM}(x)$ is approximately the same as the graph of $\Phi_M(x)$ for a sufficiently large M, if the scale of the horizontal axis of the former is n times as large as that of the latter. If we index the extrema of $\Phi_M(x)$ from the origin outward, not counting the central maximum, by $\pm 1, \pm 2, \ldots$, with positive and negative signs for $x > 0$ and $x < 0$, respectively, then for an extremum of a fixed index its function value tends to a constant but its position moves toward $x = 0$ as $M \to \infty$. This observation is of course consistent with the above conclusion that the convergence of $\Phi_M(x)$ is nonuniform in the neighborhood of $x = 0$.

From a graphical point of view, Theorem 4.4 says that, when Laurent's rule is used, the graph of $h^{(M)}(x)$ can be obtained by superimposing a series of properly scaled graphs of $\Phi_M(x)$ centered at $x_p \in U_{f,g}$ on top of the graph of $h_M(x)$. Here for ease of visualization we assume that both $f(x)$ and $g(x)$ are real functions. The effect of such a superimposition is most prominent when $h(x)$ is continuous. In that case, $h^{(M)}(x)$ will have an overshoot (if $\hat{f}_p \hat{g}_p < 0$) or an undershoot (if $\hat{f}_p \hat{g}_p > 0$) from the graph of $h_M(x)$ at $x_p \in U_{f,g}$, whose magnitude tends to $|\hat{f}_p \hat{g}_p|/8$ as $M \to \infty$. On the other hand, Theorem 4.5 says that $h^{(M)}(x)$ calculated by the inverse rule well preserves the characteristics of $h(x)$, including its continuity at $x_p \in U_{f,g}$. If we set $f(x_p + 0)/f(x_p - 0) = \alpha$ and again assume that $h(x)$ is continuous at $x_p \in U_{f,g}$, then

$$\hat{f}_p \hat{g}_p = -h(x_p) \frac{(1-\alpha)^2}{\alpha}. \tag{4.33}$$

Chapter 4. The Fourier Modal Method in Grating Theory

Thus, the magnitude of the overshoot can be arbitrarily large as $\alpha \to 0$ or $\alpha \to \pm\infty$. As an illustration, let us consider a graphical example. We choose

$$f(x) = \begin{cases} a, & |x| < \dfrac{\pi}{2} \\ \dfrac{a}{2}, & \dfrac{\pi}{2} < |x| \leq \pi \end{cases} \quad (a \neq 0), \tag{4.34}$$

and

$$g(x) = \begin{cases} b\dfrac{|x|}{\pi}, & |x| < \dfrac{\pi}{2} \\ 2b\left(1 - \dfrac{|x|}{\pi}\right), & \dfrac{\pi}{2} < |x| \leq \pi \end{cases} \quad (b \neq 0). \tag{4.35}$$

These two functions have complementary jumps at $x = \pm\pi/2$, where $h(x)$ is continuous but its derivative is not. The functions $f(x), g(x)$, and $h(x)$ are schematically shown in Figure 4.5a in decreasing line thicknesses. Figure 4.5b shows an enlarged view of $h^{(M)}(x)$, with $a = 6$ and $b = 2$, in the region enclosed by the dashed circle in Figure 4.5a. The oscillatory curve is obtained by using Laurent's rule, and the nonoscillatory curve is obtained by using the inverse rule. For both curves $M = 200$ was used. Evidently the inverse rule gives a good reconstruction of $h(x)$, but Laurent's rule renders a reconstruction that suffers from overshoot and ringing in the neighborhood of the complementary discontinuity. The amount of overshoot is approximately 3/8, as predicted by (4.29). The reader can find two other interesting graphical illustrations of Theorems 4.4 and 4.5 in [20].

For convenience, we say that a product that does not have concurrent jumps is type 1, a product that satisfies the conditions of Theorem 4.5 is type 2, and a product that is neither type 1 nor type 2 is type 3. The class of type-3 products includes those that have concurrent but noncomplementary jumps. From an operational point of view the three theorems of this subsection can be summarized as follows:

(1) A product of type 1 can be Fourier factorized by Laurent's rule.

(2) A product of type 2 can be Fourier factorized by the inverse rule, but it cannot be Fourier factorized by Laurent's rule.

(3) A product of type 3 can be Fourier factorized by neither Laurent's rule nor the inverse rule.

The above three types of the products are not without prototypes in physics. Examples are ample in grating problems, and perhaps in other physical problems. For example, the product ϵE_z in (4.1c) is type 1 because E_z is continuous and ϵ is discontinuous for lamellar gratings. On the other hand, the product ϵE_x in (4.2a) is type 2, because $\epsilon E_x = D_x$ is the normal component of the electric displacement that is always continuous across a charge-free medium discontinuity. A reader may ask the question, What about type-3 products? They do occur in physics. For example, the amplitude of an electric displacement along the diagonal of the first and second quadrants is given by a type-3 product $\epsilon(E_x + E_y)/\sqrt{2}$. However, all the grating problems that I have encountered so far can always be decomposed into

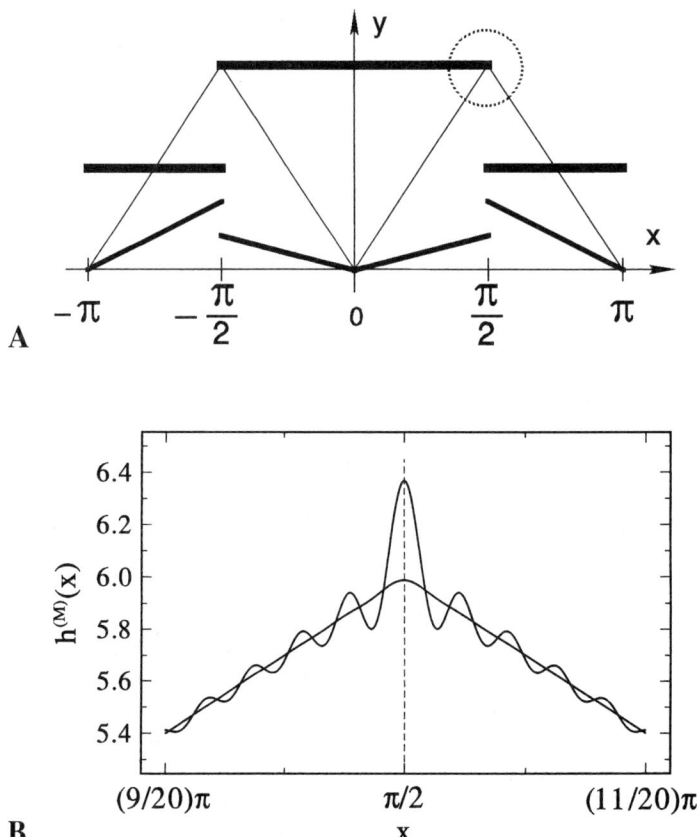

Figure 4.5: (a) Three periodic functions, in decreasing line thicknesses: $f(x), g(x)$, and $h(x) = f(x)g(x)$. (b) An enlarged view of $h^{(M)}(x)$, with $a = 6$ and $b = 2$, in the region enclosed by the dashed circle in Figure 4.5a. The oscillatory curve is obtained by using Laurent's rule, and the nonoscillatory one is obtained by using the inverse rule.

a sum of products of type 1 and type 2. The decomposition of the above example is obvious. In some cases the decomposition needs some algebraic manipulation, and in others, as in the case of [21], it requires certain approximation of the grating profiles.

It may be instructive to compare between the well-known Gibbs–Wilbraham phenomenon [35] and the phenomenon mathematically described by Theorem 4.4 and graphically illustrated in Figures 4.4 and 4.5. The reader may recall that the former refers to the overshoot of the partial sum of the Fourier series of a function in the neighborhood of a jump discontinuity. Suppose that the discontinuous function is $f(x)$ and the discontinuity is located at x_0. Then as $M \to \infty$ the amount of overshoot tends to $[\text{Si}(\pi)/\pi - \frac{1}{2}]\hat{f}_0 = 0.0895\hat{f}_0$ above the upper edge of the jump and below the lower edge of the jump. The location of the main overshoot peak (or dip) tends to x_0. In contrast, the new phenomenon here is for the partial sum of the trigonometric series of a product of two discontinuous functions with concurrent jumps that is Fourier factorized by using the finite Laurent's rule, and the number of terms in the partial sum is the same as that used in the multiplication. The sign of the main overshoot depends on the sign of the product of the two concurrent jumps. The central peak of the overshoot is fixed at the location of the concurrent jumps if the slope of $h_M(x)$ there is zero.

Prior to the work of [18, 19] some authors had suspected that the slow convergence of the FMM in TM polarization might be caused by the Gibbs–Wilbraham phenomenon or even the use of Fourier series. It is now clear that it is not the use of Fourier series but the way in which the Fourier series are used that is responsible for the slow convergence. In some sense, the Fourier series of a discontinuous function has done the best that the Fourier basis is capable of in representing the function. On the other hand, the series of a type-2 product factorized, by using Laurent's rule, has not done the best possible. Numerical evidence unmistakably suggests that the satisfaction of the electromagnetic boundary conditions is more important than the preservation of the discontinuities of the permittivity and field components. Since the Fourier series are still used to represent discontinuous functions (e.g., the function $\epsilon(x)$ in section 4.4.1), it is not the Gibbs–Wilbraham phenomenon but the avoidable phenomenon described here that is to be blamed for the historical poor performance of the FMM.

In closing the discussion, I reiterate that the mathematical results of this section are necessitated by the simultaneous matrix truncation. If computers had infinite memory and speed, then Laurent's rule could always be used. From a physics or pure mathematics point of view, the failure of $h^{(M)}(x)$ to converge to $h(x)$ at a few isolated points is insignificant. However, from a numerical point of view, with real world computers this failure can have grave consequences, and ensuring pointwise convergence is extremely important.

4.4.5 Application to Grating Problems

The application of the Fourier factorization theory to a grating problem consists of three steps:

(1) From Maxwell's equations, derive the working equations for the grating problem, i.e., the equations that define the eigenvalue problem in real space. In some cases it might be difficult to carry out step (2) below with the derived working equations. Then, one can always take the original Maxwell's equations as the working equations.

(2) Identify the continuity characteristics of the field components and arrange the working equations in such a way that the products between the field components and the scalar permittivity function or the permittivity tensor elements are either type 1 or type 2; avoid forming type-3 products.

(3) Apply Laurent's rule to the type-1 products and the inverse rule to the type-2 products.

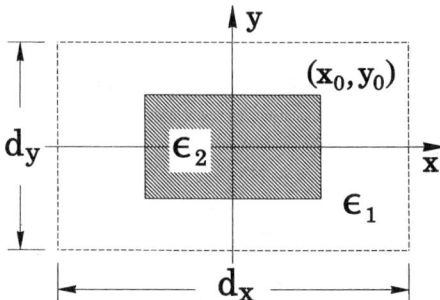

Figure 4.6: A unit cell of a rectangular crossed grating with d_x and d_y as the periods along the x- and y-axes.

The three examples to be given below are intended to illustrate how the above procedures can be applied to the analysis of gratings with the FMM. Only the Fourier analyses of the working equations will be given. The rest of the derivations are routine and can be found in many references. The accompanying numerical examples show that the improvement in convergence resulted from using the correct Fourier factorization. The following practical criterion is adopted in comparing two implementation methods. The diffraction efficiencies computed by a convergent method should stabilize as M increases, where M is the integer that we have been using for matrix truncation. For convenience, we call integer $N = 2M + 1$ the truncation order. If as N increases, the diffraction efficiencies computed by method A stabilize sooner than those by method B, then we say that method A converges faster than method B, even though mathematically they may have the same convergence rate. When considering results of convergence comparisons, the reader should keep in mind that the memory requirement and computation time to solve a grating problem is proportional to the second and the third powers of the truncation order, respectively.

Example 4.1. Rectangular Gratings in TM Polarization. This example has already been considered in section 4.4.1. Here let us justify the Fourier factorization in (4.15) and (4.16). The correctness of (4.15) is obvious because H_z is a continuous function of x. Based on physics or mathematical analysis, it is easy to see that, in (4.4), both the product $(1/\epsilon)(\partial H_z/\partial x)$ and that formed by ϵ with the term enclosed in the square brackets are continuous; therefore, the application of the inverse rule in (4.16) is also correct. This is why the combination of (4.15) and (4.16) gives the much better convergence result than that of (4.15) and (4.14). For more information about this classical grating example, the reader can read [17]–[20].

Example 4.2. Rectangular Crossed Gratings. This example is intended to show that the Fourier factorization theory given in section 4.4.2 can be used to study periodic problems in higher dimensions. Figure 4.6 depicts a unit cell of a rectangular crossed grating. The periods along x- and y-axes are d_x and d_y, respectively. The modal fields in this case are piecewise continuous, piecewise smooth, and pseu-

Chapter 4. The Fourier Modal Method in Grating Theory

doperiodic functions of x and y. The electric field components transverse to the z-axis are in general singular at the edges of the grating profile, but they should be absolutely square-integrable because the electromagnetic energy enclosed in any finite volume must be finite. We need to perform the Fourier analysis of

$$\partial_y H_z - \partial_z H_y = -\mathrm{i}k_0 \varepsilon E_x, \tag{4.36a}$$

$$\partial_z H_x - \partial_x H_z = -\mathrm{i}k_0 \varepsilon E_y, \tag{4.36b}$$

$$\partial_x H_y - \partial_y H_x = -\mathrm{i}k_0 \varepsilon E_z \tag{4.36c}$$

in the plane (x, y). Equation (4.36c) presents no challenges because the product ϵE_z is type 1 across both the vertical ($x = $ constant) and horizontal ($y = $ constant) boundaries between media 1 and 2. Applying Laurent's rule in both x- and y-directions, we have

$$(\varepsilon E_z)_{mn} = \sum_{j,l} [\![\epsilon]\!]_{mn,jl} E_{zjl}, \tag{4.37}$$

where the first integer subscripts before and after the comma are associated with the Fourier coefficients in the x-direction and the second subscripts in the y-direction, and $[\![\epsilon]\!]$ is a matrix generated by the double Fourier coefficients of $\epsilon(x,y)$ such that $[\![\epsilon]\!]_{mn,jl} = \epsilon_{m-j,n-l}$. For simplicity here, and henceforth, we do not explicitly indicate matrix truncation. At first glance, the analysis of the other two equations seems impossible. Taking the product ϵE_x as an example, it is type 2 across the vertical boundaries and type 1 across the horizontal boundaries. The difficulty vanishes once we realize that the Fourier analysis can be performed along the two directions separately, and along each direction the product is of only one type.

Let us now take a closer look at $\epsilon(x,y) E_x(x,y)$. We perform its Fourier analysis along the x-axis first. For $-y_0 < y < y_0$, the product is type 2. For $y < -y_0$ and $y > y_0$, both $\epsilon(x,y)$ and $E_x(x,y)$ are continuous. Thus, in both cases the inverse rule can be used, and we have

$$(\varepsilon E_x)_m = \sum_j \left\lceil \frac{1}{\varepsilon} \right\rceil^{-1} E_{xj}(y), \tag{4.38}$$

where $\lceil 1/\epsilon \rceil^{-1}_{mj}$ are the matrix elements of the inverse of $\lceil 1/\epsilon \rceil$, and the latter is the Toeplitz matrix generated by the Fourier coefficients of $1/\epsilon(x,y)$ with respect to variable x. We similarly define $\lfloor f \rfloor$ as the matrix generated by the Fourier coefficients of function $f(x,y)$ with respect to variable y. The matrix $\lceil 1/\epsilon \rceil^{-1}$ is clearly a discontinuous function of y. Since $E_x(x,y)$ is continuous in y and piecewise continuous in x, except at the four corners $(\pm x_0, \pm y_0)$, it can be easily shown that its Fourier coefficients $E_{xj}(y)$ are continuous everywhere. This means that each term on the right-hand side of (4.38) is a type-1 product. Therefore, the Fourier factorization of ϵE_x is given by

$$(\varepsilon E_x)_{mn} = \sum_{j,l} \lfloor\lceil\varepsilon\rceil\rfloor_{mn,jl} E_{xjl}, \qquad (4.39)$$

where

$$\lfloor\lceil\varepsilon\rceil\rfloor_{mn,jl} = \lfloor\lceil 1/\varepsilon\rceil^{-1}_{mj}\rfloor_{nl}. \qquad (4.40)$$

The symmetry of the problem allows us to write down immediately the Fourier factorization of ϵE_y

$$(\varepsilon E_y)_{mn} = \sum_{j,l} \lceil\lfloor\varepsilon\rfloor\rceil_{mn,jl} E_{yjl}, \qquad (4.41)$$

where

$$\lceil\lfloor\varepsilon\rfloor\rceil_{mn,jl} = \lceil\lfloor 1/\varepsilon\rfloor^{-1}_{nl}\rceil_{mj}. \qquad (4.42)$$

In summary, in Fourier space, (4.36) become

$$\beta_n H_{zmn} - \frac{\partial_z}{i} H_{ymn} = -k_0 \sum_{j,l} \lfloor\lceil\varepsilon\rceil\rfloor_{mn,jl} E_{xjl}, \qquad (4.43a)$$

$$\frac{\partial_z}{i} H_{xmn} - \alpha_m H_{zmn} = -k_0 \sum_{j,l} \lceil\lfloor\varepsilon\rfloor\rceil_{mn,jl} E_{yjl}, \qquad (4.43b)$$

$$\alpha_m H_{ymn} - \beta_n H_{xmn} = -k_0 \sum_{j,l} [\![\varepsilon]\!]_{mn,jl} E_{zjl}, \qquad (4.43c)$$

where $\alpha_m = \alpha_0 + 2m\pi/d_x, \beta_n = \beta_0 + 2n\pi/d_y$, and α_0 and β_0 are constants determined by the incident plane wave.

Figure 4.7 shows the convergence of the zero-order efficiency (in reflection) of a square metallic grid suspended in air. The grating parameters are normal incidence with polarization along the diagonal direction, $\epsilon_1 = (1.0 + i5.0)^2, \epsilon_2 = 1.0, d_x = d_y = 1.2\lambda_0$. Both the grid thickness and width are $0.2\lambda_0$. The data represented by the solid dots were computed using the factorization given by (4.43). The data represented by the empty circles were computed when both the matrices $\lfloor\lceil\epsilon\rceil\rfloor$ and $\lceil\lfloor\epsilon\rfloor\rceil$ in (4.43a) and (4.43b) were replaced by $[\![\epsilon]\!]$, and the matrix $[\![\epsilon]\!]$ in equation (4.43c) was replaced by $[\![1/\epsilon]\!]^{-1}$. N_x and N_y are the truncation orders along the x- and y-axes. The true truncation order for the crossed grating problem is $N_x N_y$. The maximum truncation order in the figure is 27. It is difficult to increase it much further because the demand on computing resources increases rapidly as $N_x N_y$ increases. Nonetheless, it is already evident that the correct factorization yields more stable results than the incorrect factorization does in this limited range of truncation orders.

Chapter 4. The Fourier Modal Method in Grating Theory

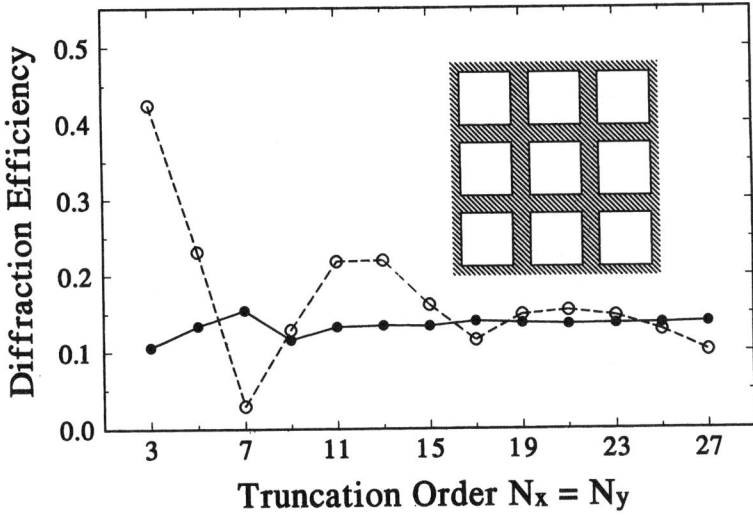

Figure 4.7: Convergence of the zero-order efficiency (in reflection) of a square metallic grid suspended in air. $\epsilon_1 = (1.0 + i5.0)^2, \epsilon_2 = 1.0, d_x = d_y = 1.2\lambda_0$. The incident plane wave is at normal incidence and its polarization is along the diagonal direction. Both the grid thickness and width are $0.2\lambda_0$. N_x and N_y are the truncation orders along the x- and y-axes. Solid dots: Correct Fourier factorization; empty circles: Incorrect Fourier factorization.

For more information on the FMM for crossed gratings, the reader may consult [15], [16], and [21]. The idea presented in this example obviously can be extended to treat problems in three-dimensional or even higher dimensional spaces. Lalanne [36] has applied this idea to the study of three-dimensional photonic crystals.

Example 4.3. Anisotropic Lamellar Gratings. This example shows that sometimes it is necessary to rearrange Maxwell's equations before Fourier factorization. As in Example 4.1, the grating grooves are along the z-direction. The periodic permittivity function is now a tensor: $\tilde{\epsilon} = \{\epsilon_{ik}(x); i, k = 1, 2, 3\}$. For convenience, here we use subscripts 1, 2, and 3 to label the x, y, and z components of the tensor, the field vectors, and the differential operator ∇. The Maxwell equation that requires factorization now takes the following form:

$$\partial_2 H_3 - \partial_3 H_2 = -ik_0(\varepsilon_{11}E_1 + \varepsilon_{12}E_2 + \varepsilon_{13}E_3), \quad (4.44a)$$

$$\partial_3 H_1 - \partial_1 H_3 = -ik_0(\varepsilon_{21}E_1 + \varepsilon_{22}E_2 + \varepsilon_{23}E_3), \quad (4.44b)$$

$$\partial_1 H_2 - \partial_2 H_1 = -ik_0(\varepsilon_{31}E_1 + \varepsilon_{32}E_2 + \varepsilon_{33}E_3). \quad (4.44c)$$

Since E_2 and E_3 are continuous across discontinuities of $\tilde{\epsilon}(x)$, the second and third terms on the right-hand sides of these equations are all type 1. However, all three first terms are type 3, which, according to the Fourier factorization theory, cannot be directly factorized. The key to overcome the difficulty is to recognize that

$(\epsilon_{11}E_1 + \epsilon_{12}E_2 + \epsilon_{13}E_3)$ is continuous because it is the x-component of the electric displacement. Rewriting equations (4.44) as

$$\partial_2 H_3 - \partial_3 H_2 = -ik_0\varepsilon_{11}\left[E_1 + \left(\frac{\varepsilon_{12}}{\varepsilon_{11}}\right)E_2 + \left(\frac{\varepsilon_{13}}{\varepsilon_{11}}\right)E_3\right], \quad (4.45a)$$

$$\partial_3 H_1 - \partial_1 H_3 = -ik_0\left[\frac{i}{k_0}\left(\frac{\varepsilon_{21}}{\varepsilon_{11}}\right)(\partial_2 H_3 - \partial_3 H_2) + \left(\frac{\varepsilon_{11}\varepsilon_{22} - \varepsilon_{21}\varepsilon_{12}}{\varepsilon_{11}}\right)E_2 \right. \quad (4.45b)$$
$$\left. + \left(\frac{\varepsilon_{11}\varepsilon_{23} - \varepsilon_{21}\varepsilon_{13}}{\varepsilon_{11}}\right)E_3\right],$$

$$\partial_1 H_2 - \partial_2 H_1 = -ik_0\left[\frac{i}{k_0}\left(\frac{\varepsilon_{31}}{\varepsilon_{11}}\right)(\partial_2 H_3 - \partial_3 H_2) + \left(\frac{\varepsilon_{11}\varepsilon_{32} - \varepsilon_{31}\varepsilon_{12}}{\varepsilon_{11}}\right)E_2 \right. \quad (4.45c)$$
$$\left. + \left(\frac{\varepsilon_{11}\varepsilon_{33} - \varepsilon_{31}\varepsilon_{13}}{\varepsilon_{11}}\right)E_3\right],$$

we have a different situation: in (4.45a), the two products inside the square brackets are type 1, and the whole right-hand side is type 2. In (4.45b) and (4.45c), every product on the right-hand sides is of type 1, provided that the combinations of permittivity tensor elements that are enclosed in parentheses are treated as solitary entities, meaning that their Fourier coefficients are calculated directly from the composite functions. Applying Laurent's rule to type-1 products and the inverse rule to type-2 products in these equations, we obtain

$$\frac{\partial_2}{i}H_{3m} - \gamma H_{2m} = -k_0 \sum_{k=1}^{3}\sum_{n}(Q_{1k})_{mn}E_{kn}, \quad (4.46a)$$

$$\gamma H_{1m} - \alpha_m H_{3m} = -k_0 \sum_{k=1}^{3}\sum_{n}(Q_{2k})_{mn}E_{kn}, \quad (4.46b)$$

$$\alpha_m H_{2m} - \frac{\partial_2}{i}H_{1m} = -k_0 \sum_{k=1}^{3}\sum_{n}(Q_{3k})_{mn}E_{kn}, \quad (4.46c)$$

where γ is the z-component of the wave vector of the incident plane wave, and Q is a 3×3 block matrix. The expressions of Q_{ik} will not be given here. The reader can easily derive them from (4.45) or find them in [22]. Equations (4.46) are the Fourier representation of (4.44).

Figure 4.8 shows the convergence of the first-order diffraction efficiencies of an anisotropic grating. This grating is the same as that of Example 4.1, except that a one-wavelength-thick anisotropic layer is added, filling the grooves and submerging the metallic grating. The permittivity tensor elements of the anisotropic material are $\epsilon_{11} = 2.25, \epsilon_{22} = 2.56, \epsilon_{33} = 2.89, \epsilon_{12} = 0.04, \epsilon_{23} = 0.16, \epsilon_{13} = 0.36$. All other parameters are the same as in Example 4.1. Because of the presence of the anisotropic layer, the TM incident plane wave excites both TE and TM diffracted orders. The curves labeled "right" were computed by using (4.46) and those labeled with "wrong" were computed with the matrices Q_{ik} replaced by $[\![\epsilon_{ik}]\!]$ in (4.46).

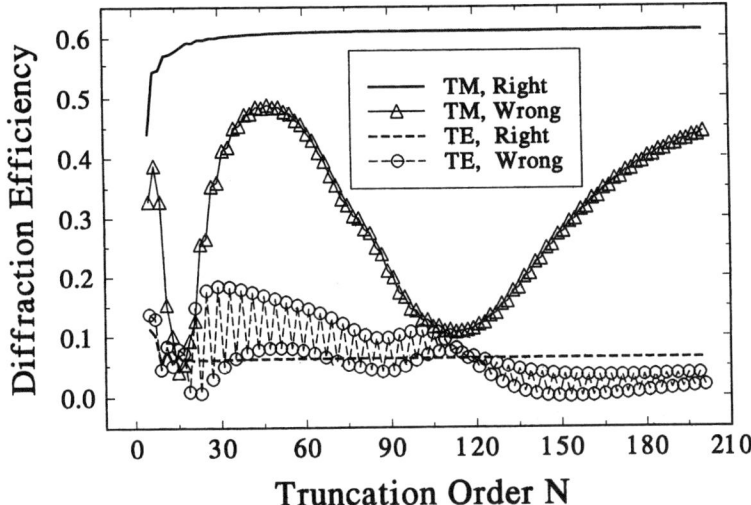

Figure 4.8: Convergence of the first-order diffraction efficiencies of an anisotropic grating that is the same as that in Example 4.1, except that a one-wavelength-thick anisotropic layer is added, filling the grooves and submerging the metallic grating. The permittivity tensor elements of the anisotropic material are $\epsilon_{11} = 2.25, \epsilon_{22} = 2.56, \epsilon_{33} = 2.89, \epsilon_{12} = 0.04, \epsilon_{23} = 0.16, \epsilon_{13} = 0.36$. All other parameters are the same as in Figures 4.2 and 4.3. The words "right" and "wrong" in the legend refer to the Fourier factorization used in the computation.

4.5 Summary

In this chapter I have discussed two mathematical aspects of the FMM for modeling diffraction gratings: convergence analysis and Fourier factorization. Both subjects belong to the domain of applied mathematics. Both studies are necessitated by the matrix truncation in numerical practice. From a theoretical point of view the investigation of the convergence of the FMM is of fundamental importance. Although so far the preliminary research has proved only the applicability of reduction method, something that researchers in the applied optics community believed in for a long time based on their numerical experience or physical intuition, continued research in this direction may yield tangible results in the future. The research on Fourier factorization, on the other hand, has already been very fruitful, as demonstrated by the numerical examples in section 4.4. Indeed, without the theoretical guidance, it would be nearly impossible for anyone to have derived the correct Fourier representations of Maxwell's equations for the grating problems in Examples 4.2 and 4.3. The theory has also been successfully applied to other methods for modeling gratings [37, 38]. With all these modeling methods, the saving in computation time offered by using the correct Fourier factorization is enormous. Because the theory of Fourier factorization is a basic result, it might also find applications in solving other problems in applied physics and engineering.

Appendix A

Proof of Theorem 4.4. Any $f(x) \in \boldsymbol{P}$ can be decomposed as follows:

$$f(x) = \tilde{f}(x) + \sum_{x_j \in U_f} \frac{\hat{f}_j}{\pi} \phi(x - x_j), \tag{A.1}$$

where $\hat{f}(x)$ is a continuous function, and

$$\phi(x) = \frac{1}{2}(\pi - x) \qquad (0 < x < 2\pi). \tag{A.2}$$

The function $h(x) = f(x)g(x)$ can then be written as

$$h(x) = Q(x) + \frac{1}{\pi} \sum_{x_j \in U_f} \hat{f}_j R(x; x_j) + \frac{1}{\pi} \sum_{x_k \in U_g} \hat{g}_k S(x; x_k)$$
$$+ \frac{1}{\pi^2} \sum_{\substack{x_j \in U_f \\ x_k \in U_g}} \hat{f}_j \hat{g}_k T(x; x_j, x_k), \tag{A.3}$$

with

$$Q(x) = \tilde{f}(x)\tilde{g}(x), \tag{A.4a}$$
$$R(x; x_j) = \phi(x - x_j)\tilde{g}(x), \tag{A.4b}$$
$$S(x; x_k) = \tilde{f}(x)\phi(x - x_k), \tag{A.4c}$$
$$T(x; x_j, x_k) = \phi(x - x_j)\phi(x - x_k). \tag{A.4d}$$

The Fourier coefficients of $\phi(x)$ are $\phi_0 = 0$ and $\phi_m = 1/(2im)$ if $m \neq 0$. Since for any function $\gamma(x) \in \boldsymbol{P}$, $\tilde{\gamma} \leq O(1/m^2)$, it is easy to show that the following estimates are valid for all $|n| \leq M$ and $0 \leq x < 2\pi$:

$$|Q_n^{(M)} - Q_n| \leq \sum_{|m|>M} |\tilde{f}_{n-m}\tilde{g}_m| \leq \frac{C}{M^2}, \tag{A.5a}$$

$$|Q^{(M)}(x) - Q_M(x)| \leq \sum_{|n| \leq M} |Q_n^{(M)} - Q_n| \leq O\left(\frac{1}{M}\right), \tag{A.5b}$$

$$|R_n^{(M)}(x_j) - R_n(x_j)| \leq \sum_{|m|>M} |\phi_{n-m}\tilde{g}_m| \leq O\left(\frac{\ln M}{M^2}\right), \tag{A.6a}$$

$$|R^{(M)}(x; x_j) - R_M(x; x_j)| \leq \sum_{|n| \leq M} |R_n^{(M)}(x_j) - R_n(x_j)| \leq O\left(\frac{\ln M}{M}\right), \tag{A.6b}$$

$$|S_n^{(M)}(x_k) - S_n(x_k)| \leq \sum_{|m|>M} |\tilde{f}_{n-m}\phi_m| \leq \frac{4C}{M(M+1-|n|)}, \tag{A.7a}$$

$$|S^{(M)}(x; x_k) - S_M(x; x_k)| \leq \sum_{|n| \leq M} |S_n^{(M)}(x_k) - S_n(x_k)| \leq O\left(\frac{\ln M}{M}\right). \tag{A.7b}$$

Chapter 4. The Fourier Modal Method in Grating Theory

By direct calculation, we have

$$T^{(M)}(x; x_j, x_k) - T_M(x; x_j, x_k)$$
$$= -\frac{1}{4} \sum_{|m|>M} \frac{e^{im(x_j - x_k)}}{m^2} + \frac{1}{4} \left[\sum_{0<|n|\leq M} \frac{e^{in(x-x_j)}}{n} \right] \left[\sum_{|m|>M} \frac{e^{im(x_j - x_k)}}{m} \right]$$
$$+ \frac{1}{4} \sum_{0<|n|\leq M} \frac{e^{in(x-x_j)}}{n} \sum_{|m|>M} \frac{e^{im(x_j - x_k)}}{n - m}. \tag{A.8}$$

The first line of the right-hand side tends to zero uniformly with respect to x. So the above equation can be rewritten as

$$T^{(M)}(x; x_j, x_k) - T_M(x; x_j, x_k) = -\frac{1}{2} \Phi_M(x - x_j, x_j - x_k) + o(1), \tag{A.9}$$

where

$$\Phi_M(x, y) = \frac{1}{2} \sum_{0<|n|\leq M} \frac{e^{inx}}{n} \sum_{|m|>M} \frac{e^{imy}}{m - n}. \tag{A.10}$$

If $y \neq 0$, then uniformly for all $0 \leq x < 2\pi$,

$$|\Phi_M(x, y)| \leq \sum_{0<|n|\leq M} \frac{1}{2|n|} \left| \sum_{|m|>M} \frac{e^{imy}}{m - n} \right| \leq O\left(\frac{\ln M}{M}\right). \tag{A.11}$$

If $y = 0$, $\Phi_M(x, y)$ becomes the function $\Phi_M(x)$ previously defined in (4.30). In any case the validity of (4.29) is established.

Next, let

$$\chi_n^{(M)} = \frac{1}{n} \sum_{|m|>M} \frac{1}{m - n} \quad (1 \leq n \leq M). \tag{A.12}$$

Then, it is elementary to show that $\chi_{n+1}^{(M)} > \chi_n^{(M)}$ for $1 \leq n < M$. Applying Abel's transformation [33] to (4.30), we have

$$\Phi_M(x) = \left(\sum_{l=1}^{M} \cos lx \right) \chi_M^{(M)} + \sum_{n=1}^{M-1} \left(\sum_{l=1}^{n} \cos lx \right) (\chi_n^{(M)} - \chi_{n+1}^{(M)}). \tag{A.13}$$

For $x \neq 0$, the sum of the cosine functions is uniformly bounded. Therefore,

$$|\Phi_M(x)| \leq C \left[\chi_M^{(M)} + \sum_{n=1}^{M-1} (\chi_{n+1}^{(M)} - \chi_n^{(M)}) \right] \leq O\left(\frac{\ln M}{M}\right). \tag{A.14}$$

If $x = 0$, then by setting $\Phi_0(0) = 0$ and recognizing that

$$\Phi_m(0) - \Phi_{m-1}(0) = \frac{3}{2m^2} \quad (m \geq 1), \tag{A.15}$$

we have

$$\Phi_M(0) = \sum_{m=1}^{M} [\Phi_m(0) - \Phi_{m-1}(0)] = \frac{3}{2} \sum_{m=1}^{M} \frac{1}{m^2}. \tag{A.16}$$

Equation (4.31) follows from (A.14) and (A.16). This completes the proof of Theorem 4.4.

Theorem 4.3 is a special case of Theorem 4.4. □

Proof of Theorem 4.5. We first show that for a function $f(x)$ satisfying the conditions stated in the theorem, the following asymptotic estimate holds:

$$\max_{|n| \leq M} \sum_{m=-M}^{M} \left| \left[\left[\frac{1}{f}\right]\right]_{nm}^{(M)-1} \right| \leq O(\sqrt{M}). \tag{A.17}$$

For convenience we set $A = [1/f]^{(M)}$. Let $\mu_{\min}$ be the smallest eigenvalue of AA^* and $\boldsymbol{u} = (u_{-M}, u_{-M+1}, \ldots, u_M)^T$ be the corresponding eigenvector. Suppose that $\boldsymbol{u}$ is normalized such that $|\boldsymbol{u}|^2 = \boldsymbol{u}^*\boldsymbol{u} = 1$, then by the Cauchy–Schwarz inequality, $|\boldsymbol{u}^*A^*\boldsymbol{u}|^2 \leq \mu_{\min}$. Since A is generated by the Fourier coefficients of $1/f$,

$$\begin{aligned}
\mu_{\min} &\geq \frac{1}{4\pi^2} \left| \int_0^{2\pi} |u_M(x)|^2 \frac{1}{f(x)} \, dx \right|^2 \\
&= \frac{1}{4\pi^2} \left\{ \left| \int_0^{2\pi} |u_M(x)|^2 \operatorname{Re}\left[\frac{1}{f}\right] dx \right|^2 + \left| \int_0^{2\pi} |u_M(x)|^2 \operatorname{Im}\left[\frac{1}{f}\right] dx \right|^2 \right\} \\
&= \frac{1}{4\pi^2} \left| \int_S |u_M(x)|^2 \operatorname{Re}\left[\frac{1}{f}\right] dx + \int_{\overline{S}} |u_M(x)|^2 \operatorname{Re}\left[\frac{1}{f}\right] dx \right|^2 \\
&\quad + \frac{1}{4\pi^2} \left| \int_S |u_M(x)|^2 \operatorname{Im}\left[\frac{1}{f}\right] dx + \int_{\overline{S}} |u_M(x)|^2 \operatorname{Im}\left[\frac{1}{f}\right] dx \right|^2,
\end{aligned} \tag{A.18}$$

where $u_M(x) = \sum_{m=-M}^{M} u_m e^{-imx}$. It is easy to verify that if either condition (a) or (b) stated in the theorem is satisfied, then there is a constant β independent of M such that $\mu_{\min} \geq \beta > 0$. Consequently, A is nonsingular and the maximum eigenvalue of $(AA^*)^{-1} = A^{*-1}A^{-1}$ is $1/\mu_{\min} < 1/\beta < \infty$. Note that $1/\mu_{\min}^{1/2}$ is $|A|^{-1}|_2$, the spectral norm of A^{-1}, and the left-hand side of (A.17) is just $|A^{-1}|_\infty$, the maximum-row-sum norm of matrix A^{-1}. Based on a relationship between these two types of matrix norms in matrix theory [39], it follows that

$$|A^{-1}|_\infty \leq \sqrt{2M+1}\, |A^{-1}|_2 \leq \sqrt{\frac{2M+1}{\beta}} = O(\sqrt{M}), \tag{A.19}$$

which is (A.17).

Now consider the product $g(x) = [1/f(x)]h(x)$. Keeping in mind that $h(x)$ is continuous, from (A.5a) and (A.6a) we have

$$\sum_{m=-M}^{M} \left(\frac{1}{f}\right)_{n-m} h_m = g_n + \Delta_n, \qquad (A.20)$$

where $\Delta_n = O[(\ln M)/M^2]$. Therefore, for $0 \leq x < 2\pi$,

$$h_n - h_n^{(M)} = \sum_{m=-M}^{M} \left[\!\!\left[\frac{1}{f}\right]\!\!\right]_{nm}^{(M)-1} \Delta_m, \qquad (A.21)$$

$$|h^{(M)}(x) - h_M(x)| \leq \sum_{n=-M}^{M} \sum_{m=-M}^{M} \left|\left[\!\!\left[\frac{1}{f}\right]\!\!\right]_{nm}^{(M)-1}\right| |\Delta_m| \leq O\left(\frac{\ln M}{\sqrt{M}}\right), \qquad (A.22)$$

where $h_n^{(M)}$ is given by the inverse rule. This completes the proof of Theorem 4.5. □

References

[1] T. Tamir, H. C. Wang, and A. A. Oliner, *Wave propagation in sinusoidally stratified dielectric media*, IEEE Trans. Microwave Theory Tech., 12 (1964), pp. 323–335.

[2] T. Tamir and H. C. Wang, *Scattering of electromagnetic waves by a sinusoidally stratified half-space, I. Formal solution and analytic approximations*, Canad. J. Phys., 44 (1966), pp. 2073–2094.

[3] T. Tamir, *Scattering of electromagnetic waves by a sinusoidally stratified half-space, II. Diffraction aspects at the Rayleigh and Bragg wavelengths*, Canad. J. Phys., 44 (1966), pp. 2461–2494.

[4] C. Yeh, K. F. Casey, and Z. A. Kaprielian, *Transverse magnetic wave propagation in sinusoidally stratified dielectric media*, IEEE Trans. Microwave Theory Tech., 13 (1965), pp. 297–302.

[5] C. B. Burckhardt, *Diffraction of a plane wave at a sinusoidally stratified dielectric grating*, J. Opt. Soc. Amer., 56 (1966), pp. 1502–1509.

[6] F. G. Kaspar, *Diffraction by thick, periodically stratified gratings with complex dielectric constant*, J. Opt. Soc. Amer., 63 (1973), pp. 37–45.

[7] S. T. Peng, T. Tamir, and H. L. Bertoni, *Theory of periodic dielectric waveguides*, IEEE Trans. Microwave Theory Tech., 23 (1975), pp. 123–133.

[8] K. Knop, *Rigorous diffraction theory for transmission phase gratings with deep rectangular grooves*, J. Opt. Soc. Amer., 68 (1978), pp. 1206–1210.

[9] M. G. Moharam and T. K. Gaylord, *Rigorous coupled-wave analysis of planar-grating diffraction*, J. Opt. Soc. Amer., 71 (1981), pp. 811–818.

[10] M. G. Moharam and T. K. Gaylord, *Diffraction analysis of dielectric surface-relief gratings*, J. Opt. Soc. Amer., 72 (1982), pp. 1385–1392.

[11] M. G. Moharam and T. K. Gaylord, *Rigorous coupled-wave analysis of grating diffraction—E-mode polarization and losses*, J. Opt. Soc. Amer., 73 (1983), pp. 451–455.

[12] M. G. Moharam and T. K. Gaylord, *Three-dimensional vector coupled-wave analysis of planar grating diffraction*, J. Opt. Soc. Amer., 73 (1983), pp. 1105–1112.

[13] M. G. Moharam and T. K. Gaylord, *Rigorous coupled-wave analysis of metallic surface-relief gratings*, J. Opt. Soc. Amer. A, 3 (1986), pp. 1780–1787.

[14] K. Rokushima and J. Yamakita, *Analysis of anisotropic dielectric gratings*, J. Opt. Soc. Amer., 73 (1983), pp. 901–908.

[15] S. T. Han, Y.-L. Tsao, R. M. Walser, and M. F. Becker, *Electromagnetic scattering of two-dimensional surface-relief dielectric gratings*, Appl. Optim., 31 (1992), pp. 2343–2352.

[16] E. Noponen and J. Turunen, *Eigenmode method for electromagnetic synthesis of diffractive elements with three-dimensional profiles*, J. Opt. Soc. Amer. A, 11 (1994), pp. 2494–2502.

[17] L. Li and C. W. Haggans, *Convergence of the coupled-wave method for metallic lamellar diffraction gratings*, J. Opt. Soc. Amer. A, 10 (1993), pp. 1184–1189.

[18] Ph. Lalanne and G. M. Morris, *Highly improved convergence of the coupled-wave method for TM polarization*, J. Opt. Soc. Amer. A, 13 (1996), pp. 779–784.

[19] G. Granet and B. Guizal, *Efficient implementation of the coupled-wave method for metallic lamellar gratings in TM polarization*, J. Opt. Soc. Amer. A, 13 (1996), pp. 1019–1023.

[20] L. Li, *Use of Fourier series in the analysis of discontinuous periodic structures*, J. Opt. Soc. Amer. A, 13 (1996), pp. 1870–1876.

[21] L. Li, *New formulation of the Fourier modal method for crossed surface-relief gratings*, J. Opt. Soc. Amer. A, 14 (1997), pp. 2758–2767.

[22] L. Li, *Reformulation of the Fourier modal method for surface-relief gratings made with anisotropic materials*, J. Modern Opt., 45 (1998), pp. 1313–1334.

[23] R. Petit, *A tutorial introduction*, in Electromagnetic Theory of Gratings, R. Petit, ed., Springer-Verlag, Berlin, 1980, pp. 1–52.

[24] L. Li, *Justification of matrix truncation in the modal methods of diffraction gratings*, J. Optics A: Pure Appl. Opt., 1 (1999), pp. 531–536.

[25] F. P. Sayer, *The eigenvalue problem for infinite systems of linear equations*, Math. Proc. Cambridge Philos. Soc., 82 (1977), pp. 269–273.

[26] E. T. Whittaker and G. N. Watson, *A Course of Modern Analysis*, 4th ed., Cambridge University Press, Cambridge, UK, 1927.

[27] A. Zygmund, *Trigonometric Series*, Vol. 1, 2nd ed., Cambridge University Press, Cambridge, UK, 1968, p. 159.

[28] F. Riesz, *Les systèmes d'èquations linéaires à une infinité d'inconnues*, Gauthier-Villars, Paris, 1913.

[29] M. Bernkopf, *A history of infinite matrices*, Arch. Hist. Exact Sci., 4 (1968), pp. 308–358.

[30] H. Poincaré, *Sur les déterminants d'ordre infini*, Bull. Soc. Math. France, XIV (1886), pp. 77–90.

[31] H. von Koch, *Sur les déterminants infinis et les équations différentielles linéaires*, Acta Math., 16 (1892), pp. 217–295.

[32] W. Magnus and S. Winkler, *Hill's Equation*, Dover, New York, 1979.

[33] K. Knopp, *Theory and Application of Infinite Series*, Dover, New York, 1990.

[34] G. H. Hardy, *Divergent Series*, 2nd ed., Chelsea, New York, 1991, pp. 227–246.

[35] E. Hewitt and R. E. Hewitt, *The Gibbs-Wilbraham phenomenon: an episode in Fourier analysis*, Arch. Hist. Exact Sci., 21 (1979), pp. 129–160.

[36] Ph. Lalanne, *Effective properties and band structures of lamellar subwavelength crystals: Plane-wave method revisited*, Phys. Rev. B, 58 (1998), pp. 9801–9807.

[37] L. Li and J. Chandezon, *Improvement of the coordinate transformation method for surface-relief gratings with sharp edges*, J. Opt. Soc. Amer. A, 13 (1996), pp. 2247–2255.

[38] E. Popov and M. Nevière, *Grating theory: New equations in Fourier space leading to fast converging results for TM polarization*, J. Opt. Soc. Amer. A, 17 (2000), pp. 1773–1784.

[39] R. A. Horn and C. R. Johnson, *Matrix Analysis*, Cambridge University Press, Cambridge, UK, 1985.

Chapter 5

Electromagnetic Models for Finite Aperiodic Diffractive Optical Elements

Dennis W. Prather, Mark S. Mirotznik, and Shouyuan Shi

5.1 Introduction

Throughout the last three decades progress in the modeling and design of diffractive optical elements (DOEs) has occurred primarily within the scalar domain [30, 12, 27, 26, 9, 59, 24, 25, 4, 60, 31]. This is due, in large part, to the relative ease with which scalar-based diffraction models can be applied to the analysis of both periodic and aperiodic DOE profiles. In scalar diffraction theory, the effect of the diffractive profile on the incident field is *assumed* to be a relative phase delay, over the extent of each feature in the profile. The phase is proportional to the relative thickness of each diffracting feature, or the optical path length through the DOE; see Figure 5.1. The result is a spatially varying phase profile that reradiates to produce the diffracted field. However, this approximation neglects the vector nature of the incident field in that it ascribes the phase values to a scalar value E, rather than the vector components, $\mathbf{E}_x, \mathbf{E}_y$, and $\mathbf{E}_z$. In addition, nonuniformities in both the amplitude and phase of the transmitted electric field that result from discontinuities in the DOE profile are neglected. These nonuniformities tend to increase as the size of the feature approaches the wavelength of illumination. In addition, as the observation plane moves closer to the DOE the effects of the nonuniformities in the diffracted field become significant. Thus, in the near-field of the DOE or as

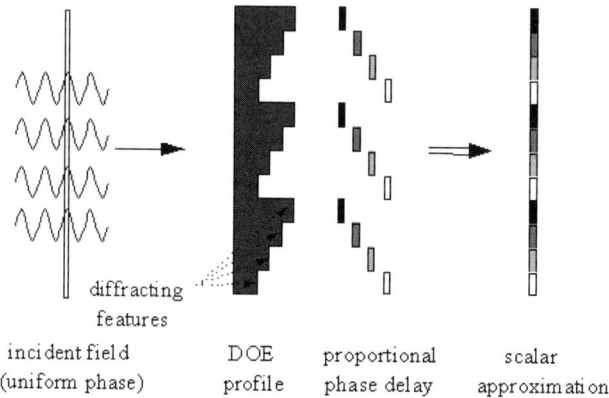

Figure 5.1: Scalar approximation for the interaction between an incident field and a diffractive structure.

the size of the feature approaches the wavelength of illumination, scalar diffraction theory becomes invalid and a complete solution to the electromagnetic boundary value problem must be used [13, 38].

Although progress has also been made in the vector analysis of DOEs [19, 37, 10, 33, 11, 36, 28, 23, 22], it has been limited primarily to infinitely periodic structures. Thus, until recently, most references to the rigorous analysis of DOEs have implicitly assumed infinitely periodic elements, i.e., gratings. The grating assumption allows wavefields to be expanded in terms of known eigenfunctions. However, for finite aperiodic elements, such as lenses and finite extent gratings, the eigenfunctions are not known, which precludes the application of such methods. Consequently, vector diffraction models for finite aperiodic DOEs need to be developed.

Recent progress in the analysis of finite aperiodic DOEs has, to a large extent, been based on well-established numerical electromagnetic techniques. Whereas the mathematical foundations for these techniques have existed for some 30 years or more, their application to the analysis of DOEs is relatively new. This is due not only to recent developments in fabrication technology [63, 32], which allows for the realization of DOEs that have feature sizes smaller than a wavelength, but also to the need for optical systems, and hence optical components, to be reduced in size. Such a reduction is necessary in order for optical technologies to keep pace with the ever decreasing scales of integration of their electronic counterparts.

Therefore, in the remainder of this section, we present an overview of recent progress in the development of rigorous, or vector-based, electromagnetic methods for the analysis of finite aperiodic DOEs.

5.2 Background

In theory, Maxwell's equations can be used to determine the exact diffracted field from any structure. However, for all but at few simple geometries, i.e., those that can be represented or conformally mapped into separable coordinate systems [56, 2, 17], closed-form analytic solutions do not exist. This is due to the fact that in order to solve Maxwell's equations one must solve the complete electromagnetic boundary value problem. That is, the electromagnetic boundary conditions must be used to match tangential and normal field components at every point along a material discontinuity. In order for this to be performed analytically it must be done in a global fashion, so as to separate the independent variables and therefore allow for application of the boundary conditions. Unfortunately, this can be done for only a small number of geometries. Consequently, for the rigorous analysis of DOEs, one must resort to numerical techniques.

During the last few years several numerical electromagnetic techniques have been successfully applied to the electromagnetic analysis of finite and aperiodic DOEs. Among these techniques are the finite-element method (FEM) [29, 34], the boundary element method (BEM) [41, 16, 42], the method of moments (MoM) [20, 57, 45, 47], the finite-difference method (FDM) [32, 14], and the finite-difference time-domain method (FDTD) [6, 40, 43]. Each of these techniques has its own unique advantage, depending on the application at hand; however, the method receiving by far the most interest lately is the FDTD. To this end, we will spend the majority of this section discussing the formulation of the FDTD method as it applies to DOE analysis. In the latter part of this section we also introduce the boundary element method for DOE analysis and refer the formulation of the finite element method to Chapter 2 in this book.

5.3 FDTD Method

Since the initial work of Yee [62], the FDTD method has become one of the most widely used numerical techniques for solving electromagnetic boundary value problems. The method has two main advantages over frequency domain techniques such as the FDM, FEM, and boundary integral methods (which includes both the BEM and MoM). First and foremost, the memory requirements of FDTD are significantly lower than those of the other methods, which permits the analysis of larger structures. Second, the method is simple to implement, even for complicated scatterers, because arbitrary dielectric parameters can be assigned to each lattice point. Consequently, structures that contain inhomogeneous, lossy, or even anisotropic material properties can be easily analyzed. In addition, the FDTD algorithm is readily implemented on multiprocessor computers because of the inherent data independencies of the difference equations, from which it is derived. This makes it possible to analyze realistic DOE profiles, i.e., $\sim 10^5$ wavelengths in extent, in reasonable time frames. In the remainder of this section we describe the FDTD method for two-dimensional, axially symmetric, and full three-dimensional DOE profiles.

5.3.1 Central Difference Expressions of Maxwell's Equations

First, consider Maxwell's time-dependent curl equations for a linear, isotropic, and source-free region,

$$\begin{cases} \mu \dfrac{\partial \mathbf{H}}{\partial t} = -\nabla \times \mathbf{E}, \\ \varepsilon \dfrac{\partial \mathbf{E}}{\partial t} = \nabla \times \mathbf{H} + \sigma \mathbf{E}, \end{cases} \quad (5.1)$$

where $\mathbf{E}$ and $\mathbf{H}$ are time-variable electric and magnetic fields (boldface is used to denote a vector quantity) and μ, ε, and σ characterize the permittivity, permeability, and conductivity, respectively. The FDTD method derives its name from a direct central finite-difference approximation to these equations. In the formulation of the FDTD method we assume that $f(u,v,w,t)$ is the component of either electric or magnetic field in an orthogonal coordinate system (u,v,w). We denote discrete points in space as $(i\Delta u, j\Delta v, k\Delta w)$ and discrete points in time as $n\Delta t$. Consequently, $f(u,v,w,t)$ can be expressed as

$$f(u,v,w,t) = f(i\Delta u, j\Delta v, k\Delta w, n\Delta t) = f_{i,j,k}^n, \quad (5.2)$$

where $\Delta u, \Delta v, \Delta w$ are the lattice increments in u, v, and w coordinate directions, respectively, and Δt is the time increment.

Yee used central finite-difference expressions for the space and time derivatives, which are second-order accurate in time and space. For example, consider the first partial space derivatives of f in the u-direction, calculated at a fixed time $t_n = n\Delta t$:

$$\frac{\partial f(i\Delta u, j\Delta v, k\Delta w, n\Delta t)}{\partial u} = \frac{f_{i+1/2,j,k}^n - f_{i-1/2,j,k}^n}{\Delta u} + O\left[(\Delta u)^2\right]. \quad (5.3)$$

We note the $\pm 1/2$ increment in the i subscript of the u coordinate as a finite difference over $\pm 1/2 \Delta u$. This notation makes the electric and magnetic fields interleaved in the space lattice at intervals Δu. Similarly, Yee's [62] expression for the time partial derivative of f, evaluated at fixed point (i,j,k):

$$\frac{\partial f(i\Delta u, j\Delta v, k\Delta w, n\Delta t)}{\partial t} = \frac{f_{i,j,k}^{n+1/2} - f_{i,j,k}^{n-1/2}}{\Delta t} + O\left[(\Delta u)^2\right], \quad (5.4)$$

where the $\pm 1/2$ increments are in the n superscript (time coordinate) of f, denoting a time finite difference over $\pm 1/2\Delta t$. Using these expressions, the electric and magnetic fields are then calculated in time at intervals of $1/2\Delta t$ in accordance with the Yee algorithm. In the formulations presented below, these difference equations are used to describe the FDTD method for the following cases: (1) two-dimensional, (2) three-dimensional axially symmetric, and (3) full three-dimensional.

5.3.2 Two-Dimensional Formulation

In the formulation of the two-dimensional FDTD method one can assume either transverse electric (TE) polarization (electric field perpendicular to the plane of incidence) or transverse magnetic (TM) polarization (magnetic field perpendicular to the plane of incidence). This results from the fact that in two dimensions, neither the fields nor the DOE profile has any variation in the z-direction, consequently, all of the partial derivatives with respect to z are zero. Therefore, Maxwell's equations, in a rectangular coordinate system, can be reduced to two decoupled sets of equations,

$$\begin{aligned}
\frac{\partial E_z}{\partial t} &= \frac{1}{\epsilon}\left(\frac{\partial H_y}{\partial x} - \frac{\partial H_x}{\partial y} - \sigma E_z\right), \\
\frac{\partial H_x}{\partial t} &= -\frac{1}{\mu}\frac{\partial E_z}{\partial y}, \\
\frac{\partial H_y}{\partial t} &= \frac{1}{\mu}\frac{\partial E_z}{\partial x}
\end{aligned} \quad (5.5)$$

and

$$\begin{aligned}
\frac{\partial E_x}{\partial t} &= \frac{1}{\epsilon}\frac{\partial H_z}{\partial y}, \\
\frac{\partial E_y}{\partial t} &= \frac{1}{\epsilon}\frac{\partial H_z}{\partial x}, \\
\frac{\partial H_z}{\partial t} &= \frac{1}{\mu}\left(\frac{\partial E_x}{\partial y} - \frac{\partial E_y}{\partial x}\right).
\end{aligned} \quad (5.6)$$

The first set includes H_x and H_y as components of the magnetic field in the x- and y-direction, respectively, and E_z as the component of electric field in the z-direction. The second set includes E_x and E_y as components of the electric field in the x- and y-direction, respectively, and H_z as the component of magnetic field in the z-direction. Applying the central difference expressions presented in section 5.3.1, the equations for the TE mode are derived to be the following:

$$E_z^n(i,j) = \frac{2\varepsilon - \sigma\Delta t}{2\varepsilon + \sigma\Delta t}E_z^{n-1}(i,j)$$

$$+ \frac{\sigma\Delta t}{2\varepsilon + \sigma\Delta t}\left[H_y^{n-1/2}(i+1/2,j) - H_y^{n-1/2}(i-1/2,j)\right]$$

$$- \frac{\sigma\Delta t}{2\varepsilon + \sigma\Delta t}\left[H_x^{n-1/2}(i,j+1/2) - H_x^{n-1/2}(i,j-1/2)\right],$$

$$H_x^{n+1/2}(i,j) = H_x^{n-1/2}(i,j) - \frac{\Delta t}{\mu\Delta y}\left[E_z^n(i,j+1/2) - E_z^n(i,j-1/2)\right],$$

$$H_y^{n+1/2}(i,j) = H_y^{n-1/2}(i,j) - \frac{\Delta t}{\mu\Delta x}\left[E_z^n(i+1/2,j) - E_z^n(i+1/2,j)\right], \quad (5.7)$$

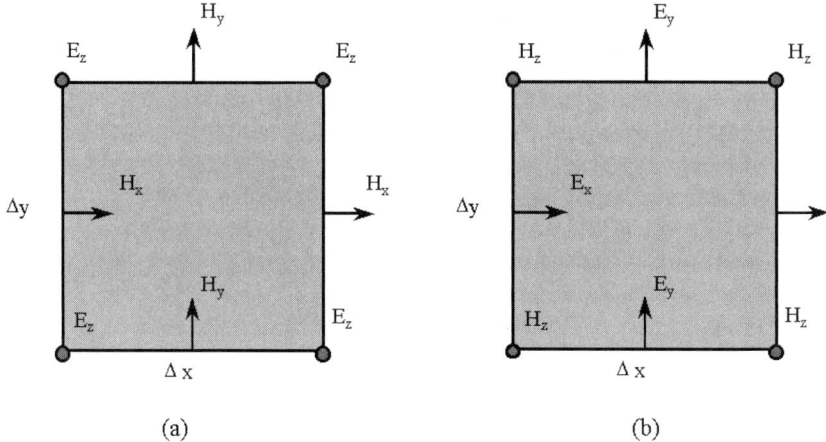

Figure 5.2: Two-dimensional FDTD lattices (a) for the TE case and (b) for the TM case.

where the electric field component is updated at time $t = n\Delta t$ and the magnetic field components are updated at time $t = (n + 1/2)\Delta t$. The spatial field components are evaluated on an interleaved grid, as shown in Figure 5.2. The difference expressions for the TM mode can be determined in a similar fashion. Whereas the two-dimensional FDTD formulation presented above is useful for many practical DOEs, e.g., cylindrical lenses and finite extent gratings, more sophisticated formulations are needed for more complex geometries. However, before presenting a general three-dimensional formulation we first present an axially symmetric formulation that exploits cylindrical symmetry within certain DOEs to reduce computational costs.

5.3.3 Three-Dimensional Axially Symmetric Formulation

Many useful DOE profiles, e.g., spherical lenses, contain axial symmetry. Therefore, it is quite useful to formulate an FDTD method that exploits this property. The primary motivation for doing so is to reduce the computational cost and memory requirements associated with DOE analysis. The formulation of an FDTD method for such structures is best represented in cylindrical coordinates:

$$\begin{aligned}
-\mu \frac{\partial H_\rho}{\partial t} &= \frac{1}{\rho} \frac{\partial E_z}{\partial \phi} - \frac{\partial E_\phi}{\partial z}, \\
-\mu \frac{\partial H_\phi}{\partial t} &= \frac{\partial E_\rho}{\partial z} - \frac{\partial E_z}{\partial \rho}, \\
-\mu \frac{\partial H_z}{\partial t} &= \frac{1}{\rho} \frac{\partial (\rho E_\phi)}{\partial \rho} - \frac{1}{\rho} \frac{\partial E_\rho}{\partial \phi}
\end{aligned} \quad (5.8)$$

and

$$\varepsilon \frac{\partial E_\rho}{\partial t} = \frac{1}{\rho} \frac{\partial H_z}{\partial \phi} - \frac{\partial H_\phi}{\partial z} + \sigma E_\rho,$$
$$\varepsilon \frac{\partial E_\phi}{\partial t} = \frac{\partial H_\rho}{\partial z} - \frac{\partial H_z}{\partial \rho} + \sigma E_\phi, \quad (5.9)$$
$$\varepsilon \frac{\partial E_z}{\partial t} = \frac{1}{\rho} \frac{\partial (\rho H_\phi)}{\partial \rho} - \frac{1}{\rho} \frac{\partial H_\rho}{\partial \phi} + \sigma E_z,$$

where E_ρ, E_ϕ, E_z and H_ρ, H_ϕ, H_z are electric and magnetic field components in cylindrical coordinates and are functions of (ρ, ϕ, z, t). For axially symmetric structures the value of each electric and magnetic field component is periodic in the azimuthal direction, consequently, their dependence on ϕ can be represented in terms of a Fourier series expansion,

$$\vec{E}_\gamma (\rho, \phi, z, t) = \sum_{k=0}^{\infty} E1_{\gamma,k} (\rho, z, t) \cos(k\phi) + \sum_{k=1}^{\infty} E2_{\gamma,k} (\rho, z, t) \sin(k\phi), \quad (5.10)$$

$$\vec{H}_\gamma (\rho, \phi, z, t) = \sum_{k=0}^{\infty} H1_{\gamma,k} (\rho, z, t) \cos(k\phi) + \sum_{k=1}^{\infty} H2_{\gamma,k} (\rho, z, t) \sin(k\phi), \quad (5.11)$$

where $\gamma = \hat{\rho}, \hat{\phi},$ and $\hat{z}$ are the vector field components, and $E1_{\gamma,k}, E2_{\gamma,k}, H1_{\gamma,k}$, and $H2_{\gamma,k}$ are the Fourier coefficients that need to be determined. Because each field component is represented using two summations, this approach appears to double the computational effort; however, for axisymmetric structures one can conveniently orient the incident field polarization to lie in the (ρ, z) plane, referred to as θ-polarization. This renders $\mathbf{E}_\rho, \mathbf{E}_z$, and $\mathbf{H}_\phi$ as even functions of ϕ and $\mathbf{H}_\rho, \mathbf{H}_z$, and $\mathbf{E}_\phi$ as odd functions. Consequently, the representation of each component is reduced to only one summation.

Because the Fourier modes in the series expansion are mutually orthogonal, one can solve for them independently. The number of modes needed in the analysis is dependent on the incident angle and the size of the object of interest. Substitution of (5.10) and (5.11) into (5.8) and (5.9) and differentiation with respect to ϕ reduces the original three-dimensional problem to an equivalent two-dimensional one. The assumption of θ-polarization thereby allows one to separate out the angular dependence from Maxwell's equations:

$$\mu \frac{\partial H2_{\rho,k}}{\partial t} = \frac{k}{\rho} E1_{z,k} + \frac{\partial E2_{\phi,k}}{\partial z},$$
$$\mu \frac{\partial H1_{\phi,k}}{\partial t} = -\frac{\partial E1_{\rho,k}}{\partial z} + \frac{\partial E1_{z,k}}{\partial \rho}, \quad (5.12)$$
$$\mu \frac{\partial H2_{z,k}}{\partial t} = -\frac{1}{\rho} \frac{\partial (\rho E2_{\phi,k})}{\partial \rho} - \frac{k}{\rho} E1_{\rho,k}$$

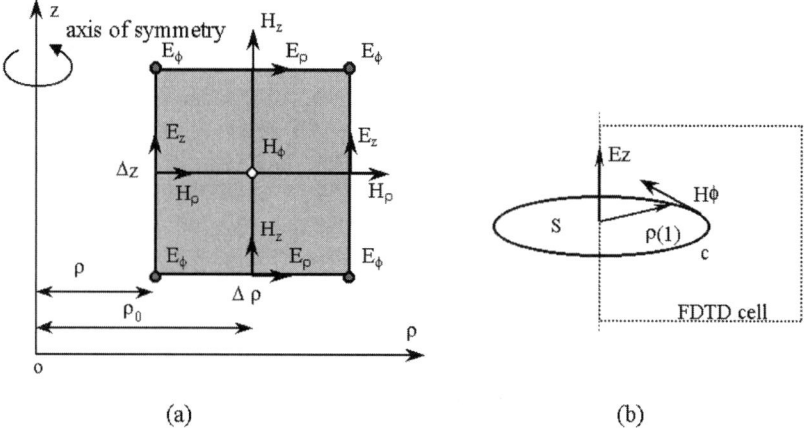

Figure 5.3: Two-dimensional grid for axially symmetric FDTD lattices in a cylindrical coordinate system: (a) off-set grid and (b) on-axis grid.

and

$$\varepsilon \frac{\partial E2_{\rho,k}}{\partial t} = -\frac{k}{\rho} H1_{z,k} - \frac{\partial H2_{\phi,k}}{\partial z} + \sigma E2_{\rho,k},$$

$$\varepsilon \frac{\partial E1_{\phi,k}}{\partial t} = \frac{\partial H1_{\rho,k}}{\partial z} - \frac{\partial H1_{z,k}}{\partial \rho} + \sigma E2_{\phi,k}, \qquad (5.13)$$

$$\varepsilon \frac{\partial E2_{z,k}}{\partial t} = \frac{1}{\rho} \frac{\partial (\rho H2_{\phi,k})}{\partial \rho} - \frac{k}{\rho} H2_{\rho,k} + \sigma E2_{z,k},$$

where $k = 0, \ldots, k_{\max}$. For convenience we simplify notation by referring to the field coefficients as $E_{\gamma,k}$ and $H_{\gamma,k}$, where the even and odd properties, denoted by 1 and 2, are implied.

The exploitation of axial symmetry can reduce the number of nodes and hence the computational costs associated with a three-dimensional analysis by projecting the three-dimensional mesh, in cylindrical coordinates, onto the two-dimensional mesh in the (ρ, z) plane, as shown in Figure 5.3(a). Even though the three-dimensional mesh is reduced to a two-dimensional mesh, the computational costs associated with the axisymmetric FDTD are actually slightly more than that of a two-dimensional problem. This is because the axisymmetric problem requires the solution of all six electromagnetic field components, as indicated in (5.12) and (5.13), whereas a true two-dimensional analysis requires the solution of only three components. Also, dependent upon the incident angle, one may have to perform k separate analyses, one for each mode. For this reason the dimensionality of an axisymmetric formulation is often referred to as 2.5.

By applying the central difference method introduced by Yee [62] to (5.12), we derive the following set of equations for the electric field components at time $t = n\Delta t$ and magnetic field components at $t = (n+1/2)\Delta t$,

$$E_{\rho,k}^{n}(i,j) = \frac{2\varepsilon - \Delta t \sigma}{2\varepsilon + \Delta t \sigma} E_{\rho,k}^{n-1}(i,j) + \frac{2k\Delta t}{\rho(i)(2\varepsilon + \Delta t \sigma)} H_{z,k}^{n-1/2}(i,j)$$
$$- \frac{2\Delta t}{\Delta z (2\varepsilon + \Delta t \sigma)} \left[H_{\phi,k}^{n-1/2}(i,j) - H_{\phi,k}^{n-1/2}(i,j-1) \right],$$

$$E_{\phi,k}^{n}(i,j) = \frac{2\varepsilon - \Delta t \sigma}{2\varepsilon + \Delta t \sigma} E_{\phi,k}^{n-1}(i,j)$$
$$+ \frac{2\Delta t}{\Delta z (2\varepsilon + \Delta t \sigma)} \left[H_{\rho,k}^{n-1/2}(i,j) - H_{\rho,k}^{n-1/2}(i,j-1) \right] \quad (5.14)$$
$$- \frac{2\Delta t}{\Delta \rho (2\varepsilon + \Delta t \sigma)} \left[H_{z,k}^{n-1/2}(i,j) - H_{z,k}^{n-1/2}(i-1,j) \right],$$

$$E_{z,k}^{n}(i,j) = \frac{2\varepsilon - \Delta t \sigma}{2\varepsilon + \Delta t \sigma} E_{z,k}^{n-1}(i,j) - \frac{2k\Delta t}{\rho_0(i)(2\varepsilon + \Delta t \sigma)} H_{\rho,k}^{n-1/2}(i,j)$$
$$- \frac{2\Delta t}{\Delta \rho \rho_0(i)(2\varepsilon + \Delta t \sigma)} \left[\rho(i) H_{\phi,k}^{n-1/2}(i,j) - \rho(i-1) H_{\phi,k}^{n-1/2}(i-1,j) \right],$$

$$H_{\rho,k}^{n+1/2}(i,j) = H_{\rho,k}^{n-1/2}(i,j) + \frac{k\Delta t}{\mu \rho_0(i)} E_{z,k}^{n}(i,j)$$
$$+ \frac{\Delta t}{\mu \Delta z} \left[E_{\phi,k}^{n}(i,j+1) - E_{\phi,k}^{n}(i,j) \right],$$

$$H_{\phi,k}^{n+1/2}(i,j) = H_{\phi,k}^{n-1/2}(i,j) - \frac{k\Delta t}{\mu \Delta z} \left[E_{\rho,k}^{n}(i,j+1) - E_{\rho,k}^{n}(i,j) \right] \quad (5.15)$$
$$+ \frac{\Delta t}{\mu \Delta \rho} \left[E_{z,k}^{n}(i+1,j) - E_{z,k}^{n}(i,j) \right],$$

$$H_{z,k}^{n+1/2}(i,j) = H_{z,k}^{n-1/2}(i,j) - \frac{k\Delta t}{\mu \rho(i)} E_{\rho,k}^{n}(i,j)$$
$$- \frac{\Delta t}{\mu \Delta \rho \rho(i)} \left[\rho_0(i+1) E_{\phi,k}^{n}(i+1,j) - \rho_0(i) E_{\phi,k}^{n}(i,j) \right],$$

where $\rho_0(i) = (i-1)\Delta \rho$ and $\rho(i) = (i - \frac{1}{2})\Delta \rho$ are the radius of the leftmost side and the center of the mesh given in Figure 5.3(a). The material parameters are assigned in each lattice given by $\mu = \mu(i,j), \varepsilon = \varepsilon(i,j)$, and $\sigma = \sigma(i,j)$.

In the solution of these equations special consideration must be given to the field values, E_ϕ, E_z, and H_ρ that lie on the axis of rotation, due to the singularity that arises as ρ approaches 0. From (5.14) and (5.15), one sees that only the components tangential to the axis, i.e., E_ϕ and E_z, are needed to update the adjacent H_ϕ and H_z fields. We note from (5.15) that, since $E_\phi(1,j)$ is multiplied by the factor $\rho = 0$, $H_z(1,j)$ can be evaluated without knowing the value of $E_\phi(1,j)$. Thus, only the $E_z(1,j)$ is needed to update all relevant magnetic field components. In addition, because $E_z(1,j)$ can be evaluated separately, it indirectly accounts for the singularity in H_ρ, which therefore need not be considered.

To solve for $E_z(1,j)$, Faraday's law is applied to the contour, c, shown in Figure 5.3(b),

$$\int_s \nabla \times \mathbf{H} ds = \int_s \left(\varepsilon \frac{\partial \mathbf{E}}{\partial t} + \sigma \mathbf{E} \right) ds = \int_c \mathbf{H} \cdot dl. \tag{5.16}$$

By substituting (5.10) and (5.11) into (5.16) and evaluating the surface and line integrals, we find that $E_{z,k}(1,j)$ is zero for all modes except $k = 0$, which is given by

$$\varepsilon \frac{\partial E_{z,0}}{\partial t} + \sigma E_{z,0} = \frac{2}{\rho(1)} H_{\phi,0}. \tag{5.17}$$

Application of the central difference discretization yields the following difference equations for $E_z(1,j)$:

$$E_{z,k}^n(1,j) = \frac{2\varepsilon - \Delta t \sigma}{2\varepsilon + \Delta t \sigma} E_{z,k}^{n-1}(1,j) + \frac{8\Delta t}{\rho(1)(2\varepsilon + \Delta t \sigma)} H_{\phi,k}^{n-1/2}(1,j), \quad k = 0, \tag{5.18}$$

$$E_{z,k}^n(1,j) = 0, \quad k \geq 1.$$

Thus, for a normally incident wave ($k = 1$), all axial components are zero. However, for an off-axis incident field, a zeroth-order analysis must be performed and (5.18) must be used. The last, and most general, FDTD formulation is that for a full three-dimensional structure, which is presented in the next section.

5.3.4 Three-Dimension Formulation

In a three-dimensional rectangular coordinate system (x, y, z), Maxwell's curl equations (5.1) can be written as a set of six coupled scalar equations given by

$$\begin{aligned}
\frac{\partial E_x}{\partial t} &= \frac{1}{\epsilon} \left[\frac{\partial H_z}{\partial y} - \frac{\partial H_y}{\partial z} - \sigma E_x \right], \\
\frac{\partial E_y}{\partial t} &= \frac{1}{\epsilon} \left[\frac{\partial H_x}{\partial z} - \frac{\partial H_z}{\partial x} - \sigma E_y \right], \\
\frac{\partial E_z}{\partial t} &= \frac{1}{\epsilon} \left[\frac{\partial H_y}{\partial x} - \frac{\partial H x}{\partial y} - \sigma E_z \right],
\end{aligned} \tag{5.19}$$

Chapter 5. Electromagnetic Models for Finite Aperiodic DOEs

$$\frac{\partial H_x}{\partial t} = \frac{1}{\mu}\left[\frac{\partial E_y}{\partial z} - \frac{\partial E_z}{\partial y}\right],$$
$$\frac{\partial H_y}{\partial t} = \frac{1}{\mu}\left[\frac{\partial E_z}{\partial x} - \frac{\partial E_x}{\partial z}\right], \quad (5.20)$$
$$\frac{\partial H_z}{\partial t} = \frac{1}{\mu}\left[\frac{\partial E_x}{\partial y} - \frac{\partial E_y}{\partial x}\right].$$

In accordance with section 5.3.1, we can approximate each partial derivative as a second-order finite-difference equation, in both space and time, which for the electric field components at time $t = n\Delta t$ are

$$E_x^n(i,j,k) = \frac{2\varepsilon - \sigma\Delta t}{2\varepsilon + \sigma\Delta t} E_x^{n-1}(i,j,k)$$
$$+ \frac{2\Delta t}{\Delta y(2\varepsilon + \sigma\Delta t)}\left[H_z^{n-1/2}(i,j,k) - H_z^{n-1/2}(i,j-1,k)\right]$$
$$- \frac{2\Delta t}{\Delta z(2\varepsilon + \sigma\Delta t)}\left[H_y^{n-1/2}(i,j,k) - H_y^{n-1/2}(i,j,k-1)\right],$$

$$E_y^n(i,j,k) = \frac{2\varepsilon - \sigma\Delta t}{2\varepsilon + \sigma\Delta t} E_y^{n-1}(i,j,k)$$
$$+ \frac{2\Delta t}{\Delta z(2\varepsilon + \sigma\Delta t)}\left[H_x^{n-1/2}(i,j,k) - H_x^{n-1/2}(i,j,k-1)\right] \quad (5.21)$$
$$- \frac{2\Delta t}{\Delta x(2\varepsilon + \sigma\Delta t)}\left[H_z^{n-1/2}(i,j,k) - H_z^{n-1/2}(i-1,j,k)\right],$$

$$E_z^n(i,j,k) = \frac{2\varepsilon - \sigma\Delta t}{2\varepsilon + \sigma\Delta t} E_z^{n-1}(i,j,k)$$
$$+ \frac{2\Delta t}{\Delta x(2\varepsilon + \sigma\Delta t)}\left[H_y^{n-1/2}(i,j,k) - H_y^{n-1/2}(i-1,j,k)\right]$$
$$- \frac{2\Delta t}{\Delta y(2\varepsilon + \sigma\Delta t)}\left[H_x^{n-1/2}(i,j,k) - H_x^{n-1/2}(i,j-1,k)\right],$$

and for the magnetic field components at time $t = (n+1/2)\Delta t$ are

$$H_x^{n+1/2}(i,j,k) = H_x^{n-1/2}(i,j,k) + \frac{\Delta t}{\mu\Delta z}\left[E_y^n(i,j,k+1) - E_y^n(i,j,k)\right]$$
$$- \frac{\Delta t}{\mu\Delta y}\left[E_z^n(i,j+1,k) - E_z^n(i,j,k)\right],$$
$$H_y^{n+1/2}(i,j,k) = H_y^{n-1/2}(i,j,k) + \frac{\Delta t}{\mu\Delta x}\left[E_z^n(i+1,j,k) - E_z^n(i,j,k)\right] \quad (5.22)$$
$$- \frac{\Delta t}{\mu\Delta z}\left[E_x^n(i,j,k+1) - E_x^n(i,j,k)\right],$$
$$H_z^{n+1/2}(i,j,k) = H_z^{n-1/2}(i,j,k) + \frac{\Delta t}{\mu\Delta y}\left[E_x^n(i,j+1,k) - E_x^n(i,j,k)\right]$$
$$- \frac{\Delta t}{\mu\Delta x}\left[E_y^n(i+1,j,k) - E_y^n(i,j,k)\right].$$

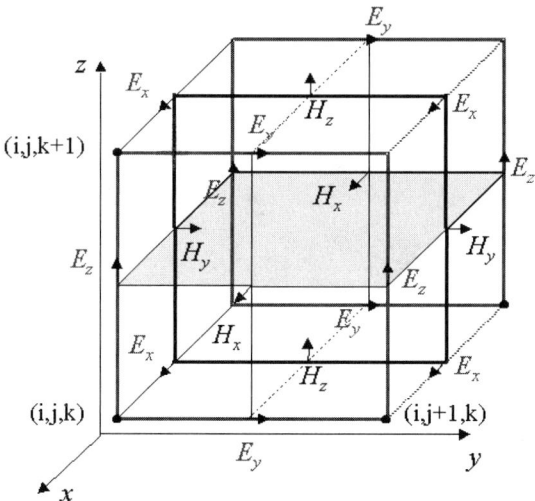

Figure 5.4: Three-dimensional FDTD lattice. Note that the gray region applies to the two-dimensional lattice.

Note that each electric field component, in three-dimensional space, is surrounded by four circulating magnetic field components, and similarly, every magnetic field component is surrounded by four circulating electric field components, as shown in Figure 5.4. Thus, the numerical solution of Maxwell's equations using the FDTD method provides for a nearly perfect correspondence with the physical propagation of electromagnetic waves. This aspect of the FDTD alone provides for great insight into the diffraction process.

In the above sections we have only discussed the discretizations of Maxwell's curl equations in terms of central difference equations. In terms of implementing an FDTD algorithm, there are several other key aspects we must discuss, namely, the stability and dispersion properties of the method, absorbing boundary conditions, source and steady-state conditions, and propagation of the electromagnetic fields from the computational region. Each of these topics is discussed in the following sections.

5.3.5 Numerical Stability and Dispersion

When implementing any numerical method it is necessary to establish the conditions under which stability of the technique can be insured. For the FDTD method stability implies that as time marching continues, according to the Yee algorithm, the electric and magnetic field values do not grow without bound. Because detailed accounts of stability can be found in [21] and [50], we present here only the criterion for which stability can be insured.

The stability condition is dependent on both the spatial increment, Δ, and the time increment, Δt. As we discuss later, to ensure the accuracy, i.e., minimize

Chapter 5. Electromagnetic Models for Finite Aperiodic DOEs   153

numerical dispersion, the spatial increment should be taken as small as possible, usually $\Delta \leq \lambda_{\min}/20$, where $\lambda_{\min}$ is the minimal wavelength of the excited source. However, one should note that as Δ is reduced, the memory requirements necessary to complete the analysis are increased. Once the spatial increment is chosen, the time increment has a corresponding specific bound, which can be shown to be [50]

$$\Delta t \leq \frac{1}{\sqrt{1/\Delta x^2 + 1/\Delta y^2}} \quad \text{for two dimensions,} \quad (5.23)$$

$$\Delta t \leq \frac{1}{\sqrt{1/\Delta x^2 + 1/\Delta y^2 + 1/\Delta z^2}} \quad \text{for three dimensions.}$$

In cylindrical coordinates it is more difficult to derive the stability condition, and as a result an empirical expression between spatial and time increments is given by [50]

$$c\Delta t \leq \frac{\min(\Delta \rho, \Delta z)}{s}, \quad (5.24)$$

where $s = \sqrt{2}$ for $k = 0$ and $s = k+1$ for $k \geq 1$.

Another factor that affects the accuracy of the FDTD method is numerical dispersion. Numerical dispersion is a nonphysical artifact that arises due to discretization of the FDTD computational space. It manifests itself by changing the phase velocity of the diffracting field and, as a result, introduces phase errors. Thus, when analyzing DOEs it is of utmost importance to reduce these errors.

In a manner similar to the analysis of stability [50], the numerical dispersion relation for the full three-dimensional FDTD algorithm can be derived by

$$\left[\frac{1}{c\Delta t}\sin\left(\frac{\omega \Delta t}{2}\right)\right]^2 = \left[\frac{1}{\Delta x}\sin\left(\frac{\widetilde{k}_x \Delta x}{2}\right)\right]^2 + \left[\frac{1}{\Delta y}\sin\left(\frac{\widetilde{k}_y \Delta y}{2}\right)\right]^2$$
$$+ \left[\frac{1}{\Delta z}\sin\left(\frac{\widetilde{k}_z \Delta z}{2}\right)\right]^2, \quad (5.25)$$

where $\widetilde{k}_x$, $\widetilde{k}_y$, and $\widetilde{k}_z$ are the x-, y-, and z-components of the numerical wave vector and ω is the wave angular frequency. Note that as Δt, Δx, Δy, and Δz all go to zero, the numerical dispersion relation tends to the analytical dispersion relation for a plane wave in a continuous lossless medium:

$$\frac{\omega^2}{c^2} = k_x^2 + k_y^2 + k_z^2.$$

5.3.6 Absorbing Boundary Conditions

The analysis of DOEs, in many cases, is defined as an "open-region" problem, where the spatial domain of the computed fields is unbounded in one or more coordinate

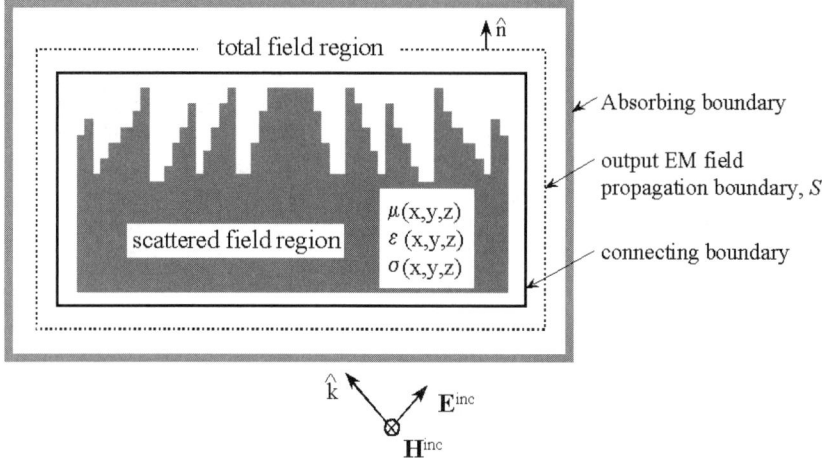

Figure 5.5: FDTD computational domain for an open-region problem.

directions. Obviously, one cannot compute the solution over such a space, due to limitations in computer memory and computational time; therefore, it is necessary to truncate the computational domain. This is achieved by enclosing the DOE and a suitable output boundary in a limited computational space, as shown in Figure 5.5. As it turns out, DOEs are especially well suited to this type of computational domain because their base is typically much larger than their height, hence the computational domain can be made into a very thin rectangle. As a result, the thickness of the computational domain in the direction of propagation can be as small as a few wavelengths. This has significant advantages when it comes to minimizing numerical dispersion, as discussed in section 5.3.5. However, because the diffracting waves are propagating out from the DOE, particular attention must be paid along the truncated boundary to insure that nonphysical, i.e., numerical, back-reflections do not occur. This is achieved by using an absorbing boundary condition (ABC), whose function is to absorb all incident radiation and to suppress spurious back-reflections of the outgoing wave.

Much of the more recent research on the FDTD method has concentrated on developing efficient ABCs. Two of the more commonly used ABCs are the Mur ABC [35] and the perfectly matched layer (PML) method [3]. Reflections from the PML ABCs are normally several orders of magnitude smaller (10^{-6} to 10^{-8} back reflections are typical) than any other ABC. In this section, we first consider the Engquist–Majda one-way wave equation, which is necessary in order to discuss the second-order Mur ABC. This is followed by a discussion of the PML ABCs for the two-dimensional FDTD algorithm.

Engquist–Majda Wave Equation

A partial differential equation that permits wave propagation in only one direction is a so-called one-way wave equation. When applied at the boundary of an FDTD

Chapter 5. Electromagnetic Models for Finite Aperiodic DOEs

grid, such an equation numerically absorbs impinging scattered waves. For purposes of illustration, consider the two-dimensional wave equation in Cartesian coordinates

$$\frac{\partial^2 U}{\partial x^2} + \frac{\partial^2 U}{\partial y^2} - \frac{1}{c^2}\frac{\partial^2 U}{\partial t^2} = 0, \qquad (5.26)$$

where U is any component of the electric or magnetic field and c is the phase velocity of light. We can define a partial differential operator

$$L = \frac{\partial^2}{\partial x^2} + \frac{\partial^2}{\partial y^2} - \frac{1}{c^2}\frac{\partial^2}{\partial t^2} = D_x^2 + D_y^2 - \frac{1}{c^2}D_t^2, \qquad (5.27)$$

from which we can express the wave equation as

$$LU = L^+ L^- U = 0, \qquad (5.28)$$

where

$$L^\pm = D_x \pm \frac{D_t}{c}\sqrt{1 - \left(c\frac{D_y}{D_t}\right)^2}. \qquad (5.29)$$

Engquist and Majda showed that on the appropriate boundary, application of L^- or L^+ performs as an exact analytical ABC for outgoing wave propagation [8]. Unfortunately, direct application of (5.29) in numerical form is not possible, due to the square root operator. Thus, approximations have to be made that unfortunately introduce error. One approximation that can be used is a two-term Taylor series expansion:

$$L^- \simeq D_x - \frac{D_t}{c} + \frac{cD_y^2}{2D_t}. \qquad (5.30)$$

Then, by performing $L^- U = 0$ we can obtain a corresponding partial differential equation that can be implemented as a second-order accurate ABC at the left boundary:

$$\frac{\partial^2 U}{\partial x \partial t} + \frac{c}{2}\frac{\partial^2 U}{\partial y^2} - \frac{1}{c}\frac{\partial^2 U}{\partial t^2} = 0. \qquad (5.31)$$

Analogous ABCs can be derived for the other grid boundaries using this method. The approximate second-order ABCs described above have been found to be useful in FDTD codes when implemented using the finite-difference scheme introduced by Mur [35], which is discussed next.

Second-Order Mur ABC

Equation (5.30) was first successfully implemented in finite-difference form by Mur [35], and for the purpose of illustration will be derived here for the left grid boundary at $x = 0$. In a two-dimensional space, i.e., TE-polarization, $U = E_z$ and (5.31) can be written as

$$\frac{\partial^2 E_z}{\partial x \partial t} - \frac{c\mu}{2}\frac{\partial^2 E_z}{\partial y \partial t} - \frac{1}{c}\frac{\partial^2 E_z}{\partial t^2} = 0.$$

By assuming the initial fields to be zero at time $t = 0$ and performing an integration with respect to t we obtain

$$\frac{\partial E_z}{\partial x} - \frac{c\mu}{2}\frac{\partial E_z}{\partial y} - \frac{1}{c}\frac{\partial E_z}{\partial t} = 0. \tag{5.32}$$

Equation (5.32) can now be used as a second-order accurate ABC, by implementing it using central difference expressions:

$$E_z^n(i,j) = E_z^{n-1}(i+1,j) + \frac{c\Delta t - \Delta x}{c\Delta t + \Delta x}\left[E_z^n(i+1,j) - E_z^{n-1}(i,j)\right]$$

$$- \frac{2\Delta x}{c\Delta t + \Delta x}\left[H_x^{n-1/2}(i,j+1/2) - H_x^{n-1/2}(i,j-1/2)\right. \tag{5.33}$$

$$\left. + H_x^{n-1/2}(i+1,j+1/2) - H_x^{n-1/2}(i+1,j-1/2)\right].$$

ABCs for the remaining boundaries can be derived in the same manner. In three dimensions, the derivation of the Mur finite-difference expressions for all six grid boundaries is similar, but the equations are obviously more complex. Numerical results using the second-order Mur ABC show reflections on the order of 10^{-2} for a board range of incident angles, which is sufficiency low enough to provide for acceptable results in the analysis of DOEs.

PML ABCs

The Mur ABC discussed above has been widely used since its introduction, but a recent advance in the development of ABCs, referred to as PMLs, is growing in popularity. As introduced by Berenger in 1994 [3], this method has been shown to reduce the reflections off ABC boundaries by several orders of magnitude as compared to any other technique.

The method is based on splitting the electric and magnetic field components in an absorbing region, where loss is assigned to individual split components. The net effect is to create a nonphysical absorbing medium adjacent to the outer FDTD mesh boundary that has a wave impedance independent of the angle of incidence and frequency of outgoing scattered waves.

We introduce its formulation by considering a modified form of Maxwell's equations in two dimensions for the TE polarization case:

$$\mu \frac{\partial H_x}{\partial t} + \sigma_x^* H_x = -\frac{\partial E_z}{\partial y},$$

$$\mu \frac{\partial H_y}{\partial t} + \sigma_y^* H_y = \frac{\partial E_z}{\partial x},$$

$$\epsilon \frac{\partial E_{zx}}{\partial t} + \sigma_x E_{zx} = \frac{\partial H_y}{\partial x}, \qquad (5.34)$$

$$\epsilon \frac{\partial E_{zy}}{\partial t} + \sigma_y E_{zy} = -\frac{\partial H_x}{\partial x},$$

$$E_z = E_{zx} + E_{zy},$$

where the parameters σ_x, σ_y, σ_x^*, and σ_y^* are electric and magnetic conductivities and satisfy the following relations:

$$\begin{cases} \dfrac{\sigma_x}{\epsilon_0} = \dfrac{\sigma_x^*}{\epsilon_0^*}, \\ \dfrac{\sigma_y}{\epsilon_0} = \dfrac{\sigma_y^*}{\epsilon_0^*}. \end{cases} \qquad (5.35)$$

The PML matching condition results in the same wave impedance of the scattered wave propagating in the PMLs as that of free space for any incident angle. As a result, the scattered wave enters the PML region with virtually no back-reflection, because the impedance of the PML is the same as that for free space. In addition, it also decays very quickly due to the lossy medium. Typically, the PML region is thick enough to absorb and completely decay the incident field by the time it reaches the outer boundary; this usually requires 8–16 PML layers. Also, the expression for the partial differential equations, (5.34), can easily be discretized using the central difference method, where E_{zx} and E_{zy} share the same spatial location as E_z.

In a three-dimensional formulation of the PML method all six field components are split, and the corresponding material in the PML layers are characterized by the electric and magnetic conductivities σ_x, σ_y, σ_z, σ_x^*, σ_y^*, and σ_z^*, which obey the same matching conditions as denoted in (5.35).

Whereas Berenger's development of the PML concept was introduced in Cartesian coordinates, more recently it has been applied to cylindrical coordinates, which are required in the axially symmetric FDTD formulation. The most common approach for formulating cylindrical PMLs is based on the complex coordinate stretching approach [53, 5, 46, 54], in which a change of variables is used to transform Maxwell's equations in the PML region into a set of equations in a complex coordinate system. Fortunately, closed-form solutions obtained in the regular media can be mapped into the PML media through a simple analytic continuation of the spatial variables into the complex coordinate system. Thus, to achieve maximum absorption of the outgoing waves, the coordinate variable transformations

$$\rho \to \widetilde{\rho} = \int_0^\rho s_\rho\left(\rho'\right) d\rho', \qquad z \to \widetilde{z} = \int_0^z s_z\left(z'\right) dz' \qquad (5.36)$$

are used to map Maxwell's equations into

$$j\omega \frac{s_\phi s_z}{s_\rho} \mu_0 H_{\rho,k} = \frac{k}{\rho} E_{z,k} + \frac{\partial E_{\phi,k}}{\partial z},$$

$$j\omega \frac{s_\rho s_z}{s_\phi} \mu_0 H_{\phi,k} = -\frac{\partial E_{\rho,k}}{\partial z} + \frac{\partial E_{z,k}}{\partial \rho}, \qquad (5.37)$$

$$j\omega \frac{s_\rho s_\phi}{s_z} \mu_0 H_{z,k} = -\frac{1}{\rho}\frac{\partial (\rho E_{\phi,k})}{\partial \rho} - \frac{k}{\rho} E_{\rho,k}$$

and

$$j\omega \frac{s_\phi s_z}{s_\rho} \varepsilon_0 E_{\rho,k} = -\frac{k}{\rho} H_{z,k} - \frac{\partial H_{\phi,k}}{\partial z},$$

$$j\omega \frac{s_\rho s_z}{s_\phi} \varepsilon_0 E_{\phi,k} = \frac{\partial H_{\rho,k}}{\partial z} - \frac{\partial H_{z,k}}{\partial \rho}, \qquad (5.38)$$

$$j\omega \frac{s_\rho s_\phi}{s_z} \varepsilon_0 E_{z,k} = \frac{1}{\rho}\frac{\partial (\rho H_{\phi,k})}{\partial \rho} - \frac{k}{\rho} H_{\rho,k},$$

where

$$\begin{aligned} s_\rho &= 1 - j\frac{\sigma_\rho}{\omega\varepsilon_0}, \\ s_\phi &= \frac{\widetilde{\rho}}{\rho} = \frac{1}{\rho}\int_0^\rho \left(1 - j\frac{\sigma_\rho}{\omega\varepsilon_0}\right) d\rho' = 1 - j\frac{\gamma_\rho}{\omega\varepsilon_0}, \quad \text{and} \\ s_z &= 1 - j\frac{\sigma_z}{\omega\varepsilon_0}. \end{aligned} \qquad (5.39)$$

Unfortunately, the electromagnetic fields in the PML region cannot be updated in the usual FDTD fashion due to their frequency dependence. However, a two-step method based on an auxiliary function,

$$B_{\rho,k} = \mu_0 \frac{s_\phi}{s_\rho} H_{\rho,k}, \qquad (5.40)$$

can be used to update the fields. By substituting (5.40) into (5.38), we have

$$j\omega B_{\rho,k} + \frac{\sigma_z}{\varepsilon_0} B_{\rho,k} = \frac{k}{\rho} E_{z,k} + \frac{\partial E_{\phi,k}}{\partial z}, \qquad (5.41)$$

$$j\omega \mu_0 H_{\rho,k} + \mu_0 \frac{\gamma_\rho}{\varepsilon_0} H_{\rho,k} = j\omega B_{\rho,k} + \frac{\sigma_\rho}{\varepsilon_0} B_{\rho,k},$$

which can be transformed back into the time domain using the Fourier transform relation $\frac{\partial}{\partial t} \Rightarrow j\omega$:

$$\frac{\partial B_{\rho,k}}{\partial t} + \frac{\sigma_z}{\varepsilon_0} B_{\rho,k} = \frac{k}{\rho} E_{z,k} + \frac{\partial E_{\phi,k}}{\partial z}, \qquad (5.42)$$

$$\mu_0 \frac{\partial H_{\rho,k}}{\partial t} + \mu_0 \frac{\gamma_\rho}{\varepsilon_0} H_{\rho,k} = \frac{\partial B_{\rho,k}}{\partial t} + \frac{\sigma_\rho}{\varepsilon_0} B_{\rho,k}.$$

One should note that this two-step procedure requires additional storage since the time history of both B_ρ and H_ρ must be known. Similar expressions for the remaining field components can be derived.

In addition to understanding the issues associated with numerical dispersion, stability, and how to implement good ABCs, one must also become knowledgable on how to implement general incident field profiles into the FDTD region. This topic is discussed in the next section.

5.3.7 Incident Wave Source Condition and Steady-State Condition

There are a variety of approaches for modeling the incident field in an FDTD calculation. One approach is to force the fields along a predefined boundary, in the FDTD space, to equal the desired incident field. The incident field will then propagate outward from this boundary and excite the scattering object some distance away. While easy to implement, this method suffers from nonphysical reflections that occur when energy reflected by the scatterer propagates back to the location of incident field boundary. Another approach, and the most widely used, is the total/scattered field formulation.

In this approach the computational region is separated into two parts, as shown in Figure 5.5, where the inner region operates on both the incident field and the scattered field (i.e., the total field) and the outer region operates on only the scattered field. The interior boundary serves to connect the inner and outer regions and is the location at which the incident field is introduced [50]. Within the FDTD space we calculate the total electric and magnetic fields, but in the scattered region we calculate only the scattered fields. Along the boundary separating the two regions we apply the simple connecting boundary condition

$$\begin{cases} E_{total} = E_{scattered} + E_{incident}, \\ H_{total} = H_{scattered} + H_{incident}. \end{cases} \qquad (5.43)$$

The incident field is then described analytically along this boundary. The formulation of this method consists of adding, or subtracting, the incident field along the connecting boundary in order to satisfy the type of field component that is being updated [50]. Arbitrary incident waveforms including plane waves and Gaussian beams can be easily implemented using this method without introducing nonphysical reflections. Propagation in this fashion is continued until a desired steady-state condition is achieved.

A steady-state condition can be achieved in one of two ways. First, the output response of the electromagnetic field components can be Fourier transformed. This method is well suited for incident fields that have a finite duration in time, e.g., a Gaussian modulated pulse. Whereas this approach exploits the inherent time-dependent nature of the FDTD method, by allowing for an entire spectral analysis from only a single computational run, it can significantly increase the memory requirements and execution time required to obtain a solution.

A second method for obtaining a steady-state condition is to extract the complex (magnitude and phase) of the field components over a complete cycle of the sinusoidal incident field. The key factor in this approach is to time-step long enough to insure that all of the transient fields have decayed away. Typically, this requires 2–5 transient periods, which is defined as the number of time steps required for one cycle of the incident field to traverse the maximum extent of the DOE [51, 49, 55, 52]. However, because DOEs are elongated structures, research has shown that a steady state condition can be reached with far fewer time steps [43]. After the FDTD calculation has reached steady state, say, after N time steps, the response for any field quantity at any lattice point within the solution space can be obtained by the following process. First, calculate the field at the current time step, N:

$$E1(\mathbf{r}) = E_0(\mathbf{r}) \sin[\omega N \Delta t + \phi(\mathbf{r})]. \tag{5.44}$$

Then, calculate it again at a time, $(\Delta N \cdot \Delta t \omega) = \pi/2$, later:

$$E2(\mathbf{r}) = E_0(\mathbf{r}) \sin[\omega(N + \Delta N)\Delta t + \phi(\mathbf{r})] = E_0(\mathbf{r}) \cos[\omega N \Delta t + \phi(\mathbf{r})]. \tag{5.45}$$

Combining (5.44) and (5.45), the magnitude $E_0(\mathbf{r})$ and phase values $\phi(\mathbf{r})$ can be determined from

$$E_0(\mathbf{r}) = \sqrt{E1^2(\mathbf{r}) + E2^2(\mathbf{r})}, \qquad \phi(\mathbf{r}) = \tan^{-1}(E2(\mathbf{r})/E1(\mathbf{r})).$$

5.3.8 Propagation of Electromagnetic Fields

Once the electromagnetic fields have reached steady-state values, we propagate them to the plane or region of interest using a propagation algorithm. In addition to reducing the number of time steps required to obtain a steady-state condition, as discussed above, this approach represents a more efficient formulation of the diffraction problem, because one can determine the diffracted fields over any region, or discrete regions, of space. In contrast, a full FDTD approach would require one to include the observation region within the computational domain, which, depending on its location and extent, can *significantly* increase the memory requirements and execution time required for a solution, not to mention its affects on numerical dispersion. To this end we introduce two propagation methods, one based on the Stratton–Chu integral method and the other on the plane wave spectrum method.

Stratton–Chu Integral Method

To determine the electromagnetic fields exterior to the FDTD computational region, we propagate the steady-state electromagnetic field components over a closed surface, S, in the scattered field region, using the Stratton–Chu [48] formula:

$$\mathbf{E}(\mathbf{r}) = -\int_{S_1} \left\{ j\omega\mu_0 G\left(\mathbf{r},\mathbf{r}'\right)\left[\widehat{n} \times \mathbf{H}\left(\mathbf{r}'\right)\right] \right.$$
$$\left. + \left[\widehat{n} \times \mathbf{E}\left(\mathbf{r}'\right)\right] \times \nabla G\left(\mathbf{r},\mathbf{r}'\right) + \left[\widehat{n} \cdot \mathbf{E}\left(\mathbf{r}'\right)\right]\nabla G\left(\mathbf{r},\mathbf{r}'\right) \right\} ds', \quad (5.46)$$

$$\mathbf{H}(\mathbf{r}) = \int_{S_1} \left\{ j\omega\varepsilon_0 G\left(\mathbf{r},\mathbf{r}'\right)\left[\widehat{n} \times \mathbf{E}\left(\mathbf{r}'\right)\right] \right.$$
$$\left. - \left[\widehat{n} \times \mathbf{H}\left(\mathbf{r}'\right)\right] \times \nabla G\left(\mathbf{r},\mathbf{r}'\right) - \left[\widehat{n} \cdot \mathbf{H}\left(\mathbf{r}'\right)\right]\nabla G\left(\mathbf{r},\mathbf{r}'\right) \right\} ds',$$

where ω is the angular frequency, $G(\mathbf{r},\mathbf{r}')$ is the free space Green's function,

$$G\left(\mathbf{r},\mathbf{r}'\right) = \frac{e^{-jk|\mathbf{r}-\mathbf{r}'|}}{4\pi|\mathbf{r}-\mathbf{r}'|}, \quad (5.47)$$

$\mathbf{E}^{scat}$ and $\mathbf{H}^{scat}$ are the scattered field components of the electric and magnetic fields, S is a closed surface, and $\widehat{n}$ is the normal unit vector directed outside of S, as shown in Figure 5.5. Because each field component is a constant over each lattice point the integrals in (5.46) are reduced to summations and are therefore easily implemented. Using this approach, the scattered fields can be determined at any location exterior to the FDTD region. Once the scattered fields are determined, the incident fields are propagated, using analytic methods, and added to the scattered fields in order to obtain the total field.

Near Field Propagation using the Plane Wave Spectrum Method

An alternative and extremely effective method for determining the propagated electromagnetic field is the plane wave spectrum (PWS) technique [12]. Whereas (5.46) is representative of an electromagnetic superposition integral, the PWS method exploits the planar relationship between the output boundary and an observation boundary by changing it to a convolution integral. This is achieved by noting that the Green's function, used in (5.46), is simply the impulse response of free space propagation. Thus, by having a planar relationship between the output and observation boundaries, the propagation process becomes shift invariant. As a result, the diffraction integral can be evaluated using a spatial convolution integral that can be determined by using a Fourier transform. Because of the efficiency of the fast Fourier transform (FFT), considerable time can be saved in the propagation of the diffracted field for large DOE profiles.

The formulation of the PWS method begins by first Fourier-transforming the known field along the output boundary. The field can then be decomposed into a PWS, which accounts for all possible directions of propagation. Once the PWS is determined, all the plane waves are propagated from the output boundary to the observation point by multiplying by an appropriate phase term.

For instance, suppose we have $E_z(x,y,0)$, as the electric field at plane $z = 0$, in a two-dimensional space. The PWS of the field is determined by using the Fourier integral

$$S(k_x, k_y, 0) = \int\int_{-\infty}^{+\infty} E_z(x, y, 0) \exp[j(k_x x + k_y y)] dx dy, \quad (5.48)$$

in which k_x and k_y are wavenumbers along the x- and y-directions, respectively. For the case when $k_x^2 + k_y^2 < k_0^2$, the propagation components are evanescent waves and do not contribute to a propagated field more than several wavelengths away. The remaining plane waves propagate in different directions and can be propagated to the observation plane, $z = z_0$, by using the relation

$$S'(k_x, k_y, z_0) = S(k_x, k_y, 0) e^{-j k_z z}, \quad (5.49)$$

where $k_z = \sqrt{k_0^2 - k_x^2 - k_y^2}$ is the wavenumber along the z-direction. In the observation plane, the propagated PWS is inverse Fourier-transformed to obtain the diffracted field at observation plane $z = z_0$

$$E_z(x, y, z_0)$$
$$= \int\int_{-\infty}^{+\infty} S(k_x, k_y, 0) \exp\left(-j\sqrt{k_0^2 - k_x^2 - k_y^2}\right) \exp[-j(k_x x + k_y y)] dk_x dk_y. \quad (5.50)$$

Although the integral extends from $-\infty$ to $+\infty$ in (5.50), the limits can be changed to $-k_0$ to k_0 if the observation point is more than a few wavelengths away from the DOE. This is due to the fact that the contribution of the evanescent waves to the diffracted field becomes negligible for distance more than a few wavelengths from the diffracted structure.

Even though the FDTD method is growing in popularity, as an analysis tool for DOEs, it does have several disadvantages as compared to alternate techniques. First, because the difference equations are implemented in a rectangular grid, a curved DOE profile must be approximated as a stair step. Second, because the solution space consists of sampled grid variations in the profile below the grid spacing are not permitted. And third, for DOE structures that are resonant, i.e., structures that are embedded in multiple dielectric layers, the number of time steps required to reach a steady state can be significant. Techniques that overcome these limitations, such as the BEM, are presented in the following sections.

5.4 Boundary Integral Methods

In this section we present the formulation of boundary integral methods (BIMs) as they apply to the diffraction analysis of DOEs of both dielectric and perfectly conducting finite aperiodic DOEs. In both cases the integral equations are solved using the BEM.

Boundary integral equations relate the interaction between an incident field and a DOE using distributions induced on the surface of the DOE by the incident field.

Chapter 5. Electromagnetic Models for Finite Aperiodic DOEs

Reradiation from the surface distribution, in turn, generates a diffracted field. Thus, the objective in applying BIMs to the analysis of DOEs is to determine accurately the surface distributions given the incident field and DOE.

Although numerical solutions of the boundary integral equations are well known and have been applied extensively within the microwave community [15, 58], their application to the analysis of DOEs has been limited [20, 57, 41, 16, 42, 45].

The formulations presented below assume a two-dimensional diffractive structure that is uniform in the z-direction. The analysis considers TE- and TM-polarized fields separately, which allows the electromagnetic fields to be represented by a single function $u^{tot}(x,y)$. For TE polarization $u^{tot}(x,y)$ denotes the total electric field $E_z(x,y)$, and for TM polarization $u^{tot}(x,y)$ denotes the total magnetic field $H_z(x,y)$. For both cases the total field, $u^{tot}(x,y)$, is the sum of the scattered field $u^{sc}(x,y)$ and the incident field $u^{inc}(x,y)$. In addition, nonmagnetic materials, i.e., $\mu_r = 0$, are also assumed.

5.4.1 Boundary Integral Formulation for Dielectrics

Boundary integral equations are formulated by defining a solution space that consists of discrete homogeneous regions. For a phase-only DOE located in free space, there are two regions: region I, which contains the DOE, and region O, which is free space, as shown in Figure 5.6. Within each region, the two-dimensional time-harmonic wave equation is used. In the interior region, the homogeneous wave equation is used; however, in the exterior region (where the incident wave is assumed to originate), an inhomogeneous wave equation must be used:

$$0 = \nabla^2 u_I^{tot}(\rho) + \beta_I^2 u_I^{tot}(\rho) \text{ for } \rho \in I, \qquad (5.51)$$

$$-f(\rho) = \nabla^2 u_O^{tot}(\rho) + \beta_O^2 u_O^{tot}(\rho) \text{ for } \rho \in O, \qquad (5.52)$$

where $u^{tot} = \mathbf{E}_z^{tot}$ for TE polarization and $u^{tot} = \mathbf{H}_z^{tot}$ for TM polarization. The wave numbers in region I and O are represented by $\beta_I = 2\pi\sqrt{\varepsilon_{r_I}}/\lambda_o$, $\beta_O = 2\pi\sqrt{\varepsilon_{r_O}}/\lambda_o$, respectively, and $f(\rho)$ represents the source term. The constants ε_{r_I} and ε_{r_O} are the permittivities of regions I and O relative to that of free space, i.e., $\varepsilon_{r_{I,O}} = \varepsilon_{I,O}/\varepsilon_0$. Equations (5.51) and (5.52) are cast into boundary integral equations by applying Green's second identity [58, 7]:

$$u_I^{tot}(\rho) = \int_C \left(G_I(\rho,\rho') v_I^{tot}(\rho') - u_I^{tot}(\rho') \frac{\partial G_I^{tot}(\rho,\rho')}{\partial \hat{n}_I} \right) dl', \; \rho \in I, \qquad (5.53)$$

$$u_O^{tot}(\rho) = u_O^{inc}(\rho) + \left(\int_C + \int_{C^\infty} \right) \left(G_O(\rho,\rho') v_O^{tot}(\rho') - u_O^{tot}(\rho') \frac{\partial G_O^{tot}(\rho,\rho')}{\partial \hat{n}_O} dl' \right),$$
$$\rho \in O,$$

where $u^{tot}(\rho)$ and

$$v_{I,O}^{tot}(\rho') = \frac{\partial u_{I,O}^{tot}(\rho')}{\partial \hat{n}_{I,O}}$$

represent the total field and its normal derivative and

$$u_O^{inc}(\rho) = \int_{C^\infty} f(\rho') G_O(\rho, \rho') dl'. \tag{5.54}$$

The vector from the origin to the observation point is $\rho(x,y)$, and the vector from the origin to the source point is $\rho'(x',y')$. Equation (5.53), which represents two equations and four unknowns, $u_{I,O}^{sc}(\rho')$ and $v_{I,O}^{sc}(\rho')$, is indeterminate. This system is solved by applying the following boundary conditions on C:

$$u_I^{tot}(\rho') = u_O^{tot}(\rho') \equiv u^{tot}(\rho'), \ \rho' \in C, \tag{5.55}$$

$$\frac{1}{p_I}\frac{\partial u_I^{tot}(\rho')}{\partial \hat{n}_I} = -\frac{1}{p_O}\frac{\partial u_O^{tot}(\rho')}{\partial \hat{n}_O} \equiv v^{tot}(\rho'), \tag{5.56}$$

where $p_{I,O} = 1$ for TE polarization and $p_{I,O} = \varepsilon_{I,O}$ for TM polarization. Equation (5.55) represents the continuity of the tangential field component across the contour, C, and (5.56) represents the normal derivative boundary condition, which is related to the boundary condition on the tangential component of the magnetic field. A minus sign is introduced in (5.56) because $\hat{n} = \hat{n}_I = -\hat{n}_O$. The contours C and C^∞ represent the contours of the DOE and a fictitious contour at infinity; see Figure 5.6. The two-dimensional free space Green's functions are

$$G_I(\rho, \rho') = \frac{1}{4j}H_o^{(2)}(\beta_I |\rho - \rho'|) = \frac{1}{4j}H_o^{(2)}\left(\beta_I \sqrt{(x-x')^2 + (y-y')^2}\right), \tag{5.57}$$

$$G_O(\rho, \rho') = \frac{1}{4j}H_o^{(2)}(\beta_O |\rho - \rho'|) = \frac{1}{4j}H_o^{(2)}\left(\beta_O \sqrt{(x-x')^2 + (y-y')^2}\right),$$

where $H_o^{(2)}(\beta|\rho-\rho'|)$ is the zeroth-order Hankel function of the second kind.

To avoid numerical errors that result when the frequency of the illumination wavelength is a resonant frequency of the structure, the boundary integral equations are better formulated in terms of the scattered fields,

$$u^{tot}(\rho) = u^{inc}(\rho) + u^{sc}(\rho). \tag{5.58}$$

Such errors are known to occur when the boundary integral equations are applied to closed boundaries [18]. By substituting (5.58) into (5.53) and incorporating the Sommerfeld radiation condition,

$$\int_{C^\infty} \left(p_O G_O(\rho, \rho') v^{sc}(\rho') - u^{sc}(\rho') \frac{\partial G_O^{tot}(\rho, \rho')}{\partial \hat{n}}\right) dl' = 0, \ \rho \in O, \tag{5.59}$$

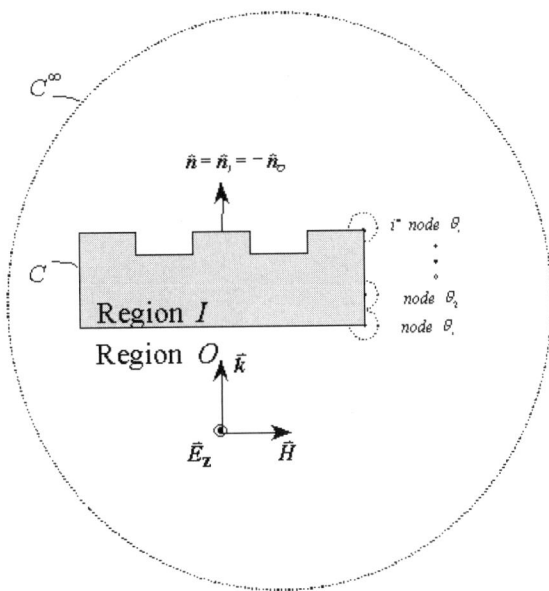

Figure 5.6: Geometry of a diffractive structure for the boundary integral equations.

and the identity,

$$\int_C \left(p_O\, G_O(\rho,\rho')\, v^{inc}(\rho') - u^{inc}(\rho') \frac{\partial G_O^{tot}(\rho,\rho')}{\partial \hat{n}} \right) dl' = 0, \; \rho \in O, \qquad (5.60)$$

we have

$$0 = u^{sc}(\rho) + \int_C \left(u^{sc}(\rho') \frac{\partial G_I^{tot}(\rho,\rho')}{\partial \hat{n}} - p_I\, G_I(\rho,\rho')\, v^{sc}(\rho') \right) dl' \qquad (5.61)$$

$$+ u^{inc}(\rho) + \int_C \left(u^{inc}(\rho') \frac{\partial G_I^{tot}(\rho,\rho')}{\partial \hat{n}} - p_I\, G_I(\rho,\rho')\, v^{inc}(\rho') \right) dl', \; \rho \in I,$$

$$0 = u^{sc}(\rho) + \int_C \left(p_O\, G_O(\rho,\rho')\, v^{sc}(\rho') - u^{sc}(\rho') \frac{\partial G_O^{tot}(\rho,\rho')}{\partial \hat{n}} \right) dl', \; \rho \in O.$$

Equation (5.60) is best understood by considering that an unperturbed incident field cannot produce a scattered field in region O.

Equation (5.61) is solved by confining the observation vector to the boundary of the DOE, i.e., $\rho=\rho_s$. However, care must be exercised when evaluating (5.61) because of singularities that exist when $\rho_s=\rho'$. These singularities are evaluated by integrating around the singularity in the limit as ρ' approaches ρ_s; see Figure 5.7.

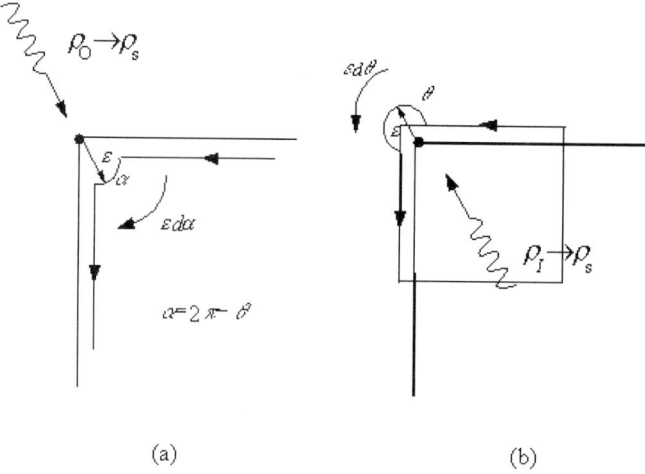

Figure 5.7: Geometry for (a) exterior and (b) interior limiting contours for the boundary integral equations.

Using the small argument approximation for the Hankel functions [1], the boundary integral equation for the outside region, O, is reduced to

$$0 = u^{sc}(\rho_s) - \int_{C-\epsilon} \!\!\!- \left(u^{sc}(\rho') \frac{\partial G_O^{tot}(\rho, \rho')}{\partial \hat{n}} - p_O\, G_O(\rho_s, \rho')\, v^{sc}(\rho') \right) dl'$$

$$- \lim_{\epsilon \to 0} \left\{ \frac{1}{4j} \int_{-\alpha/2}^{\alpha/2} \left[-u^{sc}(\rho') \left(\frac{\beta_O \epsilon}{2} + \frac{2j}{\pi \beta_O \epsilon} \right) \right. \right. \tag{5.62}$$

$$\left. \left. + p_O v^{sc}(\rho') \left(1 - \frac{2j\beta_O}{\pi} \ln\left[\left(\frac{\beta_O \epsilon}{2} \right) + \gamma \right] \right) \right] \epsilon\, d\alpha \right\},$$

where $\gamma = 0.5772$ is Euler's number and $\fint$ is Cauchy's principal value of integration. As shown in Figure 5.7(a), $u^{sc}(\rho_s)$ is the scattered field and α is the interior angle subtended by the boundary at the observation point ρ_s, and

$$\epsilon = \lim_{\rho' \to \rho_s} |\rho_s - \rho'|. \tag{5.63}$$

The singularity for the interior region I is evaluated in a similar fashion except that the exterior angle of integration θ, as shown in Figure 5.7(b), represents the complement to the angle α used in the interior integral equation. Thus, the relationship $\alpha = 2\pi - \theta$ is used. The resulting boundary integral equations in terms of θ are

Chapter 5. Electromagnetic Models for Finite Aperiodic DOEs   167

$$0 = u^{sc}(\rho_s)\left(1 - \frac{\theta}{2\pi}\right) + \oint_C \left(u^{sc}(\rho')\frac{\partial G_I^{tot}(\rho_s, \rho')}{\partial \hat{n}} - p_I\, G_I(\rho_s, \rho')\, v^{sc}(\rho')\right) dl' \quad (5.64)$$

$$+\, u^{inc}(\rho_s)\left(1 - \frac{\theta}{2\pi}\right) + \oint_C \left(u^{inc}(\rho')\frac{\partial G_I^{tot}(\rho_s, \rho')}{\partial \hat{n}} - p_I\, G_I(\rho_s, \rho')v^{inc}(\rho')\right) dl',$$

$$0 = u^{sc}(\rho_s)\left(\frac{\theta}{2\pi}\right) + \oint_C \left(p_O\, G_O(\rho_s, \rho')\, v^{sc}(\rho') - u^{sc}(\rho')\frac{\partial G_O^{tot}(\rho_s, \rho')}{\partial \hat{n}}\right) dl'.$$

Once determined, the scattered electric field $u^{sc}(\rho')$ and its normal derivative $\partial u^{sc}(\rho')/\partial \hat{n}$ are used to calculate the scattered fields anywhere in space by

$$u^{sc}(\rho) = \int_C \left[u^{sc}(\rho')\frac{\partial G_O(\rho, \rho')}{\partial \hat{n}} - p_O G_O(\rho, \rho')\, v^{sc}(\rho')\right] dl', \quad (5.65)$$

where $\rho \in O$.

In the next section, (5.64) is solved using a numerical technique called the BEM.

5.4.2 BEM for Dielectric DOEs

The BEM presented here is completely general and allows for the analysis of finite and aperiodic structures. In the application of BEM to dielectrics, the coupled integral equations in (5.64) are solved by expanding the unknown surface distributions, the scattered electric field, $u^{sc}(\rho')$, and its normal derivative, $\partial u^{sc}(\rho')/\partial \hat{n} = v^{sc}(\rho')$, in terms of a set of *shape interpolation* basis functions:

$$u^{sc}(\rho'(\xi)) = \sum_{n=1}^{N} \hat{u}_n^{sc}(\xi) = \sum_{n=1}^{N} u_n^{sc}\phi_1(\xi) + u_{n+1}^{sc}\phi_2(\xi), \quad (5.66)$$

$$v^{sc}(\rho'(\xi)) = \sum_{n=1}^{N} \hat{v}_n^{sc}(\xi) = \sum_{n=1}^{N} v_n^{sc}\phi_1(\xi) + v_{n+1}^{sc}\phi_2(\xi).$$

When $n = N$, node $N+1$ is replaced with 1. The interpolation functions $\phi_1(\xi)$ and $\phi_2(\xi)$ are defined over a single elemental segment. In this analysis, linear interpolation functions are used:

$$\phi_1(\xi) = (1-\xi)/2, \quad (5.67)$$
$$\phi_2(\xi) = (1+\xi)/2,$$

where $\xi = [-1, 1]$. However, higher-order interpolation functions can also be used [61]. In this notation the contour of the DOE is represented by a local coordinate transformation, $\rho'(\xi) = \rho'(\hat{x}_n(\xi), \hat{y}_n(\xi))$, where

$$\hat{x}_n(\xi) = x_n\phi_1(\xi) + x_{n+1}\phi_2(\xi), \quad (5.68)$$
$$\hat{y}_n(\xi) = y_n\phi_1(\xi) + y_{n+1}\phi_2(\xi).$$

The nodal values (x_n, y_n) and (x_{n+1}, y_{n+1}) represent the sample points on the boundary and $\hat{x}_n(\xi)$ and $\hat{y}_n(\xi)$ are interpolated coordinate values between nodes.

In this formulation the nodal values must be numbered in a sequential counter-clockwise fashion.

Substitution of (5.66) into (5.64) yields a system of two equations in $2N$ unknowns. To generate $2N$ equations in $2N$ unknowns, an inner product between both sides of (5.64) with a set of N weighting functions is performed. In this analysis, the weighting functions are Dirac-delta functions that sample the boundary of the DOE:

$$\omega_m = \delta\left(\rho'(\xi) - \rho'_m\right) = \begin{cases} 1, & \rho'(\xi) = \rho'_m, \\ 0, & \text{elsewhere}, \end{cases} \tag{5.69}$$

where $m = 1, 2, \ldots, N$ and ρ'_m is a set of N position vectors along the contour C of the DOE. This approach is referred to as point matching.

Substituting (5.66) and (5.67) into (5.64) and performing an inner product between ω_m yields a set of linear algebraic equations:

$$\begin{bmatrix} ZI_{n,m} & -p_I YI_{n,m} \\ ZO_{n,m} & p_O YO_{n,m} \end{bmatrix} \cdot \begin{bmatrix} u_m^{sc} \\ v_m^{sc} \end{bmatrix} = \begin{bmatrix} -ZI_{n,m} & p_I YI_{n,m} \\ 0 & 0 \end{bmatrix} \cdot \begin{bmatrix} u_m^{inc} \\ v_m^{inc} \end{bmatrix}, \tag{5.70}$$

where

$$ZI_{n,m} = \left(1 - \frac{\theta_n}{2\pi}\right) \cdot \delta_{nm} + \int_{-1}^{1} \left[\frac{\Delta l_n}{2} \phi_1(\xi) \frac{\partial G_I^{tot}\left(\rho'(\hat{x}_n, \hat{y}_n), \rho'_m\right)}{\partial \hat{n}_I} \right. \tag{5.71}$$

$$\left. + \frac{\Delta l_{n-1}}{2} \phi_2(\xi) \frac{\partial G_I^{tot}\left(\rho'(\hat{x}_{n-1}, \hat{y}_{n-1}), \rho'_m\right)}{\partial \hat{n}_I} \right] d\xi,$$

$$YI_{n,m} = \int_{-1}^{1} \left[\frac{\Delta l_n}{2} \phi_1(\xi) G_I\left(\rho'(\hat{x}_n, \hat{y}_n), \rho'_m\right) \right.$$

$$\left. + \frac{\Delta l_{n-1}}{2} \phi_2(\xi) G_I\left(\rho'(\hat{x}_{n-1}, \hat{y}_{n-1}), \rho'_m\right) \right] d\xi,$$

$$ZO_{n,m} = \left(\frac{\theta_n}{2\pi}\right) \cdot \delta_{nm} - \int_{-1}^{1} \left[\frac{\Delta l_n}{2} \phi_1(\xi) \frac{\partial G_O^{tot}\left(\rho'(\hat{x}_n, \hat{y}_n), \rho'_m\right)}{\partial \hat{n}_O} \right.$$

$$\left. + \frac{\Delta l_{n-1}}{2} \phi_2(\xi) \frac{\partial G_O^{tot}\left(\rho'(\hat{x}_{n-1}, \hat{y}_{n-1}), \rho'_m\right)}{\partial \hat{n}_O} \right] d\xi,$$

$$YO_{n,m} = \int_{-1}^{1} \left[\frac{\Delta l_n}{2} \phi_1(\xi) G_O\left(\rho'(\hat{x}_n, \hat{y}_n), \rho'_m\right) \right.$$

$$\left. + \frac{\Delta l_{n-1}}{2} \phi_2(\xi) G_O\left(\rho'(\hat{x}_{n-1}, \hat{y}_{n-1}), \rho'_m\right) \right] d\xi.$$

In addition, $d\xi = 2dl'/\Delta l_n$, $u_m^{inc} = u^{inc}(x_m, y_m)$, $v_m^{inc} = \partial u_m^{inc}(\rho')/\partial \hat{n}$, and Δl_n represents the length of the corresponding line segment for element n. The integral equations in (5.71) specify a unique relationship between the electric field and its normal derivative at each node on the boundary to each and every other node on the boundary; i.e., they impose the boundary value problem.

Once the solution to (5.70) is determined, (5.65) is used to calculate the scattered field values at any observation point. To obtain the total diffracted field, the incident field is added to the scattered field in the observation plane. In our analysis, the incident wave was propagated to the observation plane using Rayleigh–Sommerfeld diffraction [12]. In the next section we extend this formulation of the BEM to include perfectly conducting DOEs.

5.4.3 Boundary Integral Formulation for Perfect Conductors

The analysis presented above is applicable to both dielectrics and imperfect conductors, in which case the permittivity of the interior region ε_I is complex, $\varepsilon_I = \varepsilon_{rI}\varepsilon_o + j\sigma_I/\omega$, where σ_I denotes electrical conductivity. For the analysis of a TE incident field on a perfect conductor, (5.53) reduces to a single integral equation in the exterior region,

$$\mathbf{E}^{sc}(\rho_s) = -\int_C v^{tot}(\rho') G_O(\rho_s, \rho') \, dl'. \tag{5.72}$$

Equation (5.72) can be cast into the more familiar electric field integral equation for perfect conductors,

$$\mathbf{E}^{inc}(\rho_s) = \frac{\omega\mu}{4} \int_C \mathbf{J}(\rho') H_o^2\left(\beta_O \sqrt{[x_s - x']^2 + [y_s - y']^2}\right) dl', \ \rho_s \in C, \tag{5.73}$$

where the relationships $\mathbf{E}^{inc}(\rho_s) = -\mathbf{E}^{sc}(\rho_s)$ and $\mathbf{J}(\rho') = (1/j\omega\mu)[\partial\mathbf{E}^{tot}(\rho')/\partial\hat{n}]$ have been used. $\mathbf{J}(\rho')$ denotes the surface current density and is related to the normal derivative of the electric field. Equation (5.73) is solved in the next section using the BEM.

5.4.4 BEM for Perfectly Conducting DOEs

To determine the current distribution on the contour of the DOE, the unknown surface current distribution is expanded in terms of interpolation functions,

$$\hat{J}_n(\xi) = \sum_{n=1}^{N} J_n \phi_1(\xi) + J_{n+1}\phi_2(\xi), \tag{5.74}$$

where, again, $\phi_1(\xi)$ and $\phi_2(\xi)$ are continuous functions of $\xi = [-1, 1]$ defined over a single line segment as in (5.67).

The substitution of (5.68) and (5.74) into (5.73) yields

$$u^{inc}(x_s, y_s) = \frac{\omega\mu}{8} \sum_{n=1}^{N} \Delta l_n \int_{-1}^{1} \hat{J}_n H_o^{(2)}\left(\beta\sqrt{[x_s - \hat{x}_n(\xi)]^2 + [y_s - \hat{y}_n(\xi)]^2}\right) d\xi, \tag{5.75}$$

where again $d\xi = 2dl'/\Delta l_n$ and Δl_n represents the length of the corresponding line segment for element n.

To yield a discrete system of equations, we again use point matching:

$$[E_m] = [L_{m,n}] \cdot [J_n], \tag{5.76}$$

where

$$E_m = u^{inc}(x_m, y_m), \tag{5.77}$$

$$L_{m,n} = \frac{\omega\mu}{8} \int_{-1}^{1} \left\{ \Delta l_n \phi_1(\xi) H_o^{(2)}\left(\beta\sqrt{[x_m - \hat{x}_n(\xi)]^2 + [y_m - \hat{y}_n(\xi)]^2}\right) \right. \tag{5.78}$$
$$\left. + \Delta l_{n-1} \phi_2(\xi) H_o^{(2)}\left(\beta\sqrt{[x_m - \hat{x}_{n-1}(\xi)]^2 + [y_m - \hat{y}_{n-1}(\xi)]^2}\right) \right\} d\xi.$$

The values J_n are determined through matrix inversion of (5.76), and the diffracted field values anywhere in space are determined by

$$u^{tot}(x,y) = u^{inc}(x,y) - \frac{\omega\mu}{8} \sum_{n=1}^{N} \Delta l_n \tag{5.79}$$
$$\times \int_{-1}^{1} [J_n \phi_1(\xi) + J_{n+1}\phi_2(\xi)] H_o^{(2)}\left(\beta\sqrt{[x - \hat{x}_n(\xi)]^2 + [y - \hat{y}_n(\xi)]^2}\right) d\xi.$$

5.4.5 Semi-infinite and Symmetric BEM Formulation

The traditional BEMs presented in sections 5.4.2 and 5.4.4 represent accurate techniques for the analysis of dielectric and conducting DOEs. However, their application to the analysis of DOEs that are realistic in size requires significant computational effort and excessive memory requirements.

In this section we present a more efficient implementation of the BEM, referred to as the semi-infinite symmetric BEM (SSBEM), that uses specific DOE properties to reduce the computational and memory requirements needed in the analysis of DOEs. For example, most DOEs are fabricated on the front side of a substrate whose front and back sides are planar. As a result, the contribution of the back side of the DOE to its diffractive behavior is insignificant as compared to the contributions of the front side. In addition, some DOEs have twofold symmetry about the axis of propagation. As presented below, these two properties can be used to formulate an efficient BEM model of diffraction.

Semi-infinite BEM

In the traditional formulation of the BEM each point on the boundary of the DOE is included in the formulation. However, if the back side of the DOE is planar, its diffractive effects can be accounted for by scaling the amplitude of the incident field

Chapter 5. Electromagnetic Models for Finite Aperiodic DOEs

by the Fresnel reflection coefficient of the planar dielectric interface [39]. Thus, one can formulate a semi-infinite BEM by accounting for the back side of the DOE with an amplitude scale.

A semi-infinite formulation of the BEM considers only the side of the DOE boundary that contains the diffractive structure, in which case the other side is not included. As a result, the computations on the front side of the DOE must be truncated at some point beyond the region that contains the diffractive profile. The regions beyond the diffractive profile are also neglected because within these regions the incident field is zero. Consequently, in these regions no surface distributions are induced and reradiation does not occur.

However, in this formulation the incident field must go to zero before the truncation point, otherwise a discontinuity will exist and produce errors in the analysis. Unfortunately, a truncated incident field no longer satisfies the homogeneous wave equation (which is implicitly assumed in the formulation); however, errors that arise due to this truncation can be minimized by decreasing the amplitude of the incident field gradually before the truncation point [16]. This is achieved by multiplying the incident field by a window function,

$$w(x) = \begin{cases} 1, & 0 \leq |x| < D/2, \\ \cos^2\left(\frac{|x| - D/2}{2d}\pi\right), & D/2 \leq |x| \leq d, \\ 0, & d < |x| < \infty, \end{cases} \qquad (5.80)$$

where D is the diameter of the DOE and d is the transition region from the incident field to zero. Thus, an incident plane wave is approximated as a windowed incident field. In alternate incident fields, such as a Gaussian beam, the window function may be inherently accounted for.

This formulation reduces the number of nodes in the traditional formulation, N, by approximately a factor of 2 (depending on the topology of the DOE profile), and, because the size of the system of equations is proportional to N^2, represents a nearly fourfold reduction in computational effort and memory requirements. In the next section planar symmetry is considered to reduce the computational effort further.

Planar Symmetric BEM

If a DOE has planar symmetry about the axis of propagation, the induced fields on the DOE are symmetric. Thus, one can formulate a BEM that incorporates this property to further reduce the computational effort and memory requirements needed in analysis. Note, that in order to take advantage of these reductions the incident field must also have planar symmetry.

In the BEM formulation, the induced field at each point on the boundary is coupled to every other point on the boundary through a Green's function and its normal derivative, as represented in (5.71). To apply a symmetric boundary condition to the BEM one must still account for the contribution of each symmetric node.

Table 5.1: Table of properties for the finite-difference, finite-element, boundary element, and hybrid FEBE methods.

Numerical technique	FDTD	BEM	FEM	Hybrid FEBEM
Material properties	Inhomogen. anisotropic lossy	Homogeneous isotropic lossy	Inhomogen. anisotropic lossy	Inhomogen. anisotropic lossy
Arbitrary contours	Poor	Good	Good	Good
Memory requirements	Good	Poor	Good	Moderate
Sampling rate for $\leq 5\%$ error	$\lambda/20$	$\lambda/10$	$\lambda/20$	$\lambda/20$
Type of system of equations	None	Full	Sparse	Partial
Require absorbing boundaries	Yes	No	Yes	No

This is accomplished by renumbering the nodes from $-N/2$ to $N/2$, calculating the terms in (5.71), applying symmetric boundary conditions,

$$\begin{aligned} u_m^{sc} &= u_{-m}^{sc}, \\ u_m^{inc} &= u_{-m}^{inc}, \\ v_m^{sc} &= v_{-m}^{sc}, \\ v_m^{inc} &= v_{-m}^{inc} \end{aligned} \quad (5.81)$$

to (5.70), and combining like terms. The result is an additional factor of 2 reduction in the size of the system of equations and thereby a net factor of 16 in memory requirements. Thus, a semi-infinite and symmetric formulation to the BEM reduces the order of the system of equations by a factor of 4 and the memory requirements by a factor of 16, as compared to the traditional formulation.

The BEM as presented in sections 5.4.2, 5.4.4, and 5.4.5 represent accurate and efficient models of diffraction provided the DOE is homogeneous. For inhomogeneous DOEs either the FDTD method, as discussed in section 5.3, or a variational approach based on the FEM can be used, as discussed elsewhere in this book.

5.5 Summary of Numerical Methods

For the analysis of a specific DOE profile one has to choose the numerical method that best suits the problem at hand, based on the attributes of each method and their suitability to the problem. To this end we discussed two numerical electromagnetic techniques, as they apply to the analysis of finite aperiodic DOEs. To aid in the decision process we conclude this section with a summary of these techniques as well as others, based on generally accepted appraisals of them, in Table 5.1. Current work in this field is focused on streamlining computational costs in order to improve algorithm efficiency. To this end variations of the above methods that incorporate the unique properties of DOEs, such as semi-infinite extent and field stitching, have been presented in the literature [44].

References

[1] M. Abramowitz and I. A. Stegun, *Handbook of Mathematical Functions*, Dover, New York, 1970, p. 360.

[2] C. A. Balanis, *Advanced Engineering Electromagnetics*, Wiley, New York, 1989.

[3] J. P. Berenger, *A perfectly matched layer for the absorption of electromagnetic waves*, J. Comp. Phys., 114 (1994), pp. 185–200.

[4] O. Bryngdahl and F. Wyrowski, *Progress in Optics*, Vol. XXVIII, North Holland, New York, 1990.

[5] W. C. Chew and W. H. Weedon, *A 3d perfectly matched medium from modified Maxwell's equations with stretched coordinates*, Microwave Opt. Tech. Lett., 7 (1994), pp. 599–604.

[6] D. Davidson and R. Ziolkowski, *Body-of-revolution finite-difference time-domain modeling of space-time focusing by a three-dimensional lens*, J. Opt. Soc. Amer. A, 11 (1994), pp. 1471–1490.

[7] W. El-Mikati and J. B. Davies, *Improved boundary element techniques for two-dimensional scattering problems with circular boundaries*, IEEE Trans. Antennas and Propagation, AP-35 (1987), pp. 539–544.

[8] B. Engquist and A. Majda, *Absorbing boundary conditions for the numerical simulation of waves*, Math. Comp., 31 (1977), pp. 629–651.

[9] N. C. Gallagher and B. Liu, *Method for computing kinoforms that reduces image reconstruction error*, Appl. Opt., 12 (1973), pp. 2328–2335.

[10] T. K. Gaylord and M. G. Moharam, *Planar dielectric grating diffraction theories*, Appl. Phys. B, 28 (1982), pp. 1–14.

[11] T. K. Gaylord and M. G. Moharam, *Analysis and applications of optical diffraction by gratings*, Proc. IEEE, 73 (1985), pp. 894–937.

[12] J. W. Goodman, *Introduction to Fourier Optics*, McGraw-Hill, San Francisco, 1968.

[13] D. A. Gremaux and N. C. Gallagher, *Limits of scalar diffraction theory for conducting gratings*, Appl. Opt., 32 (1993), pp. 1948–1953.

[14] G. R. Hadley, *Numerical simulation of reflecting structures by solution of the two-dimensional Helmholtz equation*, Opt. Lett., 19 (1994), pp. 84–86.

[15] R. F. Harrington, *Field Computation by Moment Methods*, Krieger, Malabar, FL, 1968.

[16] K. Hirayama, E. N. Glytsis, T. K. Gaylord, and D. W. Wilson, *Rigorous electromagnetic analysis of diffractive cylindrical lenses*, J. Opt. Soc. Amer. A, 13 (1996), pp. 2219–2231.

[17] A. Ishimaru, *Electromagnetic Wave Propagation, Radiation, and Scattering*, Prentice-Hall, Englewood Cliffs, NJ, 1991.

[18] J. Jin, *The Finite Element Method in Electromagnetics*, Wiley-Interscience, New York, 1993.

[19] K. Knop, *Rigorous diffraction theory for transmission phase gratings with deep rectangular grooves*, J. Opt. Soc. Amer. A, 68 (1978), pp. 1206–1210.

[20] G. Koppelmann and M. Totzeck, *Diffraction near fields of small phase objects: comparison of 3-cm wave measurements with moment-method calculations*, J. Opt. Soc. Amer. A, 8 (1991), pp. 554–558.

[21] K. S. Kunz and R. J. Luebbers, *The Finite Difference Time Domain Method for Electromagnetics*, CRC Press, Boca Raton, FL, 1993.

[22] P. Lalanne, *Improved formulation of the coupled-wave method for two-dimensional gratings*, J. Opt. Soc. Amer. A, 14 (1997), pp. 1592–1598.

[23] P. Lalanne and G. Morris, *Highly improved convergence of the coupled-wave method for TM polarization*, J. Opt. Soc. Amer. A, 13 (1996), pp. 779–784.

[24] S. H. Lee, ed., *Selected Papers on Computer-Generated Holograms and Diffractive Optics*, SPIE Milestone Series MS 33, SPIE Press, Bellingham, WA, 1992.

[25] S. H. Lee, *Diffractive and miniaturized optics*, in Critical Reviews of Optical Science and Technology, CR 49, SPIE Press, Bellingham, WA, 1993.

[26] W. H. Lee, *Sampled Fourier transform hologram generated by computer*, Appl. Opt., 9 (1970), pp. 639–643.

[27] L. B. Lesem, P. M. Hirsch, and J. A. Jordan, *The kinoform: a new wavefront reconstruction device*, IBM J. Res. Develop., 13 (1969), pp. 150–155.

[28] L. Li, *Multilayer modal method for diffraction gratings of arbitrary profile, depth, and permittivity*, J. Opt. Soc. Amer. A, 10 (1993), pp. 2581–2591.

[29] B. Lichtenberg and N. C. Gallagher, *Numerical modeling of diffractive devices using the finite element method*, Opt. Eng., 33 (1994), pp. 3518–3526.

[30] A. W. Lohmann and D. P. Paris, *Binary Fraunhofer holograms, generated by computer*, Appl. Opt., 6 (1967), pp. 1739–1748.

[31] J. N. Mait, *Understanding diffractive optic design in the scalar domain*, J. Opt. Soc. Amer. A, 12 (1995), pp. 2145–2158.

[32] P. D. Maker, D. W. Wilson, and R. E. Muller, *Fabrication and performance of optical interconnect analog phase holograms made by electron beam lithography*, in Optoelectronic Interconnects and Packaging, R. T. Chen and P. S. Guilfoyle, eds., Proc. SPIE CR 62, Bellingham, WA, 1996, pp. 415–430.

[33] D. Maystre, *Progress in Optics*, Vol. XXI, North Holland, New York, 1984.

[34] M. S. Mirotznik, D. W. Prather, and J. N. Mait, *A hybrid finite-boundary element method for the analysis of diffractive elements*, J. Mod. Opt., 43 (1996), pp. 1309–1322.

[35] G. Mur, *Absorbing boundary conditions for the finite-difference approximation of the time-domain electromagnetic-field equations*, IEEE Trans Electromagn. Compat., EMC-23 (1981), pp. 377–382.

[36] E. Noponen, J. Turunen, and A. Vasara, *Electromagnetic theory and design of diffractive-lens arrays*, J. Opt. Soc. Amer. A, 10 (1993), pp. 434–443.

[37] R. Petit, ed., *Electromagnetic Theory of Gratings*, Springer-Verlag, Berlin, 1980.

[38] D. A. Pommet, M. G. Moharam, and E. B. Grann, *Limits of scalar diffraction theory for diffractive phase elements*, J. Opt. Soc. Amer. A, 11 (1994), pp. 1827–1834.

[39] D. W. Prather, *Analysis and Synthesis of Finite Aperiodic Diffractive Optical Elements Using Rigorous Electromagnetic Models*, Ph.D. thesis, University of Maryland, College Park, 1997.

[40] D. Prather and S. Shi, *Electromagnetic analysis of axially-symmetric DOEs using the FDTD method*, in Diffractive Optics and Micro-Optics, OSA Technical Digest, 1998.

[41] D. W. Prather, M. S. Mirotznik, and J. N. Mait, *Boundary element method for vector modeling diffractive optical elements*, in Diffractive and Holographic Optics Technology II, I. Cindrich and S. H. Lee, eds., Proc. SPIE 2404, Bellingham, WA, 1995, pp. 28–39.

[42] D. W. Prather, M. S. Mirotznik, and J. N. Mait, *Boundary integral methods applied to the analysis of diffractive optical elements*, J. Opt. Soc. Amer. A, 14 (1997), pp. 34–43.

[43] D. W. Prather and S. Shi, *Formulation and application of the finite-difference time-domain method for the analysis of axially-symmetric DOEs*, J. Opt. Soc. Amer. A, 16 (1999), pp. 1131–1142.

[44] D. W. Prather, S. Shi, and J. S. Bergey, *A field stitching algorithm for the analysis of electrically large finite aperiodic diffractive optical elements*, Opt. Lett., 24 (1999), pp. 273–275.

[45] D. W. Prather, S. Shi, M. S. Mirotznik, and J. N. Mait, *Vector-based analysis of axially-symmetric DOEs using the method of moments*, 1998 OSA Topical Meeting on Diffractive Optics, Kona, HI.

[46] C. M. Rappaport, *Interpreting and improving the PML absorbing boundary condition using anisotropic lossy mapping of space*, IEEE Trans. Magnetics, 32 (1996), pp. 968–974.

[47] S. Shi and D. Prather, *Vector-based plane-wave spectrum method for the propagation of cylindrical electromagnetic fields*, Opt. Lett., 24 (1999), pp. 1445–1447.

[48] J. A. Stratton, *Electromagnetic Theory*, McGraw-Hill, New York, 1941.

[49] A. Taflove, *Application of the finite-difference time-domain method to sinusoidal steady-state electromagnetic-penetration problems*, IEEE Trans. Electromagn. Compat., EMC-22 (1980), pp. 191–202.

[50] A. Taflove, *Computational Electromagnetics: The Finite-Difference Time Domain Method*, Artech House, Boston, MA, 1995.

[51] A. Taflove and M. E. Brodwin, *Numerical solution of steady-state electromagnetic scattering problems using the time-dependent Maxwell's equations*, IEEE Trans. Microwave Theory Tech., MTT-23 (1975), pp. 623–630.

[52] A. Taflove and K. R. Umashankar, *The finite-difference time-domain (FD-TD) method for numerical modeling of electromagnetic scattering*, IEEE Trans. Magnetics, 25 (1989), pp. 3086–3091.

[53] F. L. Teixeira and W. C. Chew, *Systematic derivation of anisotropic PML absorbing media in cylindrical and spherical coordinates*, IEEE Microwave Guided Wave Lett., 7 (1997), pp. 371–373.

[54] F. L. Teixeira and W. C. Chew, *PML-FDTD in cylindrical and spherical grids*, IEEE Microwave Guided Wave Lett., 7 (1997), pp. 285–287.

[55] K. Umashankar and A. Taflove, *A novel method to analyze electromagnetic scattering of complex objects*, IEEE Trans. Electromagn. Compat., EMC-24 (1982), pp. 397–405.

[56] H. C. van de Hulst, *Light Scattering by Small Particles*, Dover, New York, 1981.

[57] A. Wang and A. Prata, *Lenslet analysis by rigorous vector diffraction theory*, J. Opt. Soc. Amer. A, 12 (1995), pp. 1161–1169.

[58] J. J. Wang, *Generalized Moment Methods in Electromagnetics*, Wiley, New York, 1991.

[59] F. Wyrowski and O. Bryngdahl, *Iterative Fourier-transform algorithm applied to computer holography*, J. Opt. Soc. Amer. A, 5 (1988), pp. 1058–1065.

[60] F. Wyrowski and O. Bryngdahl, *Digital holography as part of diffractive optics*, Rep. Prog. Phys., 54 (1991), pp. 1481–1571.

[61] K. Yashiro and S. Ohkawa, *Boundary element method for electromagnetic scattering from cylinders*, IEEE Trans. Antennas and Propagation, AP-33 (1985), pp. 383–389.

[62] K. S. Yee, *Numerical solution of initial boundary value problems involving Maxwell's equations in isotropic media*, IEEE Trans. Antennas and Propagation, AP-14 (1966), pp. 302–307.

[63] D. Zaleta, W. Daschner, M. Larsson, B. C. Kress, J. Fan, K. S. Urquhart, and S. H. Lee, *Diffractive Optics Fabricated by Electron-Beam Direct Write Methods*, CR 49, SPIE Press, Bellingham, WA, 1993, pp. 117–137.

Chapter 6

Analysis of the Diffraction from Chiral Gratings

Habib Ammari and Jean-Claude Nédélec

6.1 Introduction

The optical activity appearing in some biological or crystalline substances was discovered at the beginning of the 19th century, and by the middle of the century, this phenomenon was explained by postulating that it was due to the chirality of their molecules. A three-dimensional chiral object is one that cannot be brought into congruence with its mirror image by rotation or translation, so that a collection of chiral objects will form a material that is characterized by right-handedness or right-handedness. Such materials have been manufactured in several laboratories [58] by placing either left-handed or right-handed metal chiral objects (helices) randomly in an isotropic host material. Helices are perhaps the most common chiral objects found in nature. It appears that the helix as well as the spiral, its two-dimensional counterpart, may be considered as the canonical chiral structures. All other manifestations of geometric chirality stem from the helix.

Chiral media are isotropic, reciprocal, and, more important, circularly birefringent. The optical activity in a chiral medium has the property of rotating the plane of polarization (containing the electric field vector and the direction of propagation) of an electromagnetic wave. In general, electromagnetic wave propagation in a chiral medium is governed by Maxwell's equations and a set of constitutive equations known as *the Drude–Born–Fedorov constitutive equations*, in which the electric and magnetic fields are coupled. The coupling is responsible for the chirality of the medium. It is measured by the magnitude of the chirality admittance β, which, along with the dielectric coefficient ε and the magnetic permeability constant μ, characterizes completely the electromagnetic properties of the medium. The parameter β measures the degree of handedness of the material, and for media of both handedness in equal number this parameter vanishes. A change in the sign of β means taking the mirror image of the medium. The case $\beta = 0$ corresponds to the propagation of electromagnetic waves in a standard achiral medium.

In most of the papers dealing with chiral materials, the frequency dependence of the chirality parameter due to the nature of the chiral material is not taken into account and neither is the medium permittivity or the permeability dependence on the chirality parameter itself. In general, for the sake of simplicity, the parameters β, ε, and μ are taken to be constant over the frequency range of interest. It should be noted that different expressions exist in the literature for the constitutive relations in chiral media. But they are in fact equivalent to the Drude–Born–Fedorov equations after appropriate redefinitions of the field vectors and the material parameters.

The Drude–Born–Fedorov equations have a long history. In 1890, Drude claimed that chirality can be predicted by the Maxwell postulates, provided the polarization has an additional term proportional to the curl of the electric field. Born used similar equations as the constitutive equations in chiral media. Drude's proposal for isotropic chirality was modified by Fedorov in 1959 to the Drude–Born–Fedorov equations which are symmetric under time-reversality and duality transformations. In 1920, Lidman was the first to synthesize artificial isotropic chiral composite materials by immersing small conducting helices with random orientation in a cardboard box [58]. Recently, there has been considerable interest in the study of scattering and diffraction by chiral media. Antennas coated with chiral materials have significant radiation characteristics due to the extra degree of freedom offered by the presence of the chirality parameter [39], [74]. Chirality must also be recognized as an important aspect of the future of waveguides [59], [76], [71], microwave devices [27], [68], band-gap structures, and liquid crystals. The liquid crystals community is fascinated with the possibility of creating polar fluids [38]. Numerous reviews on chiral media have appeared in the engineering literature in the last few years [58], [48], [47], [50], [31], [27], and a comprehensive list of references on chiral materials has been issued electronically [73].

To the best of our knowledge, few mathematical contributions have been published in this relatively young applied mathematics field. Recently, we have obtained some mathematical results on the scattering of electromagnetic waves in chiral media, e.g., mathematical justification of the Drude–Born–Fedorov model [14], representation theorems [5], the radiation condition for the Drude–Born–Fedorov equations [5], existence and uniqueness of a solution [5], convergence to Maxwell's equations [7], equivalent boundary conditions for thin chiral coatings [6], low-frequency asymptotics [12], and chiral gratings [2]–[4]. In a joint work with Hamdache [14], we have provided the mathematical foundation of the Drude–Born–Fedorov equations and connected the permittivity, the permeability, and the chirality admittance of a chiral sample to the dimensions of the chiral inclusions (the metallic helices). More precisely, we have derived the Drude–Born–Fedorov constitutive relations from the standard constitutive relations for a homogeneous, isotropic medium by embedding spaced, randomly oriented helical conductors, each modeled as a dipole, and estimate the chirality admittance, which is a function of the length, the radius, and the width of the single helix. The first step of the method is in the spirit of the point interaction method [36]. It consists in replacing, by using the Born approximation, the metallic helices along with the perfect conductor boundary condition on them by a collection of equivalent electric and magnetic dipoles in the constitutive relations which are then together with the Maxwell equations hold everywhere. The second step consists in analyzing, by the integral equations method, these new constitutive relations with the Maxwell equations as the number of metallic inclusions becomes large. The Drude–Born–Fedorov model appears then as a continuum limit of the Maxwell equations in the presence of a collection of electric dipoles. The chirality parameter in this model vanishes if the wave length differs greatly from the dimension of the helix [14]. If the wavelength in the medium is comparable to the dimension of the helix, the Born approximation does not hold. However, in general, the width of the single helix is very small compared to its length and its radius, and an asymptotic analysis may be carried out. To the best of our knowledge, there is no mathematical description of the interaction of a collection of helices with electromagnetic waves at this band of high frequencies. The problem seems to be very difficult, but the challenge is rewarding.

In [5], a combined Hilbert space and integral equations method is employed to prove the existence and uniqueness of a solution to the scattering problem by an inhomogeneous object in a chiral medium. It is shown that the classical Silver–Müller radiation condition remains valid in chiral media.

It is known that for solutions of the Maxwell equations, the Silver–Müller radiation condition is equivalent to the Sommerfeld radiation condition for the Cartesian components [26]. We have proved that this classical result is not valid in chiral media [5].

The diffraction problem of the plane electromagnetic field by a chiral curved thin layer covering a perfectly conducting object is considered in [6]. Approximative impedance conditions that can be used, along with the radiation condition, to compute the scattered field without requiring detailed modeling of the field quantities inside the chiral coating are derived, and optimal error estimates are obtained. It should be noted that these approximate boundary conditions, unlike those for achiral coatings [6], may generate nonphysical surface waves.

In [7], the boundary integral method together with a saddle point formulation is used to prove the convergence of the Drude–Born–Fedorov equations with variable coefficients, possibly nonsmooth, to Maxwell's equations as the chirality admittance tends to zero.

The low-frequency behavior of solutions to electromagnetic scattering problems in chiral media is carried out in a joint work with Laouadi [12]. The full asymptotic expansions of the electric and magnetic fields with respect to the frequency are given and the asymptotic expansion's dependence on the chirality parameter is shown.

For further significant mathematical results on the electromagnetic scattering in chiral media, we also refer to Kong [46], Lakhtakia, Varadan, and Varadan [49], Ola [64], Rojas [69], [70], Karlsson and Kristensson [44], Lindell and Shivola [56], [57], Lindell and Olyslager [54], Lindell and Weiglhofer [55], Athanasiadis and Stratis [15], Athanasiadis, Martin, and Stratis [16], and McDowall [60].

In this chapter, we present some recent mathematical aspects of the diffraction problem in periodic chiral structures. Much of our present discussion on the solution of periodic scattering problems in chiral media is centered around the recent work by Ammari and Bao [2]–[4]. We have attempted, however, to make this chapter self-contained and, in this spirit, our discussion contains a number of remarks on the Drude–Born–Fedorov equations and provides complete proofs of the main results.

Periodic structures (gratings) have received increasing attention through the years because of important applications in integrated optics, optical lenses, antireflective structures, holography, lasers, communication, and computing. Significant mathematical results on periodic achiral structures may be found in Bao and Dobson [18], Bao, Dobson, and Cox [19], Chen and Friedman [24], Nédélec and Starling [63], Dobson and Friedman [29], Dobson [28], Abboud [1], Bruno and Reitich [23], and Bao [17].

Chiral gratings provide an exciting combination of the medium and structure. The combination gives rise to new features and applications. For instance, chiral gratings are capable of converting a linearly polarized incident field into two nearly circularly polarized diffracted modes in different directions.

For various physical and computational aspects of the electromagnetic wave propagation inside periodic chiral media, we refer to Jaggar et al. [41], Lakhtakia, Varadan, and Varadan [51], and Yueh and Kong [77]. Jaggar et al. [41] have investigated the electromagnetic properties of a structure with sinusoidally periodic permittivity, permeability, and chirality admittance by using coupled-mode equations linking forward and backward propagating waves of opposite circular polarizations. Lakhtakia, Varadan, and Varadan [51] have investigated a similar problem with a different approach. They have obtained coupled first-order differential equations for reduced fields and analyzed a piecewise constant case and a constant impedance case. Lakhtakia, Varadan, and Varadan [51] have solved wave scattering at an interface with a singly periodic geometry separating a chiral medium from an achiral one by a fully vectorial treatment and have studied the reflection and transmission characteristics of these gratings. Yueh and Kong [77] have analyzed the diffraction of waves by chiral gratings placed over a dielectric substrate, for arbitrary angles of incidence and polarizations, by a generalization of the coupled wave theory. Numerical examples have been given to illustrate the effects of chirality on the polarization states of waves diffracted by gratings with rectangular grooves. It has been found that the chiral grating is able to more evenly distribute the power between Floquet modes and also to make modes nearly circularly.

In this chapter, we consider a time-harmonic electromagnetic plane wave incident on a very general biperiodic structure in $\mathbf{R}^3$. By biperiodic structure or doubly periodic structure, we mean that the structure is periodic in two orthogonal directions. The periodic structure separates two *chiral homogeneous* regions. The medium inside the structure is *chiral* and *nonhomogeneous*. The study of the propagation of the reflected and transmitted waves away from the structure is the diffraction problem.

The purpose of this chapter is to introduce a variational formulation of the diffraction problem by chiral gratings. The main result is concerned with the well-posedness of the model problem. It is shown that for all but possibly a discrete set of frequencies, there is a unique quasi-periodic weak solution to the diffraction problem. The approach is based on a *Hodge decomposition* and a *compact imbedding result*.

An important step in studying diffraction from periodic structures is to reduce the diffraction problem into a bounded domain. This can be done by introducing

Chapter 6. Analysis of the Diffraction from Chiral Gratings

a pair of transparent boundary conditions. For the diffraction problem by a chiral grating, the derivation of these boundary conditions uses the *Bohren decomposition* of fields inside homogeneous chiral media.

Another approach for reducing the diffraction problem into a bounded domain consists in deriving exact radiation conditions on the boundary of the heterogeneous chiral medium. This method couples a finite-element method (FEM) in the nonhomogeneous chiral media with a method of integral equations or boundary element method (BEM) on the periodic interfaces. The fact that these boundary conditions are formulated on the surface of the periodic structure implies that no mesh of the surrounding medium would be needed. Coupling FEM/BEM variational formulations for the diffraction problem by periodic structures were derived by Ammari and Bao in [4].

We emphasize that the variational approach discussed in this chapter is very general; in particular, the material parameters ε, μ, and β are assumed only to be bounded functions. The incident angles and grating shapes may be arbitrary. The geometry can be extremely general as well. Existence results for the diffraction from periodic chiral structures with nonsmooth interfaces may also be established. Moreover, the present formulation can be used successfully to study certain inverse problems and to develop efficient and reliable numerical methods for solving direct and optimal design problems. A class of FEMs can be formulated based on the variational approach and on the family of Nédélec's finite elements (tetrahedra) in $H(\text{curl})$. We also emphasize that the variational formulations coupling FEMs in the heteregeneous media with a method of integral equations on the periodic interfaces [4] may also be applicable to the present diffraction problem. However, the derivation and the analysis of coupling variational formulations are much more complicated than the approach adapted in this chapter for solving the diffraction problem.

The chapter is outlined as follows. In section 6.2, the Maxwell equations and the constitutive Drude–Born–Fedorov equations are presented. Section 6.3 is devoted to a variational formulation of the model problem. Section 6.4 is concerned with some auxiliary results. A version of the Hodge decomposition is studied. We also prove a compact imbedding result. The well-posedness of the model is investigated in section 6.5. Results on existence and uniqueness of the weak quasi-periodic solutions are proved. In section 6.6, uniform convergence for the discrete variational approximation is obtained. Some interesting future directions of the diffraction problem from chiral gratings are discussed in section 6.7.

6.2 The Drude–Born–Fedorov Equations

6.2.1 Physical Background

The electromagnetic fields are governed by the time-harmonic Maxwell equations (time dependence $e^{-i\omega t}$):

$$\nabla \times E - i\omega B = 0 , \quad (6.1)$$

$$\nabla \times H + i\omega D = 0 , \quad (6.2)$$

where E, H, D, and B denote the electric field, the magnetic field, and the electric and magnetic displacement vectors in $\mathbf{R}^3$, respectively.

In addition, the following Drude–Born–Fedorov constitutive equations hold:

$$D = \varepsilon(x)\Big(E + \beta(x)\nabla \times E\Big), \tag{6.3}$$

$$B = \mu(x)\Big(H + \beta(x)\nabla \times H\Big), \tag{6.4}$$

where ε is the electric permittivity, μ is the magnetic permeability, and β is the chirality admittance.

It is easily seen that the following equations are equivalent to the constitutive equations (6.3)–(6.4):

$$\Big(1 - (k(x)\beta(x))^2\Big) D = \varepsilon(x)E + \frac{i\beta(x)}{\omega}(k(x))^2 H, \tag{6.5}$$

$$\Big(1 - (k(x)\beta(x))^2\Big) B = \mu(x)H - \frac{i\beta(x)}{\omega}(k(x))^2 E, \tag{6.6}$$

where

$$k(x) = \omega\sqrt{\varepsilon(x)\mu(x)}.$$

Similarly, the Maxwell equations may be rewritten as

$$\nabla \times E = (\gamma(x))^2 \beta(x) E + i\omega\mu(x)\left(\frac{\gamma(x)}{k(x)}\right)^2 H, \tag{6.7}$$

$$\nabla \times H = (\gamma(x))^2 \beta(x) H - i\omega\varepsilon(x)\left(\frac{\gamma(x)}{k(x)}\right)^2 E. \tag{6.8}$$

In these equations, the parameter $\gamma(x)$ is defined as

$$(\gamma(x))^2 = \frac{(k(x))^2}{1 - (k(x)\beta(x))^2}.$$

Throughout, we always assume that

$$(k(x)\beta(x))^2 \neq 1, \quad x \in \mathbf{R}^3.$$

Moreover, the above system may be shown to be equivalent in a weak sense to

$$\nabla \times \left(\frac{1 - \omega^2\beta^2\varepsilon\mu}{\mu}\right)\nabla \times E - \omega^2\nabla \times (\varepsilon\beta E) - \omega^2\varepsilon\beta\nabla \times E - \omega^2\varepsilon E = 0, \tag{6.9}$$

$$\nabla \times E = \gamma^2 \beta E + i\omega\mu\left(\frac{\gamma}{k}\right)^2 H. \tag{6.10}$$

Hence, in order to solve the Maxwell equations, it suffices to solve (6.9) for E. Once the electric field E is determined, the magnetic field H follows from (6.10).

Chapter 6. Analysis of the Diffraction from Chiral Gratings

Standard jump conditions may be deduced from the above system. In fact, the tangential parts of the electric and magnetic fields are continuous across an interface. Let ν denote the unit normal to the interface. We then have

$$[\nu \times E] = 0,$$
$$\left[\nu \times \frac{1 - \beta^2 k^2}{i\omega\mu} \nabla \times E - \frac{\gamma^2 \beta(1 - k^2\beta^2)}{i\omega\mu} \nu \times E\right] = 0.$$

Note that when the structure is only assumed to be periodic in one dimension, the diffraction problem cannot be reduced to two-dimensional transverse electric (TE) and transverse magnetic (TM) diffraction problems; this is also the case for Maxwell's equations. This is due to the structure of the Drude–Born–Fedorov equations.

6.2.2 Plane Waves

We next consider a free space example of plane waves propagating in a chiral medium, which is assumed to be homogeneous, say $\varepsilon = \varepsilon_j$, $\mu = \mu_j$, and $\beta = \beta_j$. Here ε_j and μ_j are positive constants. Let

$$\begin{cases} \gamma_j^\pm = \dfrac{\omega\sqrt{\varepsilon_j\mu_j}}{1 \pm \omega\sqrt{\varepsilon_j\mu_j}\beta_j}, \\ \omega_j^\pm = \dfrac{\omega}{1 \pm \omega\sqrt{\varepsilon_j\mu_j}\beta_j}. \end{cases} \qquad (6.11)$$

It is easy to verify that the following time-harmonic electromagnetic plane wave satisfies the equations

$$E_I(x) = q^- e^{i\gamma_j^- p^- \cdot x} + q^+ e^{i\gamma_j^+ p^+ \cdot x},$$
$$H_I(x) = -i\sqrt{\frac{\varepsilon_j}{\mu_j}} \left(q^- e^{i\gamma_j^- p^- \cdot x} + q^+ e^{i\gamma_j^+ p^+ \cdot x} \right),$$

where the complex vectors $p^\pm$ and $q^\pm$ satisfy

$$p^\pm \cdot q^\pm = 0, \quad p^\pm \times q^\pm = \pm i q^\pm. \qquad (6.12)$$

From (6.12), we have

$$\nabla \times (q^\pm e^{i\gamma_j^\pm p^\pm \cdot x}) = \mp \gamma_j^\pm q^\pm e^{i\gamma_j^\pm p^\pm \cdot x}.$$

Thus the field E_I is a combination of left-circularly polarized plane wave $q^- e^{i\gamma_j^- p^- \cdot x}$ with a positive helicity γ_j^-, and right-circularly polarized one $q^+ e^{i\gamma_j^+ p^+ \cdot x}$ with a negative helicity $-\gamma_j^+$, and it solves the Drude–Born–Fedorov equations in $\mathbf{R}^3$.

Finally, note that in an unbounded chiral media, the left-circularly and right-circularly polarized plane waves (the Beltrami fields) can propagate without interfering with each other. However, when a wave of either polarization encounters a boundary or interface, a mode conversion takes place. In general, the scattered and transmitted fields then consist of waves of both circular polarization states.

6.3 Diffraction Problem

6.3.1 Geometry of the Problem

We first specify the geometry of the problem. Let (x_1, x_2, x_3) be the Cartesian coordinate system equipped with an orthonormal basis (ν_1, ν_2, ν_3). Let Λ_1 and Λ_2 be two positive constants such that the material functions ε, μ, and β satisfy, for any $n_1, n_2 \in Z = \{0, \pm 1, \pm 2, \ldots\}$,

$$\varepsilon(x_1 + n_1\Lambda_1, x_2 + n_2\Lambda_2, x_3) = \varepsilon(x_1, x_2, x_3),$$
$$\mu(x_1 + n_1\Lambda_1, x_2 + n_2\Lambda_2, x_3) = \mu(x_1, x_2, x_3),$$
$$\beta(x_1 + n_1\Lambda_1, x_2 + n_2\Lambda_2, x_3) = \beta(x_1, x_2, x_3).$$

In addition, it is assumed that, for some fixed positive constant b and sufficiently small $\delta > 0$,

$$\varepsilon(x) = \varepsilon_1, \quad \mu(x) = \mu_1, \quad \beta(x) = \beta_1 \text{ for } x_3 > b - \delta, \quad (6.13)$$
$$\varepsilon(x) = \varepsilon_2, \quad \mu(x) = \mu_2, \quad \beta(x) = \beta_2 \text{ for } x_3 < -b + \delta, \quad (6.14)$$

where ε_1, ε_2, μ_1, and μ_2 are positive constants and β_1 and β_2 are real constants.

We make the following general assumptions:

- $\varepsilon(x)$, $\mu(x)$, and $\beta(x)$ are all real-valued L^∞ functions, $\varepsilon(x) \geq \varepsilon_0$, $\mu(x) \geq \mu_0$, and $\beta \geq 0$, where ε_0 and μ_0 are positive constants;

- $d(x) = \dfrac{1 - \omega^2 \beta^2 \varepsilon \mu}{\mu} \geq d_0 > 0$ for some positive constant d_0.

The first assumption is a technical one. Analogous results hold for materials that absorb energy, i.e., if $\text{Im}(\varepsilon) > 0$ or $\text{Im}(\mu) > 0$. The second assumption appears to be common in the literature and justifiable since the chirality parameter β is generally small.

Let

$$\Omega = \{x \in \mathbf{R}^3 : -b < x_3 < b\}, \qquad \Omega_1 = \{x \in \mathbf{R}^3 : x_3 > b\},$$

$$\Omega_2 = \{x \in \mathbf{R}^3 : x_3 < -b\},$$

and

$$d_j = \frac{1 - \omega^2 \beta_j^2 \varepsilon_j \mu_j}{\mu_j}.$$

Consider a plane wave in Ω_1:

$$E_I(x) = q^- e^{i\gamma_1^- p^- \cdot x}, \qquad H_I(x) = -i\sqrt{\frac{\varepsilon_j}{\mu_j}} q^- e^{i\gamma_1^- p^- \cdot x}, \quad (6.15)$$

incident on Ω. The vectors q^- and p^- satisfy (6.12).

Chapter 6. Analysis of the Diffraction from Chiral Gratings

Let
$$\alpha = (\alpha_1, \alpha_2, 0) = \gamma_1^-(p_1^-, p_2^-, 0).$$

We are interested in quasi-periodic solutions, i.e., solutions E and H such that the fields E_α, H_α defined by

$$E_\alpha = e^{-i\alpha \cdot x} E(x_1, x_2, x_3), \qquad (6.16)$$
$$H_\alpha = e^{-i\alpha \cdot x} H(x_1, x_2, x_3) \qquad (6.17)$$

are periodic in x_1 with period Λ_1 and in x_2 with period Λ_2.

Denote
$$\nabla_\alpha = \nabla + i\alpha = \nabla + i(\alpha_1, \alpha_2, 0).$$

It is easy to see from (6.9) and (6.10) that E_α and H_α satisfy

$$\nabla_\alpha \times \left(d \nabla_\alpha \times E_\alpha \right) - \omega^2 \nabla_\alpha \times (\varepsilon \beta E_\alpha) - \omega^2 \varepsilon \beta \nabla_\alpha \times E_\alpha - \omega^2 \varepsilon E_\alpha = 0, \qquad (6.18)$$

$$\nabla_\alpha \times E_\alpha = \gamma^2 \beta E_\alpha + i\omega\mu \left(\frac{\gamma}{k}\right)^2 H_\alpha. \qquad (6.19)$$

6.3.2 The Bohren Transform and the Periodic Radiation Condition

In order to solve the system of differential equations, we need boundary conditions in the x_3-direction. These conditions may be derived by the Bohren transform and the periodic radiation condition for Maxwell's equations. In the homogeneous regions Ω_1 and Ω_2, the electric field E can be expanded as follows:

$$E|_{\Omega_1}(x) = E_{I,\alpha}(x) + \sum_{n \in Z^2} E_{1,n}(x_3) e^{i\alpha_n \cdot x}, \qquad (6.20)$$

$$E|_{\Omega_2}(x) = \sum_{n \in Z^2} E_{2,n}(x_3) e^{i\alpha_n \cdot x}, \qquad (6.21)$$

where
$$E_{I,\alpha}(x) = e^{-i\alpha \cdot x} E_I(x).$$

Now, if we define the fields $(\mathcal{E}_j^+, \mathcal{E}_j^-)$ in the homogeneous domain Ω_j by

$$\begin{pmatrix} \mathcal{E}_j^+ \\ \mathcal{E}_j^- \end{pmatrix} = \begin{pmatrix} 1 & i\sqrt{\frac{\mu_j}{\varepsilon_j}} \\ 1 & -i\sqrt{\frac{\mu_j}{\varepsilon_j}} \end{pmatrix} \begin{pmatrix} E_\alpha \\ H_\alpha \end{pmatrix}, \qquad (6.22)$$

and the fields $(\mathcal{H}_j^+, \mathcal{H}_j^-)$ by

$$\mathcal{H}_j^\pm = \frac{1}{i\omega_j^\pm \mu_j} \nabla \times \mathcal{E}_j^\pm,$$

we can easily verify that
$$\nabla \times \mathcal{H}_j^\pm = -i\omega_j^\pm \varepsilon_j \mathcal{E}_j^\pm.$$

The fields $(\mathcal{E}_j^\pm, \mathcal{H}_j^\pm)$ are then solutions of the Maxwell equations
$$\nabla \times \mathcal{E}_j^\pm = i\omega_j^\pm \mu_j \mathcal{H}_j^\pm \quad \text{in } \Omega_j,$$
$$\nabla \times \mathcal{H}_j^\pm = -i\omega_j^\pm \varepsilon_j \mathcal{E}_j^\pm \quad \text{in } \Omega_j.$$

Furthermore, we have
$$\nabla \times \mathcal{E}_j^\pm = \mp \gamma_j^\pm \mathcal{E}_j^\pm,$$
and then
$$\mathcal{H}_j^\pm = \pm i \sqrt{\frac{\varepsilon_j}{\mu_j}} \mathcal{E}_j^\pm \quad \text{in } \Omega_j.$$

Observe that inside Ω_j $(j = 1, 2)$
$$\nabla_\alpha \cdot \mathcal{E}_j^\pm = 0;$$
the following Maxwell equations then imply
$$(\Delta_\alpha + (\omega_j^\pm)^2 \varepsilon_j \mu_j)\mathcal{E}_j^\pm = 0 \quad \text{in } \Omega_j, \tag{6.23}$$
where
$$\Delta_\alpha = \Delta + 2i\alpha \cdot \nabla - |\alpha|^2.$$

The decomposition (6.22) is known in the literature as the Bohren transform. Since $\mathcal{E}_j^\pm$ is $\Lambda = (\Lambda_1, \Lambda_2)$ periodic, we can expand $\mathcal{E}_j^\pm$ in a Fourier series:
$$\mathcal{E}_j^\pm(x) = \mathcal{E}_{I,j}(x) + \sum_{n \in Z^2} e_{j,n}^\pm(x_3) e^{i\alpha_n \cdot x}, \tag{6.24}$$
where
$$\mathcal{E}_{I,1}(x) = \left(E_I(x) + i\sqrt{\frac{\mu_j}{\varepsilon_j}} H_I(x) \right) e^{-i\alpha \cdot x},$$

$$\mathcal{E}_{I,2}(x) = 0,$$

$$e_{j,n}^\pm(x_3) = \frac{1}{\Lambda_1 \Lambda_2} \int_0^{\Lambda_1} \int_0^{\Lambda_2} (\mathcal{E}_j^\pm(x) - \mathcal{E}_{I,j}(x)) e^{-i\alpha_n \cdot x} dx_1 dx_2,$$
and
$$\alpha_n = (2\pi n_1/\Lambda_1, 2\pi n_2/\Lambda_2, 0).$$

Denote
$$\Gamma_1 = \{x \in \mathbf{R}^3 : x_3 = b\} \text{ and } \Gamma_2 = \{x_3 = -b\}.$$

Define for $j = 1, 2$ the coefficients

$$\beta_{j,n}^{\pm}(\alpha) = \begin{cases} \sqrt{(\omega_j^{\pm})^2 \varepsilon_j \mu_j - |\alpha_n + \alpha|^2}, & (\omega_j^{\pm})^2 \varepsilon_j \mu_j > |\alpha_n + \alpha|^2, \\ i\sqrt{|\alpha_n + \alpha|^2 - (\omega_j^{\pm})^2 \varepsilon_j \mu_j}, & (\omega_j^{\pm})^2 \varepsilon_j \mu_j < |\alpha_n + \alpha|^2. \end{cases} \quad (6.25)$$

We make the following general assumptions:
- $(\omega_j^{\pm})^2 \varepsilon_j \mu_j \neq |\alpha_n + \alpha|^2$ for all $n \in Z^2$, $j = 1, 2$;
- $\beta_{j,n}^{+}(\alpha) \cdot \beta_{j,n}^{-}(\alpha) \in \mathbf{R}$ for all $n \in Z^2$, $j = 1, 2$.

The first assumption excludes "resonances." The second assumption appears to be justifiable since the parameters β_j are generally small.

For convenience, we also introduce the following notation:
$$\Lambda_j^{+} = \{n \in Z^2 : \operatorname{Im}(\beta_{j,n}^{\mp}) = 0\},$$
$$\Lambda_j^{-} = \{n \in Z^2 : \operatorname{Im}(\beta_{j,n}^{\mp}) \neq 0\}.$$

Since the medium in Ω_j ($j = 1, 2$) is homogeneous, the method of separation of variables implies that $\mathcal{E}_j^{\pm}$ can be expressed as a sum of plane waves:

$$\mathcal{E}_j^{\pm}|_{\Omega_j} = \mathcal{E}_{I,j}(x) + \sum_{n \in Z^2} A_{j,n}^{\pm} e^{\pm i \beta_j^{(n)} x_3 + i \alpha_n \cdot x}, \quad j = 1, 2, \quad (6.26)$$

where the $A_{j,n}^{\pm}$ are constant (complex) vectors.

We next impose a radiation condition on $\mathcal{E}_j^{\pm}$. Due to the (infinite) periodic structure, the usual Sommerfeld or Silver–Müller radiation condition is no longer valid. Instead, the following radiation condition based on the diffraction theory is employed: Since $\beta_{j,n}^{\pm}$ is real for at most finitely many n, there are only a finite number of propagating plane waves in the sum (6.26); the remaining waves are exponentially decaying (or unbounded) as $|x_3| \to \infty$. We will insist that $\mathcal{E}_j^{\pm}$ is composed of bounded outgoing plane waves in Ω_1 and Ω_2, plus the incident (incoming) wave $\mathcal{E}_{I,1}$ in Ω_1.

From (6.24) and (6.25) we deduce

$$e_{j,n}^{\pm}(x_3) = \begin{cases} e_{j,n}^{\pm}(b) e^{i \beta_{1,n}^{\pm}(x_3 - b)} & \text{in } \Omega_1, \\ e_{j,n}^{\pm}(-b) e^{-i \beta_{2,n}^{\pm}(x_3 + b)} & \text{in } \Omega_2. \end{cases} \quad (6.27)$$

By matching the two expansions (6.24) and (6.26), we get

$$A_{1,n}^{\pm} = e_{1,n}^{\pm}(b) e^{-i \beta_{1,n}^{\pm} b} \quad \text{on } \Gamma_1, \quad (6.28)$$
$$A_{2,n}^{\pm} = e_{2,n}^{\pm}(-b) e^{-i \beta_{2,n}^{\pm} b} \quad \text{on } \Gamma_2. \quad (6.29)$$

Further, since in the regions $\{x : x_3 > b - \delta\} \cup \{x : x_3 < -b + \delta\}$,

$$\nabla_\alpha \cdot \mathcal{E}_j^{\pm} = 0, \quad \nabla_\alpha \cdot \mathcal{E}_{I,j} = 0,$$

we have from (6.26) that

$$(\alpha_n + \alpha) \cdot e^{\pm}_{1,n}(b) + \beta^{\pm}_{1,n} e^{\pm}_{1,n}(b) \cdot \nu_3 = 0 \quad \text{on } \Gamma_1, \tag{6.30}$$

$$(\alpha_n + \alpha) \cdot e^{\pm}_{2,n}(-b) - \beta^{\pm}_{2,n} e^{\pm}_{1,n}(-b) \cdot \nu_3 = 0 \quad \text{on } \Gamma_2, \tag{6.31}$$

where ν_3 is the outward normal to Γ_1.

LEMMA 6.1. *There exist boundary pseudodifferential operators $\mathcal{B}^{\pm}_j$ ($j = 1, 2$) of order 1, such that*

$$\nu_3 \times (\nabla_\alpha \times (\mathcal{E}^{\pm}_1 - \mathcal{E}_{I,1})) = \mathcal{B}^{\pm}_1(P(\mathcal{E}^{\pm}_1 - \mathcal{E}_{I,1})) \quad \text{on } \Gamma_1, \tag{6.32}$$

$$\nu_3 \times (\nabla_\alpha \times \mathcal{E}^{\pm}_2) = \mathcal{B}^{\pm}_2(P(\mathcal{E}^{\pm}_2)) \quad \text{on } \Gamma_2, \tag{6.33}$$

where the operator $\mathcal{B}^{\pm}_j$ is defined by

$$\mathcal{B}^{\pm}_j f = -i \sum_{n \in Z^2} \frac{1}{\beta^{\pm}_{j,n}} \left\{ (\beta^{\pm}_{j,n})^2 (f^{(n)}_1, f^{(n)}_2, 0) + ((\alpha + \alpha_n) \cdot f^{(n)})(\alpha + \alpha_n) \right\} e^{i\alpha_n \cdot x}, \tag{6.34}$$

where P is the projection onto the plane orthogonal to ν_3, i.e.,

$$Pf = -\nu_3 \times (\nu_3 \times f)$$

and

$$f^{(n)} = \Lambda_1^{-1} \Lambda_2^{-1} \int_0^{\Lambda_1} \int_0^{\Lambda_2} f(x) e^{-i\alpha_n \cdot x} dx_1 dx_2 \;.$$

Proof. The proof may be given by using the expansion (6.26) together with (6.28)–(6.31), and some simple calculation. □

Remark 6.1. In order to define the Dirichlet to Neumann operator $\mathcal{B}^{\pm}_j$, it is crucial to assume that $\beta^{\pm}_{j,n}$ is nonzero.

From (6.22) and (6.26) we deduce that expansion (6.20)–(6.21) can be rewritten as a sum of plane waves in the homogeneous regions Ω_1 and Ω_2:

$$E|_{\Omega_1}(x) = E_{I,\alpha}(x) \tag{6.35}$$
$$+ \frac{1}{2} \sum_{n \in Z^2} \left(e^+_{1,n}(b) e^{i\beta^+_{1,n}(x_3-b)} + e^-_{1,n}(b) e^{i\beta^-_{1,n}(x_3-b)} \right) e^{i\alpha_n \cdot x},$$

$$E|_{\Omega_2}(x) = \frac{1}{2} \sum_{n \in Z^2} \left(e^+_{2,n}(b) e^{i\beta^+_{2,n}(x_3-b)} + e^-_{2,n}(b) e^{i\beta^-_{2,n}(x_3-b)} \right) e^{i\alpha_n \cdot x}, \tag{6.36}$$

where $e^{\pm}_{j,n}$ are unknown coefficients. Introduce for $j = 1, 2$ the coefficients

$$\gamma_{j,n} = \frac{i}{\omega \sqrt{\varepsilon_j \mu_j}} \left(\frac{1}{\beta^+_{j,n}} - \frac{1}{\beta^-_{j,n}} \right) (1 - \omega^2 \varepsilon_j \mu_j \beta_j^2)$$

Chapter 6. Analysis of the Diffraction from Chiral Gratings

and

$$\lambda_{j,n} = \frac{i}{\omega\sqrt{\varepsilon_j\mu_j}}(\beta_{j,n}^+ - \beta_{j,n}^-)(1 - \omega^2\varepsilon_j\mu_j\beta_j^2),$$

and the matrix $M_{j,n}$ given by

$$\begin{pmatrix} \dfrac{1 + \lambda_{j,n}\gamma_{j,n}|\alpha_n+\alpha|^2 + \lambda_{j,n}^2\gamma_{j,n}^2(\alpha_n+\alpha)_2^2(\alpha_n+\alpha)_1^2}{1+\lambda_{j,n}\gamma_{j,n}(\alpha_n+\alpha)_2^2(1+\lambda_{j,n}\gamma_{j,n}|\alpha_n+\alpha|^2+\lambda_{j,n}^2)} & \dfrac{\lambda_{j,n}+\lambda_{j,n}\gamma_{j,n}(\alpha_n+\alpha)_2(\alpha_n+\alpha)_1}{1+\lambda_{j,n}\gamma_{j,n}|\alpha_n+\alpha|^2+\lambda_{j,n}^2} \\[2mm] \dfrac{-\lambda_{j,n}+\lambda_{j,n}\gamma_{j,n}(\alpha_n+\alpha)_2(\alpha_n+\alpha)_1}{1+\lambda_{j,n}\gamma_{j,n}|\alpha_n+\alpha|^2+\lambda_{j,n}^2} & \dfrac{1+\lambda_{j,n}\gamma_{j,n}(\alpha_n+\alpha)_2^2}{1+\lambda_{j,n}\gamma_{j,n}|\alpha_n+\alpha|^2+\lambda_{j,n}^2} \end{pmatrix},$$

where

$$(\alpha_n + \alpha)_j = \alpha_j + \frac{2\pi n_j}{\Lambda_j}.$$

The following holds.

LEMMA 6.2. *There exist boundary pseudodifferential operators B_j ($j = 1, 2$) of order 1, such that*

$$\nu_3 \times (\nabla_\alpha \times (E_\alpha - E_{I,\alpha})) = B_1(P(E_\alpha - E_{I,\alpha})) \quad on \ \Gamma_1, \tag{6.37}$$
$$\nu_3 \times (\nabla_\alpha \times E_\alpha) = B_2(P(E_\alpha)) \quad on \ \Gamma_2, \tag{6.38}$$

where the operator B_j is defined by

$$B_j f = -\frac{i}{2}\sum_{n\in Z^2}\left\{\left(\frac{1}{\beta_{j,n}^+}(1-\omega\sqrt{\varepsilon_j\mu_j}) + \frac{1}{\beta_{j,n}^-}(1+\omega\sqrt{\varepsilon_j\mu_j})\right)((\alpha_n+\alpha)\cdot f^{(n)})(\alpha_n+\alpha)\right.$$
$$+(\beta_{j,n}^+(1-\omega\sqrt{\varepsilon_j\mu_j}) + \beta_{j,n}^-(1+\omega\sqrt{\varepsilon_j\mu_j}))(g_{j,1}^{(n)}, g_{j,2}^{(n)}, 0)$$
$$\left.+\gamma_{j,n}(\beta_{j,n}^+(1-\omega\sqrt{\varepsilon_j\mu_j}) + \beta_{j,n}^-(1+\omega\sqrt{\varepsilon_j\mu_j}))(-g_{j,2}^{(n)}, g_{j,1}^{(n)}, 0)\right\}e^{i\alpha_n\cdot x},$$

where

$$f^{(n)} = \Lambda_1^{-1}\Lambda_2^{-1}\int_0^{\Lambda_1}\int_0^{\Lambda_2} f(x)e^{-i\alpha_n\cdot x}dx_1 dx_2$$

and

$$g_j^{(n)} = M_{j,n}(f^{(n)}).$$

Remark 6.2. In the case $\beta_j = 0$, we have $M_{j,n} = I$ for all $n \in Z^2$ and then $B_j = \mathcal{B}_j^+ = \mathcal{B}_j^-$.

We introduce the L^2 scalar product

$$(f, g) = \int_D f\bar{g},$$

where D is the domain.

Denote by B_j^* the adjoint of B_j, that is,

$$(B_j f, g) = (f, B_j^* g).$$

It is easily seen that the adjoint operator of B_j in the above lemma is given by

$$\begin{aligned} B_j^* f = \frac{i}{2} \sum_{n \in Z^2} &\left\{ \left(\frac{1}{\overline{\beta}_{j,n}^+}(1 - \omega\sqrt{\varepsilon_j \mu_j}) + \frac{1}{\overline{\beta}_{j,n}^-}(1 + \omega\sqrt{\varepsilon_j \mu_j}) \right) (\alpha_n + \alpha) \cdot f^{(n)}(\alpha_n + \alpha) \right. \\ &+ (\overline{\beta}_{j,n}^+(1 - \omega\sqrt{\varepsilon_j \mu_j}) + \overline{\beta}_{j,n}^-(1 + \omega\sqrt{\varepsilon_j \mu_j}))(\overline{g}_{j,1}^{(n)}, \overline{g}_{j,2}^{(n)}, 0) \quad (6.39) \\ &\left. + \overline{\gamma}_{j,n}(\overline{\beta}_{j,n}^+(1 - \omega\sqrt{\varepsilon_j \mu_j}) + \overline{\beta}_{j,n}^-(1 + \omega\sqrt{\varepsilon_j \mu_j}))(-\overline{g}_{j,2}^{(n)}, \overline{g}_{j,1}^{(n)}, 0) \right\} e^{i\alpha_n \cdot x}, \end{aligned}$$

where

$$\overline{g}_j^{(n)} = \overline{M}_{j,n}(f^{(n)}).$$

Remark 6.3. If we assume that $\beta_{j,n}^+ \cdot \beta_{j,n}^- \in \mathbf{R}$ for all $n \in Z^2, j = 1, 2$, then the matrix $M_{j,n}$ is real. Otherwise, it is real for n large enough. A similar result holds for the coefficients $\gamma_{j,n}$.

Define

$$\Lambda = \Lambda_1 Z \times \Lambda_2 Z \times \{0\} \subset \mathbf{R}^3.$$

Since the fields E_α are Λ-periodic, we can move the problem from $\mathbf{R}^3$ to the quotient space $\mathbf{R}^3/\Lambda$. For the remainder of the paper, we shall identify Ω with the cylinder Ω/Λ, and similarly for the boundaries $\Gamma_j \equiv \Gamma_j/\Lambda$. Thus from now on,

all functions defined on Ω and Γ_j are implicitly Λ-periodic.

Define div_α by

$$\text{div}_\alpha u = \nabla_\alpha \cdot u = (\partial_{x_1} + i\alpha_1)u_1 + (\partial_{x_2} + i\alpha_2)u_2.$$

Let H^m be the mth-order L^2-based Sobolev spaces of complex-valued functions. We denote by $H_p^m(\Omega)$ the subset of all functions in $H^m(\Omega)$ which are the restrictions to Ω of the functions in $H_{loc}^m(\mathbf{R}^2 \times (-b, b))$ that are Λ-periodic. Similarly we define $H_p^m(\Omega_j)$ and $H_p^m(\Gamma_j)$. In the future, for simplicity, we shall drop the subscript p. We shall also drop the subscript α from E_α, $E_{I,\alpha}$, ∇_α, and div_α.

Introduce the usual Sobolev spaces

$$H(\text{curl}, \Omega) = \left\{ v \in (L^2(\Omega))^3, \nabla \times v \in (L^2(\Omega))^3 \right\}$$

and

$$H(\text{div}, \Omega) = \left\{ v \in (L^2(\Omega))^3, \nabla \cdot v \in L^2(\Omega) \right\}.$$

Chapter 6. Analysis of the Diffraction from Chiral Gratings

Recall that the imbeddings of $H(\text{curl}, \Omega)$ and $H(\text{curl}, \Omega) \cap H(\text{div}, \Omega)$ into $(L^2(\Omega))^3$ are not compact. To see this, it suffices to consider a sequence $\{h_m\}_{m \in \mathbb{N}}$ which tends to 0 weakly but not strongly in $H^{1/2}(\Gamma_1)$ and the unique Λ-periodic solution ξ_m of the problem

$$\nabla \cdot (\varepsilon \nabla \xi_m) = 0 \quad \text{in } \Omega,$$
$$\xi_m = h_m \quad \text{on } \Gamma_1,$$
$$\xi_m = 0 \quad \text{on } \Gamma_2.$$

Then, the sequence $\{\xi_m\}_{m \in \mathbb{N}}$ is bounded in $H^1(\Omega)$, and it tends to 0 weakly but not strongly in $H^1(\Omega)$. Finally, for any m, the function $u_m = \nabla \xi_m$ satisfies $\nabla \cdot (\varepsilon u_m) = 0$ and $\nabla \times u_m = 0$ in Ω. Then, it tends to 0 in $(L^2(\Omega))^3$ weakly but not strongly. This example is due to Murat [61]. It is also known that homogeneous normal or tangential boundary conditions are sufficient to enforce compactness. The following compactness result due to Weber [75] holds. Let

$$\tilde{\Omega} = \{-b' \leq x_3 \leq b', \ 0 < x_1 < \Lambda_1, \ 0 < x_2 < \Lambda_2\} \text{ with } b' > b.$$

Introduce

$$\tilde{W}_D(\tilde{\Omega}) = \left\{ v : v \in H(\text{curl}, \tilde{\Omega}), \ \nabla \cdot (\varepsilon v) = 0, \ \nu_3 \times v = 0 \text{ on } x_3 = \pm b' \right\}$$

and

$$\tilde{W}_N(\tilde{\Omega}) = \left\{ v : v \in H(\text{curl}, \tilde{\Omega}), \ \nabla \cdot (\varepsilon v) = 0, \ \nu_3 \cdot v = 0 \text{ on } x_3 = \pm b' \right\}.$$

LEMMA 6.3. *The imbeddings from $\tilde{W}_D(\tilde{\Omega})$ and $\tilde{W}_N(\tilde{\Omega})$ into $(L^2(\tilde{\Omega}))^3$ are compact.*

Therefore, the diffraction problem can be reformulated as follows:

$$\nabla \times (d \nabla \times E) - \omega^2 \nabla \times (\varepsilon \beta E) - \omega^2 \varepsilon \beta \nabla \times E - \omega^2 \varepsilon E = 0 \text{ in } \Omega, \quad (6.40)$$
$$\nu_3 \times (\nabla \times E) = B_1 P(E) - f \text{ on } \Gamma_1, \quad (6.41)$$
$$\nu_3 \times (\nabla \times E) = B_2 P(E) \text{ on } \Gamma_2, \quad (6.42)$$

where

$$f = d_1 \left(B_1 P(E_I)|_{\Gamma_1} - \nu_3 \times (\nabla \times E_I)|_{\Gamma_1} \right). \quad (6.43)$$

The weak form of the above boundary value problem is to find $E \in H(\text{curl}, \Omega)$, such that for any $F \in H(\text{curl}, \Omega)$

$$\int_\Omega d \nabla \times E \cdot \overline{\nabla \times F} - \int_\Omega \omega^2 \beta \varepsilon E \cdot \overline{\nabla \times F} - \int_\Omega \omega^2 \varepsilon \beta \nabla \times E \cdot \overline{F} \quad (6.44)$$
$$- \int_\Omega \omega^2 \varepsilon E \cdot \overline{F} + \int_{\Gamma_1} d_1 B_1 P(E) \cdot \overline{F} + \int_{\Gamma_2} d_2 B_2 P(E) \cdot \overline{F} = \int_{\Gamma_1} f \cdot \overline{F}.$$

6.4 Auxiliary Results

We present a version of the Hodge decomposition and compactness lemma. The results are crucial in the proof of our theorem on existence and uniqueness. We also state a useful trace regularity estimate. For simplicity, no attempt is made to present the most general forms of these results.

Hodge decomposition is now a well-known method for the study of Maxwell's equations in bounded or unbounded domains. The trouble with the Maxwell system is that it is neither elliptic nor semibounded. The use of a Hodge decomposition is one way to cope with both of these problems. In Birman and Solomyak [21], Leis [53], Kirsch and Monk [45], and our recent papers [8], [9], and [13], the Hodge decomposition has proven to be very useful.

Let us begin with a simple property of the operator B_j. From now on, we define div_Γ as the surface divergence. Its meaning should be clear from the contexts. In particular, if the surface is Γ_j ($j = 1$ or 2), then $\text{div}_\Gamma = P(\text{div})$. We also define curl_Γ as the scalar surface curl on Γ. From [2]-[4], the following holds.

PROPOSITION 6.1.
- For $j = 1, 2$ and $q \in H^1(\Omega)$

$$-\text{Re} \int_{\Gamma_j} B_j P(\nabla q) \cdot \overline{\nabla q} \geq 0 \,.$$

- There exists a positive constant C such that

$$\text{Re} \int_{\Gamma_j} B_j P(u) \cdot \overline{u} \geq -C \|\nu_3 \times u\|^2_{(H^{-1/2}(\Gamma_j))^3}$$

for any $u \in (H^{-1/2}(\Gamma_j))^3$, $u \cdot \nu_3 = 0$ such that $\text{curl}_{\Gamma_j} u \in H^{-1/2}(\Gamma_j)$.
- The pseudodifferential operator $B_j - B_j^*$ is of order $-\infty$.

We also need the following result.

LEMMA 6.4. *For any function $f \in (H^1(\Omega))'$ that is smooth near Γ_1 and Γ_2, the boundary value problem*

$$\nabla \cdot (\varepsilon \nabla p) = f \text{ in } \Omega, \tag{6.45}$$

$$\varepsilon_1 \frac{\partial p}{\partial \nu_3} = -d_1 \text{div}_{\Gamma_1} B_1 P(\nabla p) \text{ on } \Gamma_1, \tag{6.46}$$

$$\varepsilon_2 \frac{\partial p}{\partial \nu_3} = -d_2 \text{div}_{\Gamma_2} B_2 P(\nabla p) \text{ on } \Gamma_2, \tag{6.47}$$

has a unique solution in $H^1_a(\Omega) = \{q : q \in H^1(\Omega), \int_\Omega q = 0\}$.

Proof. We examine the weak form of the boundary value problem (6.45)–(6.47). For any $q \in H^1_a(\Omega)$, multiplying both sides of (6.45) by $\overline{q}$ and integrating over Ω yield

$$\int_\Omega \nabla \cdot (\varepsilon \nabla p) \cdot \overline{q} = \int_\Omega f \cdot \overline{q} \,.$$

Chapter 6. Analysis of the Diffraction from Chiral Gratings

By using the boundary conditions, integration by parts gives

$$\int_\Omega \varepsilon \nabla p \cdot \overline{\nabla q} + \int_{\Gamma_1} d_1 \mathrm{div}_{\Gamma_1} B_1 P(\nabla p) \cdot \bar{q} + \int_{\Gamma_2} d_2 \mathrm{div}_{\Gamma_2} B_2 P(\nabla p) \cdot \bar{q} = -\int_\Omega f \cdot \bar{q}. \quad (6.48)$$

Denote the left-hand side of (6.48) by $b(p,q)$. From integration by parts on the boundary, we obtain

$$b(p,q) = \int_\Omega \varepsilon \nabla p \cdot \overline{\nabla q} - \int_{\Gamma_1} d_1 B_1 P(\nabla p) \cdot \overline{P(\nabla q)} - \int_{\Gamma_2} d_2 B_2 P(\nabla p) \cdot \overline{P(\nabla q)}.$$

The variational problem takes the following form: Find $p \in H_a^1(\Omega)$ such that

$$b(p,q) = -\int_\Omega f \cdot \bar{q} \quad \text{for all } q \in H_a^1(\Omega).$$

It is now obvious from Proposition 6.1 that

$$\mathrm{Re}\, b(p,p) = \int_\Omega \varepsilon |\nabla p|^2 - \mathrm{Re}\Big\{ \int_{\Gamma_1} d_1 B_1 P(\nabla p) \cdot \overline{P(\nabla p)} \quad (6.49)$$

$$+ \int_{\Gamma_2} d_2 B_2 P(\nabla p) \cdot \overline{P(\nabla p)} \Big\} \quad (6.50)$$

$$\geq C \|\nabla p\|_{(L^2(\Omega))^3}^2. \quad (6.51)$$

Therefore, by a version of Poincaré's inequality ($\int_\Omega p = 0$), we obtain

$$\mathrm{Re}\, b(p,p) \geq C \|p\|_{H^1(\Omega)}^2.$$

The proof is complete by a direct application of the Lax–Milgram lemma. □

Next, we present an imbedding result. Let $W(\Omega)$ be a functional space defined by

$$W(\Omega) = \Big\{ u : u \in H(\mathrm{curl}, \Omega), \ \nabla \cdot (\varepsilon u) = 0 \text{ in } \Omega, \text{ and}$$

$$\omega^2 \varepsilon_j u \cdot \nu_3 = -d_j \mathrm{div}_{\Gamma_j} B_j P(u) \text{ on } \Gamma_j, \ j = 1,2 \Big\}.$$

LEMMA 6.5. *The imbedding from $W(\Omega)$ to $(L^2(\Omega))^3$ is compact.*

Proof. Let u be a function in $W(\Omega)$. Define an extension of u by

$$\tilde{u} = \begin{cases} u_1 \text{ in } \Omega_1, \\ u \text{ in } \Omega, \\ u_2 \text{ in } \Omega_2, \end{cases}$$

where u_j ($j = 1, 2$) satisfies

$$(1 - \omega^2 \beta_j^2 \varepsilon_j \mu_j) \nabla \times \nabla \times u_j - 2\omega^2 \varepsilon_j \mu_j \beta_j \nabla \times u_j - \omega^2 \varepsilon_j \mu_j u_j = 0 \text{ in } \Omega_j, \quad (6.52)$$

$$u_j \times \nu_3 = u \times \nu_3 \text{ on } \Gamma_j, \quad (6.53)$$

and the radiation condition at the infinity. $\quad (6.54)$

Since the medium in Ω_j is homogeneous, it may be shown that

$$\omega^2 \varepsilon_j u_j \cdot \nu_3 = -d_j \text{div}_{\Gamma_j} B_j P(u) \text{ on } \Gamma_j, \quad j = 1, 2. \tag{6.55}$$

In the following, we outline the proof of (6.55). In fact, it is easy to see that the function u_j satisfies the boundary condition

$$\nu_3 \times \nabla \times u_j = B_j(P(u_j)).$$

Hence

$$\text{div}_{\Gamma_j}(\nu_3 \times \nabla \times u_j) = \text{div}_{\Gamma_j}(B_j(P(u_j))). \tag{6.56}$$

But

$$\text{div}_{\Gamma_j}(\nu_3 \times \nabla \times u_j) = -\nabla \times \nabla \times u_j \cdot \nu_3,$$

which, together with the homogeneous Drude–Born–Fedorov system for u_j yield that

$$-\omega^2 \varepsilon_j \mu_j u_j \cdot \nu_3 = \text{div}_{\Gamma_j} B_j P(u_j). \tag{6.57}$$

From (6.56), (6.57), the boundary identity (6.55) follows.

Therefore, from $[\tilde{u} \times \nu_3] = 0$, it follows that $[\tilde{u} \cdot \nu_3] = 0$ on Γ_j and hence

$$\nabla \cdot (\varepsilon \tilde{u}) = 0 \text{ in } \overline{\Omega} \cup \Omega_1 \cup \Omega_2.$$

It follows from $[\tilde{u} \times \nu_3] = 0$ on Γ_j and the radiation condition that $\tilde{u} \in H(\text{curl}, D)$ for any compact domain $D \subset \overline{\Omega} \cup \Omega_1 \cup \Omega_2$.

Now let $\{\tilde{u}_j\}$ be a sequence of functions in W that converges weakly to zero in $W(\Omega)$. Construct a cutoff function χ with the properties that χ is supported in $\tilde{\Omega} \supset\supset \Omega$ and $\chi = 1$ in Ω. Here

$$\tilde{\Omega} = \{-b' \leq x_3 \leq b', \ 0 < x_1 < \Lambda_1, \ 0 < x_2 < \Lambda_2\} \text{ with } b' > b.$$

Hence

$$\{\chi \tilde{u}_j\} \subset \tilde{W}_D(\tilde{\Omega}) = \left\{ v : \ v \in H(\text{curl}, \tilde{\Omega}), \ \nabla \cdot (\varepsilon v) = 0, \ \nu_3 \times v = 0 \text{ on } x_3 = \pm b' \right\}.$$

It follows from Lemma 6.3 that the imbedding from $\tilde{W}(\tilde{\Omega})$ to $(L^2(\tilde{\Omega}))^3$ is compact. Therefore, the sequence $\{\tilde{u}_j\}$ converges strongly to zero in $(L^2(\Omega))^3$, which completes the proof. □

Finally, we state a useful trace regularity result.

LEMMA 6.6. *Let D be a bounded domain. Let ν denote the unit normal to the boundary ∂D. Let u be in $H(\text{curl}, D)$. For any $\eta > 0$, there is a constant $C(\eta)$ such that the estimate*

$$\|\nu \times u\|_{(H^{-1/2}(\partial D))^3} \leq \eta \|\nabla \times u\|_{(L^2(D))^3} + C(\eta) \|u\|_{(L^2(D))^3}$$

holds.

Chapter 6. Analysis of the Diffraction from Chiral Gratings

6.5 Existence and Uniqueness

In this section, we investigate questions of existence and uniqueness for the model problem. Our main result is as follows.

THEOREM 6.1. *For all but possibly a discrete set of frequencies ω, the variational problem (6.44) admits a unique weak solution E in $H(\mathrm{curl}, \Omega)$.*

Proof. The proof is based on the Lax–Milgram lemma. We first decompose the field E into two parts

$$E = u + \nabla p, \quad u \in H(\mathrm{curl}, \Omega), \quad p \in H^1(\Omega).$$

By choosing $E = u + \nabla p$, $F = v$ in (6.44), we arrive at

$$\int_\Omega d\nabla \times u \cdot \overline{\nabla \times v} - \omega^2 \int_\Omega \varepsilon \beta u \cdot \overline{\nabla \times v} \quad (6.58)$$

$$-\omega^2 \int_\Omega \varepsilon \beta \nabla \times u \cdot \overline{v} - \omega^2 \int_\Omega \varepsilon u \cdot \overline{v} + \int_{\Gamma_1} d_1 B_1 P(u) \cdot \overline{v} + \int_{\Gamma_2} d_2 B_2 P(u) \cdot \overline{v}$$

$$-\omega^2 \int_\Omega \varepsilon \beta \nabla p \cdot \overline{\nabla \times v} - \omega^2 \int_\Omega \varepsilon \nabla p \cdot \overline{v} + \int_{\Gamma_1} d_1 B_1 P(\nabla p) \cdot \overline{v}$$

$$+ \int_{\Gamma_2} d_2 B_2 P(\nabla p) \cdot \overline{v} = \int_{\Gamma_1} f \cdot \overline{v}.$$

Similarly, by choosing $E = u + \nabla p$, $F = \nabla q$ in (6.44), we get

$$-\omega^2 \int_\Omega \varepsilon \beta \nabla \times u \cdot \overline{\nabla q} - \omega^2 \int_\Omega \varepsilon u \cdot \overline{\nabla q} + \int_{\Gamma_1} d_1 B_1 P(u) \cdot \overline{\nabla q} \quad (6.59)$$

$$+ \int_{\Gamma_2} d_2 B_2 P(u) \cdot \overline{\nabla q} - \omega^2 \int_\Omega \varepsilon \nabla p \cdot \overline{\nabla q} + \int_{\Gamma_1} d_1 B_1 P(\nabla p) \cdot \overline{\nabla q}$$

$$+ \int_{\Gamma_2} d_2 B_2 P(\nabla p) \cdot \overline{\nabla q} = \int_{\Gamma_1} f \cdot \overline{\nabla q}.$$

We use the following Hodge decomposition:

$$E = u + \nabla p,$$

where $p \in H^1(\Omega)$ and $u \in W(\Omega)$. The functional space $W(\Omega)$ consists of all functions $u \in H(\mathrm{curl}, \Omega)$ that satisfy

$$\nabla \cdot (\varepsilon u) = 0 \text{ in } \Omega, \quad (6.60)$$

$$\omega^2 \varepsilon_1 u \cdot \nu_3 = -d_1 \mathrm{div}_{\Gamma_1} B_1 P(u) \text{ on } \Gamma_1, \quad (6.61)$$

$$\omega^2 \varepsilon_2 u \cdot \nu_3 = -d_2 \mathrm{div}_{\Gamma_2} B_2 P(u) \text{ on } \Gamma_2. \quad (6.62)$$

The fact that this decomposition is valid follows from Lemma 6.4. Actually, it is obvious to see that for any given E, Lemma 6.4 implies that there is a function p, such that $\nabla \cdot (\varepsilon \nabla p) = \nabla \cdot (\varepsilon E)$ and the suitable boundary conditions hold. Therefore, $u = E - \nabla p$ solves the problem (6.60)–(6.62).

Moreover, according to Lemma 6.5, the imbedding from $W(\Omega)$ to $(L^2(\Omega))^3$ is compact. Recall that the imbedding from $H(\mathrm{curl}, \Omega)$ to $(L^2(\Omega))^3$ is not compact. Denote the left-hand sides of (6.58), (6.59) by $a_1(u,v)$, $a_2(p,q)$, respectively. After some simple calculation, we obtain for $u, v \in W(\Omega)$, $p, q \in H^1(\Omega)$ that

$$a_1(u,v) = \int_\Omega d\nabla \times u \cdot \overline{\nabla \times v} - \omega^2 \int_\Omega \varepsilon\beta u \cdot \overline{\nabla \times v} - \omega^2 \int_\Omega \varepsilon\beta \nabla \times u \cdot \overline{v} \quad (6.63)$$
$$-\omega^2 \int_\Omega \varepsilon u \cdot \overline{v} + d_1 \int_{\Gamma_1} B_1 P(u) \cdot \overline{v} + d_2 \int_{\Gamma_2} B_2 P(u) \cdot \overline{v}$$
$$-\omega^2 \int_\Omega \varepsilon\beta \nabla p \cdot \overline{\nabla \times v} - \int_{\Gamma_1} d_1 p \, \mathrm{div}_{\Gamma_1}(\overline{(B_1^* - B_1)P(v)})$$
$$- \int_{\Gamma_2} d_2 p \, \mathrm{div}_{\Gamma_2}(\overline{(B_2^* - B_2)P(v)})$$

and

$$a_2(p,q) = -\omega^2 \int_\Omega \varepsilon\beta \nabla \times u \cdot \overline{\nabla q} - \omega^2 \int_\Omega \varepsilon \nabla p \cdot \overline{\nabla q} \quad (6.64)$$
$$+ d_1 \int_{\Gamma_1} B_1 P(\nabla p) \cdot \overline{\nabla q} + \int_{\Gamma_2} d_2 B_2 P(\nabla p) \cdot \overline{\nabla q}.$$

By taking $v = u$, $q = p$, we deduce from (6.63), (6.64) that

$$a_1(u,u) - a_2(p,p) = \int_\Omega d|\nabla \times u|^2 \quad (6.65)$$
$$-\omega^2 \int_\Omega \varepsilon\beta u \cdot \overline{\nabla \times u} - \omega^2 \int_\Omega \varepsilon\beta \nabla \times u \cdot \overline{u}$$
$$-\omega^2 \int_\Omega \varepsilon |u|^2 + d_1 \int_{\Gamma_1} B_1 P(u)\overline{u} + d_2 \int_{\Gamma_2} B_2 P(u) \cdot \overline{u} - \omega^2 \int_\Omega \varepsilon\beta \nabla p \cdot \overline{\nabla \times u}$$
$$- \int_{\Gamma_1} d_1 p \, \mathrm{div}_{\Gamma_1}(\overline{(B_1^* - B_1)P(v)}) - \int_{\Gamma_2} d_2 p \, \mathrm{div}_{\Gamma_2}(\overline{(B_2^* - B_2)P(v)})$$
$$+ \omega^2 \int_\Omega \varepsilon\beta \nabla \times u \cdot \overline{\nabla p} + \omega^2 \int_\Omega \varepsilon|\nabla p|^2,$$
$$- d_1 \int_{\Gamma_1} B_1 P(\nabla p) \cdot \overline{\nabla p} - \int_{\Gamma_2} d_2 B_2 P(\nabla p) \cdot \overline{\nabla p} = \int_{\Gamma_1} f \cdot \overline{(u - \nabla p)}.$$

By noting that

$$\mathrm{Re}\Big\{-\omega^2 \int_\Omega \varepsilon\beta \nabla p \cdot \overline{\nabla \times u} + \omega^2 \int_\Omega \varepsilon\beta \nabla \times u \cdot \overline{\nabla p}\Big\} = 0,$$

Chapter 6. Analysis of the Diffraction from Chiral Gratings

we have

$$\text{Re}\left\{a_1(u,u) - a_2(p,p)\right\} \geq d_0 \|\nabla \times u\|^2_{(L^2(\Omega))^3} \quad (6.66)$$

$$-\omega^2 \int_\Omega \varepsilon \beta (u \cdot \overline{\nabla \times u} + \nabla \times u \cdot \overline{u})$$

$$-\omega^2 \int_\Omega \varepsilon |u|^2 + \text{Re}\left\{d_1 \int_{\Gamma_1} B_1 P(u)\overline{u} + d_2 \int_{\Gamma_2} B_2 P(u) \cdot \overline{u}\right\}$$

$$-\text{Re}\left\{\int_{\Gamma_1} d_1 p \, \text{div}_{\Gamma_1}(\overline{(B_1^* - B_1)P(v)}) + \int_{\Gamma_2} d_2 p \, \text{div}_{\Gamma_2}(\overline{(B_2^* - B_2)P(v)})\right\}$$

$$+\omega^2 \int_\Omega \varepsilon |\nabla p|^2 - \text{Re}\left\{d_1 \int_{\Gamma_1} B_1 P(\nabla p) \cdot \overline{\nabla p} - \int_{\Gamma_2} d_2 B_2 P(\nabla p) \cdot \overline{\nabla p}\right\}.$$

We now estimate the terms on the right-hand side of (6.66) one by one.

By the Cauchy–Schwarz inequality, for any $\eta > 0$,

$$\int_\Omega \varepsilon \beta (u \cdot \overline{\nabla \times u} + \nabla \times u \cdot \overline{u}) \leq \eta \|\nabla \times u\|^2_{(L^2(\Omega))^3} + C(\eta) \|u\|^2_{(L^2(\Omega))^3}.$$

It follows from Proposition 6.1 that

$$\text{Re} \int_{\Gamma_j} d_j B_j P(u)\overline{u} \geq -C\|\nu_3 \times u\|^2_{(H^{-1/2}(\Gamma_j))^3},$$

where to get the last estimate, we have used the expression (6.25). An application of Lemma 6.6 leads to

$$\text{Re}\left\{\int_{\Gamma_1} d_1 B_1 P(u)\overline{u} + \int_{\Gamma_2} d_2 B_2 P(u)\overline{u}\right\} \geq -\eta \|\nabla \times u\|^2_{(L^2(\Omega))^3} - C(\eta) \|u\|^2_{(L^2(\Omega))^3}.$$

We next estimate the term

$$-\text{Re}\left\{\int_{\Gamma_j} d_j p \, \text{div}_{\Gamma_j}(\overline{(B_j^* - B_j)P(v)})\right\}.$$

From Proposition 6.1, it follows that

$$-\text{Re}\left\{\int_{\Gamma_j} d_j p \, \text{div}_{\Gamma_j}(\overline{(B_j^* - B_j)P(v)})\right\} \leq C\|p\|_{H^{1/2}(\Gamma_j)} \|\nu_3 \times v\|_{(H^{-1/2}(\Gamma_j))^3}.$$

Hence Lemma 6.6 and the trace theorem may be used once again to yield

$$-\sum_{j=1,2} \text{Re}\left\{\int_{\Gamma_j} d_j p \, \text{div}_{\Gamma_j}(\overline{(B_j^* - B_j)P(v)})\right\} \leq \eta \|p\|^2_{H^1(\Omega)} + \eta \|\nabla \times v\|_{(L^2(\Omega))^3}$$

$$+ C(\eta) \|v\|_{(L^2(\Omega))^3}.$$

Finally, by Proposition 6.1,

$$-\text{Re} \int_{\Gamma_j} d_j B_j P(\nabla p) \cdot \overline{\nabla p} = \text{Re} \int_{\Gamma_j} d_j \text{div}_{\Gamma_j} B_j P(\nabla p) \cdot \overline{p} \geq 0.$$

Combining the above estimates, we have shown that for any $u \in W(\Omega)$ and $p \in H^1(\Omega)$

$$\mathrm{Re}\Big\{a_1(u,u) - a_2(p,p)\Big\} \geq C_1 \|u\|^2_{H(curl,\Omega)} + C_2 \|p\|^2_{H^1(\Omega)} \quad (6.67)$$
$$- C_3 \Big(\|u\|^2_{(L^2(\Omega))^3} + \|p\|^2_{L^2(\Omega)} \Big).$$

Since the imbedding from $W(\Omega)$ to $(L^2(\Omega))^3$ is compact, the Fredholm alternative holds. The proof is complete. □

Remark 6.4. In the special case $\beta_1 = \beta_2 = 0$, Theorem 6.1 was proved by Ammari and Bao [2], [3]. A different approach for solving the diffraction problem, which couples an FEM with a method of integral equations or BEMs on the periodic interfaces, was presented by Ammari and Bao in [4]. In the case $\beta = 0$ everywhere, i.e., if the medium is achiral, Theorem 6.1 was shown in Abboud [1] by using a saddle point formulation. In that case, our present method provides a different proof of the result.

6.6 Convergence Analysis

We discretize our variational formulation using a family of finite-element subspaces $\mathcal{H}_h$ in $H(\mathrm{curl}, \Omega)$, where the parameter h measures the maximum diameter of the elements in the associated finite-element mesh. The domain Ω can be meshed using curvilinear tetrahedra. We assume that the family $\mathcal{H}_h$ satisfies the Hodge decomposition used for the continuous problem. A vector $E_h \in \mathcal{H}_h$ can be written as $E_h = u_h + \nabla p_h$, where $u \in W(\Omega)$ and $p \in H^1(\Omega)$. We also require that

$$\frac{1}{\|E_h\|_{H(curl,\Omega)}} \|R_h u_h - u_h\|_{(L^2(\Omega))^3} \to 0, \quad h \to 0,$$

where $R_h : (H^2(\Omega))^3 \mapsto \mathcal{H}_h$ is an interpolation operator. The family used to discretize the vector unknown u is then the projection of $\mathcal{H}_h$ to $W(\Omega)$. The family of Nédélec's finite element in $H(\mathrm{curl})$ [62] satisfies the required assumptions. It has the nice property that $\nabla S_h \subset \mathcal{H}_h$, where S_h is the usual P_1-Lagrange finite-element approximation.

Next, let

$$\mathcal{X}_h \subset \mathcal{X} = W(\Omega) \times H^1(\Omega)$$

be the discretized subspace associated with h.

By arguments essentially the same as in [4], the following theorem holds.

THEOREM 6.2. *There exists $h_0 > 0$ such that the discrete solution*

$$E_h = u_h + \nabla p_h$$

is well defined provided $0 < h < h_0$ and E_h satisfies the error estimate

$$\|E - E_h\|_{H(curl,\Omega)} \leq C \inf_{F_h \in \mathcal{X}_h} \|E - F_h\|_{H(curl,\Omega)}, \quad (6.68)$$

where C is a positive constant independent of h.

Chapter 6. Analysis of the Diffraction from Chiral Gratings

Remark 6.5. Note that the estimate (6.68) may not be improved. This is essentially due to the fact that the solution is only in $H(\text{curl}, \Omega)$ for bounded measurable material parameters ε, μ, and β. Nevertheless, with additional smoothness, one could expect to improve (6.68) and get the following optimal error estimate:

$$\|E - E_h\|_{H(curl,\Omega)} \leq C\,h\,\|E\|_{(H^2(\Omega))^3}.$$

An interesting open problem is to determine quantitatively how additional smoothness of the geometry could improve the convergence result.

Remark 6.6. Consider the so-called interface problem where the medium Ω is divided by the interfaces into a finite number of regions $\Omega_1, \ldots, \Omega_N$ with constant material parameters in each region. If the interfaces are of class $C^{1,1}$, it can be shown that the bilinear form $a_1(u,v)$ given by (6.63) is of Fredholm type on $(H^1(\Omega_1))^3 \times (H^1(\Omega_2))^3 \times \cdots \times (H^1(\Omega_N))^3$.

6.7 Future Directions

It seems worthwhile to intensify research on chiral gratings to further explore their use in microstrip and narrow strip gratings, band-gap structures [25], [37], [43], and as absorbers.

Engheta and Pelet [33]–[35] have shown that for a line source located above a chiral substrate, on the whole, the surface wave power decreases as the chirality of the substrate increases, thus increasing the radiation efficiency of the antenna, improving its bandwidth, and reducing the mutual coupling among the elements of an array.

More work is needed in order to investigate the effect of chirality parameter on resonant frequencies for narrow strip and microstrip gratings.

A microstrip grating is constituted by a parallel-plane waveguide in which there are periodically opened narrow slots, infinite in one direction, while the narrow strip grating is the dual structure: the small dimension of the periodic structure is the width of conducting strips. Significant mathematical results on the existence and distribution of the resonances for achiral narrow strip and microstrip gratings are obtained in [10] and [11].

Other interesting future directions are to develop computational schemes for solving the design problems and maximazing band gap in periodic chiral structures. The chirality provides an additional degree of freedom that cannot be achieved using conventional dielectric materials. More work is needed in order to investigate the prospects of maximizing band gaps and improving absorption by periodic chiral structures.

References

[1] T. Abboud, *Formulation variationnelle des équations de Maxwell dans un réseau bipériodique de R^3*, C. R. Acad. Sci. Paris Sér. I, 317 (1993), pp. 245–248.

[2] H. Ammari and G. Bao, *Analysis of the diffraction from periodic chiral structures*, C. R. Acad. Sci. Paris Sér. I Math., 326 (1998), pp. 1371–1376.

[3] H. Ammari and G. Bao, *Maxwell's equations in periodic chiral structures*, preprint, 1999.

[4] H. Ammari and G. Bao, *Coupling of finite element and boundary element methods for the diffraction from a periodic chiral structure*, in Res. Notes Math., A K Peters, Wellesley, MA, to appear.

[5] H. Ammari and J.-C. Nédélec, *Time-harmonic electromagnetic fields in chiral media*, in Modern Mathematical Methods in Diffraction Theory and Its Applications in Engineering, Methoden Verfahren Math. Phys., 42 Lang, Frankfurt am Main, 1997, pp. 174–202.

[6] H. Ammari and J.-C. Nédélec, *Time-harmonic electromagnetic fields in thin chiral curved layers*, SIAM J. Math. Anal., 29 (1998), pp. 395–423.

[7] H. Ammari and J.-C. Nédélec, *Small chirality behavior of solutions to electromagnetic scattering problems in chiral media*, Math. Methods Appl. Sci., 21 (1998), pp. 327–359.

[8] H. Ammari and J.-C. Nédélec, *Propagation d'ondes électromagnétiques à basses fréquences*, J. Math. Pures Appl., 77 (1998), pp. 839–849.

[9] H. Ammari and J.-C. Nédélec, *Low-frequency electromagnetic scattering*, SIAM J. Math. Anal., 31 (2000), pp. 836–861.

[10] H. Ammari, N. Béreux, and J.-C. Nédélec, *Resonant frequencies for a narrow strip grating*, Math. Methods Appl. Sci., 22 (1999), pp. 1121–1152.

[11] H. Ammari, N. Béreux, and J.-C. Nédélec, *Resonances for Maxwell's equations in a periodic structure*, Japan J. Appl. Math., 17 (2000), pp. 149–198.

[12] H. Ammari, M. Laouadi, and J.-C. Nédélec, *Low frequency behavior of solutions to electromagnetic scattering problems in chiral media*, SIAM J. Appl. Math., 58 (1998), pp. 1022–1042.

[13] H. Ammari, C. Latiri-Grouz, and J.-C. Nédélec, *The Leontovich boundary value problem for the time-harmonic Maxwell equations in an inhomogeneous medium: A singular perturbation problem*, SIAM J. Appl. Math., 59 (1999), pp. 1322–1334.

[14] H. Ammari, K. Hamdache, and J.-C. Nédélec, *Chirality in the Maxwell equations by the dipole approximation*, SIAM J. Appl. Math., 59 (1999), pp. 2045–2059.

[15] C. Athanasiadis and I. G. Stratis, *Electromagnetic scattering by a chiral obstacle*, IMA J. Appl. Math., 58 (1997), pp. 83–91.

[16] C. Athanasiadis, P. A. Martin, and I. G. Stratis, *Electromagnetic scattering by a homogeneous chiral obstacle: Boundary integral equations and low-chirality approximations*, SIAM J. Appl. Math., 59 (1999), pp. 1745–1762.

[17] G. Bao, *Variational approximation of Maxwell's equations in biperiodic structures*, SIAM J. Appl. Math., 57 (1997), pp. 364–381.

[18] G. Bao and D. Dobson, *On the scattering by a periodic structure*, Proc. Amer. Math. Soc., 128 (2000), pp. 2715–2723.

[19] G. Bao, D. Dobson, and J. A. Cox, *Mathematical studies of rigorous grating theory*, J. Opt. Soc. Amer. A, 12 (1995), pp. 1029–1042.

[20] S. Bassiri, C. H. Papas, and N. Engheta, *Electromagnetic wave propagation through a dielectric-chiral interface and through a chiral slab*, J. Opt. Soc. Amer. A, 5 (1988), pp. 1450–1459.

[21] M. Sh. Birman and M. Z. Solomyak, L_2-*Theory of the Maxwell operator in arbitrary domains*, Russian Math. Surveys, 42 (1987), pp. 75–96.

[22] C. F. Bohren, *Light scattering by an optically active sphere*, Chem. Phys. Lett., 29 (1974), pp. 458–462.

[23] O. P. Bruno and F. Reitich, *Numerical solution of diffraction problems: A method of variation of boundaries* III. *Doubly-periodic gratings*, J. Opt. Soc. Amer., 10 (1993), pp. 2551–2562.

[24] X. Chen and A. Friedman, *Maxwell's equations in a periodic structure*, Trans. Amer. Math. Soc., 323 (1991), pp. 465–507.

[25] J. Chongjun, Q. Bai, Y. Miao, and Q. Ruhu, *Two-dimensional photonic band structure in the chiral medium-transfer matrix method*, Opt. Commun., 142 (1997), pp. 179–183.

[26] D. Colton and R. Kress, *Inverse Acoustic and Electromagnetic Scattering Theory*, Appl. Math. Sci., 93, Springer-Verlag, New York, 1992.

[27] H. Cory, *Chiral devices—an overview of canonical problems*, J. Electro. Waves Appl., 9 (1995), pp. 805–829.

[28] D. Dobson, *A variational method for electromagnetic diffraction in biperiodic structures*, RAIRO Modél. Math. Anal. Numér., 28 (1993), pp. 321–340.

[29] D. Dobson and A. Friedman, *The time-harmonic Maxwell equations in a doubly periodic structure*, J. Math. Anal. Appl., 166 (1992), pp. 507–528.

[30] C. Eftimiu and L. W. Pearson, *Guided electromagnetic waves in chiral media*, Radio Sci., 24 (1989), pp. 351–359.

[31] N. Engheta, guest ed., *Wave interation with chiral and complex media*, Special Issue, J. Electro. Waves Appl., 6 (1992), pp. 537–798.

[32] N. Engheta and D. L. Jaggar, *Electromagnetic chirality and its application,* IEEE Trans. Antennas and Propagat. Soc. Newsletter, 30 (1988), pp. 6–12.

[33] N. Engheta and P. Pelet, *Surface waves in chiral layers,* Opt. Lett., 16 (1991), pp. 723–725.

[34] N. Engheta and P. Pelet, *Reduction of surface waves in chirostrip antennas,* Electron. Lett., 27 (1991), pp. 5–7.

[35] N. Engheta and P. Pelet, *Chirostrip antenna: line source problem,* J. Electro. Waves Appl., 6 (1992), pp. 771–793.

[36] R. Figari, G. Papanicolaou, and J. Rubinstein, *Remarks on the point interaction approximation,* in Hydrodynamic Behavior and Interacting Particle Systems, IMA Vol. Math. Appl. 9, Springer-Verlag, New York, 1987, pp. 45–56.

[37] K. M. Flood and D. L. Jaggard, *Band-gap structure for periodic chiral media,* J. Opt. Soc. Amer. A, 13 (1996), pp. 1395–1406.

[38] G. Heppke and D. Moro, *Chiral order from achiral molecules,* Science, 279 (1998), pp. 1872–1873.

[39] D. J. Hoppe and Y. Rahmat-Samii, *Higher order impedance boundary conditions revisited: Application to chiral coatings,* J. Electro. Waves Appl., 8 (1994), pp. 1303–1329.

[40] D. L. Jaggar, A. R. Michelson, and C. H. Papas, *On electromagnetic waves in chiral media,* Appl. Phys., 18 (1979), pp. 211–216.

[41] D. L. Jaggar, N. Engheta, M. Kowarz, P. Pelet, J. Liu, and Y. Kim, *Periodic chiral structures,* IEEE Trans. Antennas and Propagation, 37 (1989), pp. 1447–1452.

[42] D. L. Jaggar and J. C. Liu, *Chiral layers on curved surfaces,* J. Electro. Waves Appl., 6 (1992), pp. 669–694.

[43] V. Karathanos, N. Stefanou, and A. Modinos, *Optical activity of photonic crystals,* J. Mod. Opt., 42 (1995), pp. 619–626.

[44] A. Karlsson and G. Kristensson, *Constitutive relations, dissipation, and reciprocity for the Maxwell equations in the time domain,* J. Electro. Waves Appl., 6 (1992), pp. 537–551.

[45] A. Kirsch and P. Monk, *Corrigendum to "A finite element/spectral method for approximating the time-harmonic Maxwell system in R^3, SIAM J. Appl. Math., 55 (1995), 1324–1344,"* SIAM J. Appl. Math., 58 (1998), pp. 2024–2028.

[46] J. A. Kong, *Theorems of bianistropic media,* Proc. IEEE, 60 (1972), pp. 1036–1046.

[47] A. Lakhtakia, ed., *Selected Papers on Natural Optical Activity,* SPIE Milestone Series MS15, SPIE Optical Engineering Press, Bellingham, WA, 1990.

[48] A. Lakhtakia, *Beltrami Fields in Chiral Media,* World Sci. Ser. Contemp. Chem. Phys. 2, World Scientific, River Edge, NJ, 1994.

[49] A. Lakhtakia, V. K. Varadan, and V. V. Varadan, *Field equations, Huygens's principle, integral equations, and theorems for radiation and scattering of electromagnetic waves in sotropic chiral media,* J. Opt. Soc. Amer. A, 5 (1988), pp. 175–184.

[50] A. Lakhtakia, V. K. Varadan, and V. V. Varadan, *Time-Harmonic Electromagnetic Fields in Chiral Media,* Lecture Notes in Phys. 355, Springer-Verlag, New York, 1989.

[51] A. Lakhtakia, V. K. Varadan, and V. V. Varadan, *Scattering by periodic achiral-chiral interfaces,* J. Opt. Soc. Amer. A, 6 (1989), pp. 1675–1681.

[52] A. Lakhtakia, V. K. Varadan, and V. V. Varadan, *On electromagnetic fields in a periodically inhomogeneous chiral medium,* Z. Naturforsch., 459 (1990), pp. 639–644.

[53] R. Leis, *Initial Boundary Value Problems in Mathematical Physics,* Teubner and Wiley, Stuttgart, 1986.

[54] I. V. Lindell and F. Olyslager, *Generalized decomposition of electromagnetic fields in bi-anistropic media,* IEEE Trans. Antennas and Propagation, 46 (1998), pp. 1584–1585.

[55] I. V. Lindell and W. S. Weiglhofer, *Green dyadic for an unixial bianistropic medium,* IEEE Trans. Antennas and Propagation, 42 (1994), pp. 1013–1016.

[56] I. V. Lindell and A. Shivola, *Quasi-static analysis of scattering from a chiral sphere,* J. Electro. Waves Appl., 4 (1990), pp. 1223–1231.

[57] I. V. Lindell and A. Shivola, *Analysis on chiral mixtures,* J. Electro. Waves Appl., 6 (1992), pp. 553–572.

[58] I. V. Lindell, A. Shivola, S. A. Tretyakov, and A. J. Viitanen, *Electromagnetic Waves in Chiral and Bi-Isotropic Media,* Artech House, Boston, 1994.

[59] F. Mariotte and N. Engheta, *Effect of chiral material loss on guided electromagnetic modes in parallel-plate chirowaveguides,* J. Electro. Waves Appl., 7 (1993), pp. 1307–1321.

[60] S. R. McDowall, *An electromagnetic inverse problem in chiral media,* Trans. Amer. Math. Soc., 352 (2000), pp. 2993–3013.

[61] F. Murat, *Compacité par compensation* II, in Proc. Intern. Meeting on Recent Methods in Nonlinear Analysis, E. De Giorgi, E. Magenes, and U. Mosco, eds., Pitagora Editrice, Bologna, Italy, 1979, pp. 245–256.

[62] J.-C. Nédélec, *A new family of mixed finite elements in* $\mathbf{R}^3$, Numer. Math., 50 (1986), pp. 57–81.

[63] J.-C. Nédélec and F. Starling, *Integral equation methods in a quasi-periodic diffraction problem for the time harmonic Maxwell's equations*, SIAM J. Math. Anal., 22 (1991), pp. 1679–1701.

[64] P. Ola, *Boundary integral equations for the scattering of electromagnetic waves by a homogeneous chiral obstacle*, J. Math. Phys., 35 (1994), pp. 3969–3680.

[65] P. Pelet and N. Engheta, *The theory of chirowaveguides*, IEEE Trans. Antennas and Propagation, 38 (1990), pp. 90–98.

[66] D. M. Pozar, *Microstrip antennas and arrays on chiral substrates*, IEEE Trans. Antennas and Propagation, 40 (1992), pp. 1260–1263.

[67] I. E. Psarobas, N. Stefanou, and A. Modinos, *Photonic crystals of chiral spheres*, J. Opt. Soc. Amer. A, 16 (1999), pp. 343–347.

[68] R. Ro, V. V. Varadan, and V. K. Varadan, *Electromagnetic activity and absorption in microwave chiral composites*, IEE Proceedings-H, 139 (1992), pp. 441–448.

[69] R. G. Rojas, *Integral equations for the scattering by three dimensional inhomogeneous chiral bodies*, J. Electro. Waves Appl., 6 (1992), pp. 733–750.

[70] R. G. Rojas, *Integral equations for EM scattering by homogeneous/ inhomogeneous two-dimensional chiral bodies*, IEE Proceedings-H, 141 (1994), pp. 385–392.

[71] A. Toscano and L. Vegni, *Effects of chirality admittance on the propagating modes in a parallel-plate waveguide partially filled with a chiral slab*, Microwave Opt. Tech. Lett., 6 (1993), pp. 806–809.

[72] S. A. Tretyakov and A. A. Sochava, *Proposed composite material for nonreflecting shields and antenna radomes*, Electron. Lett., 29 (1993), pp. 1048–1049.

[73] U. Unrau, *A Bibiliograhy on Chiral and Bi(an)isotropic Microwave Materials*, Hochfrequenztechnik, Tech. Univers. Bramschweig, Germany.

[74] P. L. E. Uslenghi, *Scattering by an impedance sphere coated with a chiral sphere*, Electromagnetics, 10 (1990), pp. 201–211.

[75] Ch. Weber, *A local compactness theorem for Maxwell's equations*, Math. Methods Appl. Sci., 2 (1980), pp. 12–25.

[76] Y. Wenyan and L. Pao, *Effects of chirality admittance on the cutoff wavelength of propagating modes in a chiral coaxial line*, Microwave Opt. Tech. Lett., 7 (1994), pp. 810–812.

[77] S. Y. Yueh and J. A. Kong, *Analysis of diffraction from chiral gratings*, J. Electro. Waves Appl., 5 (1991), pp. 701–714.

Chapter 7

The Mathematics of Photonic Crystals

Peter Kuchment

7.1 Introduction

A photonic crystal, or photonic band gap (PBG) optical material, is an artificially created periodic low-loss dielectric medium in which electromagnetic waves of certain frequencies cannot propagate. The range of the prohibited frequencies is called the complete band gap. A simple example of such a medium is a dielectric background material with a periodic array of air bubbles. The reason why the band gap arises (if it does) is the coherent multiple scattering of waves and destructive interference. To put it simply, if a wave of a prohibited frequency somehow managed to propagate in the medium, it would reflect and self-interfere in such a way that it would cancel itself completely. It is expected that industrial manufacturing of photonic crystals will bring about a new technological revolution in optics, computing, information transmission, and other areas. The idea of photonic crystals was coined in [107, 192], though simpler versions of such materials like layered media and optical gratings have been known for a long time. We will not dwell much on the physics aspects of this field of research, since the reader can refer to the surveys and proceedings [33, 101, 108, 134, 154, 156, 167, 168, 182, 189] devoted to this topic, and especially to the lovely book [106]. The bibliography [153] and the collection of photonic crystal links [190] are also very useful.

The area of photonic crystal research presents a bonanza of beautiful, important, and hard problems for a mathematician, most of which are still unexplored or explored only tangentially. The number of mathematics publications dealing with PBG materials is growing (see [3, 7, 8, 21, 41, 42, 44, 45, 50, 51], [61]–[81], [86, 87, 100, 126, 127, 155, 188]) but is probably still not sufficient. We hope that this survey will play some role in publicizing the topic. One of the big attractions and advantages of the PBG research is that the mathematical model one studies is considered to be practically precise in most circumstances (see, for instance, [106]), a luxury not very often enjoyed by applied mathematicians.

In this article the author tries to expose some basic analytic ideas and techniques, to collect the recent mathematical results on PBG materials and their acoustic analogs, to present some basic problems that still await their resolution, and to indicate analogies with research in other areas (mostly related to solid-state physics) that could provide some leads for the PBG studies. Due to the limited space, the reader is referred to the corresponding literature for the details, complete formulations of the results, precise conditions on the coefficients, or exact definitions of some operators. Since the surveys [101, 152, 189] and collections cited before do a good job describing the numerical techniques that are commonly used, this paper addresses only a few recent, less standard numerical approaches that are not covered by these surveys.

There are many areas of the photonic crystal research that deserve and have not yet enjoyed close mathematical attention, but which we were not able to include in this survey. Among these are effects of losses, finiteness of samples, surface waves, nonlinear effects, magnetic effects, effects of metallic inclusions, gap solitons, and many others. The reader can find discussion of all of these and many other exciting topics in the surveys and bibliography quoted above and also in the papers [3, 41, 63, 66, 67, 86, 87, 188]. Regretfully, the important topic of Anderson localization of classical waves, where crucial results have recently been achieved, is just briefly mentioned due to the space limitations. We provide references to the relevant publications on this subject in section 7.6.6.

The theory of PBG materials as an area of mathematics is still in its childhood. As a result, there is no common choice of topics, approaches, etc. This article, therefore, reflects the author's views and interests and would probably be written in a totally different (maybe even orthogonal) manner by other researchers.

7.2 The Maxwell Operator

The reader has probably already seen the Maxwell equations many times in this book. We need, however, to briefly address them again. Our goal is to summarize the information we need and to mention some specific mathematical questions relevant to the theory of photonic crystals and to optics in general. Good general references concerning the Maxwell equations are [103, 135]. A mathematician can also be interested in the discussion of these equations presented in [53].

The macroscopic Maxwell equations that govern the light propagation in a photonic crystal in absence of free charges and currents look as follows:

$$\begin{cases} \nabla \times \mathbf{E} = -\frac{1}{c}\frac{\partial \mathbf{B}}{\partial t}, & \nabla \cdot \mathbf{B} = 0, \\ \nabla \times \mathbf{H} = \frac{1}{c}\frac{\partial \mathbf{D}}{\partial t}, & \nabla \cdot \mathbf{D} = 0. \end{cases} \quad (7.1)$$

Here c is the speed of light, $\mathbf{E}$ and $\mathbf{H}$ are the macroscopic electric and magnetic fields, and $\mathbf{D}$ and $\mathbf{B}$ are the displacement and magnetic induction fields, respectively.

Chapter 7. The Mathematics of Photonic Crystals

All these fields are vector-valued functions from $\mathbb{R}^3$ (or a subset of $\mathbb{R}^3$) into $\mathbb{R}^3$. We denote such fields with boldface letters. The standard vector notations $\nabla\times$ (or $\nabla^\times$), $\nabla\cdot$, and ∇ are used for the curl, divergence, and gradient, although we will also use curl, div, and grad. The system (7.1) is incomplete until we add the so-called constitutive relations that describe how the fields $\mathbf{D}$ and $\mathbf{H}$ depend on $\mathbf{E}$ and $\mathbf{B}$. Although in general these relations are nonlinear and even nonlocal, in materials other than ferroelectrics and ferromagnets and when the fields are weak enough, the following linear approximations to the constitutive relations work:

$$\mathbf{D} = \varepsilon \mathbf{E}, \quad \mathbf{B} = \mu \mathbf{H}. \tag{7.2}$$

Here ε and μ are the so-called material tensors. We will mostly address the case of isotropic media, where ε and μ can be considered as scalar time-independent functions called electric permittivity (or dielectric constant) and magnetic permeability, correspondingly. In most photonic crystals considerations it is assumed that the material is nonmagnetic, and hence $\mu = 1$.

After introducing the above assumptions, the Maxwell system reduces to the form

$$\begin{cases} \nabla \times \mathbf{E} = -\frac{1}{c}\mu(x)\frac{\partial \mathbf{H}}{\partial t}, & \nabla \cdot \mu\mathbf{H} = 0, \\ \nabla \times \mathbf{H} = \frac{1}{c}\varepsilon(x)\frac{\partial \mathbf{E}}{\partial t}, & \nabla \cdot \varepsilon\mathbf{E} = 0, \end{cases} \tag{7.3}$$

or, in the nonmagnetic case,

$$\begin{cases} \nabla \times \mathbf{E} = -\frac{1}{c}\frac{\partial \mathbf{H}}{\partial t}, & \nabla \cdot \mathbf{H} = 0, \\ \nabla \times \mathbf{H} = \frac{1}{c}\varepsilon(x)\frac{\partial \mathbf{E}}{\partial t}, & \nabla \cdot \varepsilon\mathbf{E} = 0. \end{cases} \tag{7.4}$$

These linear partial differential equations have time-independent coefficients, so the Fourier transform in the time domain reduces considerations to the case of monochromatic waves $\mathbf{E}(x,t) = e^{i\omega t}\mathbf{E}(x)$, $\mathbf{H}(x,t) = e^{i\omega t}\mathbf{H}(x)$. This leads from (7.3) to

$$\begin{cases} \nabla \times \mathbf{E} = -\frac{i\omega}{c}\mu(x)\mathbf{H}, & \nabla \cdot \mu\mathbf{H} = 0, \\ \nabla \times \mathbf{H} = \frac{i\omega}{c}\varepsilon(x)\mathbf{E}, & \nabla \cdot \varepsilon\mathbf{E} = 0, \end{cases} \tag{7.5}$$

which can be rewritten in the matrix form as

$$\begin{pmatrix} 0 & -\frac{i}{\varepsilon}\nabla^\times \\ \frac{i}{\mu}\nabla^\times & 0 \end{pmatrix} \begin{pmatrix} \mathbf{E} \\ \mathbf{H} \end{pmatrix} = \frac{\omega}{c}\begin{pmatrix} \mathbf{E} \\ \mathbf{H} \end{pmatrix} \tag{7.6}$$

on the subspace of vectors $\begin{pmatrix} \mathbf{E} \\ \mathbf{H} \end{pmatrix}$ satisfying

$$\nabla \cdot \varepsilon\mathbf{E} = 0, \quad \nabla \cdot \mu\mathbf{H} = 0. \tag{7.7}$$

We use in (7.6) and in the rest of the text the notation $\nabla^\times$ for the curl operator.

We are now facing the spectral problem for the Maxwell operator

$$M = \begin{pmatrix} 0 & -\frac{i}{\varepsilon}\nabla^\times \\ \frac{i}{\mu}\nabla^\times & 0 \end{pmatrix} \tag{7.8}$$

on the subspace (7.7). One of the principal tasks of the photonic crystals theory is to choose periodic functions $\varepsilon(x) \geq 1$ and μ (although μ is usually assumed to be equal to 1) such that the corresponding spectrum has a gap. Existence of such a gap would mean that electromagnetic waves with a frequency ω in the gap cannot propagate in the material.

7.2.1 Defining a Self-Adjoint Maxwell Operator

Before studying the spectrum of the problem (7.8), one needs to define the corresponding self-adjoint operator. It is not hard to define the operator in the case when the material tensors are smooth and the operator is considered either in the whole space or in a smooth domain with conducting boundaries. In fact, smoothness of the material tensors is also not a big issue when one deals with the whole space (or with a torus, as one often does in the photonic crystal theory). However, if the domain has nonsmooth conducting boundaries, the solutions can develop singularities. Although we will constrain ourselves to the case when no metallic inclusions are present, PBG materials with nonsmooth metallic inclusions and/or boundaries are actually considered. In such cases one should consult with a comprehensive study of the Maxwell operator done by Birman and Solomyak in [22, 23].

Let us assume that $\mu(x)$ and $\varepsilon(x)$ are positive measurable functions uniformly bounded by positive constants from below and from above. In most of our discussion μ is equal to 1 and $\varepsilon(x) \geq 1$ is periodic and piecewise constant. We want to define the Maxwell operator as a self-adjoint operator in appropriate spaces. We will use notation $L^2(\mathbb{R}^3, w(x)dx)$ for the weighted L^2-space with the norm

$$||f||^2 = \int |f(x)|^2 w(x)dx$$

and $L^2(\mathbb{R}^3, w(x)dx; \mathbb{C}^3)$ for the corresponding space of vector fields.

Consider now the subspace J of the space

$$L^2(\mathbb{R}^3, \varepsilon(x)dx; \mathbb{C}^3) \oplus L^2(\mathbb{R}^3, \mu(x)dx; \mathbb{C}^3)$$

that consists of all pairs of vector fields (u_1, u_2) such that

$$\nabla \cdot \varepsilon u_1 = \nabla \cdot \mu u_2 = 0. \tag{7.9}$$

On the space J we can define the Maxwell operator M with the matrix (7.8) and the domain consisting of pairs (u_1, u_2) such that

$$\nabla \times u_j \in L^2(\mathbb{R}^3), \qquad j = 1, 2.$$

The derivatives here are understood in the distributional sense.

THEOREM 7.1 (Lemma 2.2 in [22]). *The Maxwell operator M defined this way is self-adjoint.*

7.2.2 Ellipticity

The trouble with the Maxwell operator is that it is neither elliptic nor semi-bounded (so its spectrum extends to both positive and negative infinity). There are ways to cope with both of these problems. Squaring the operator produces a new operator

$$M^2 = \begin{pmatrix} \frac{1}{\varepsilon}\nabla\times\frac{1}{\mu}\nabla\times & 0 \\ 0 & \frac{1}{\mu}\nabla\times\frac{1}{\varepsilon}\nabla\times \end{pmatrix},$$

which is already positive definite. Thus, as is customarily done in photonic crystals theory, we can consider either one of the following positive definite spectral problems:

$$\begin{cases} \frac{1}{\varepsilon}\nabla\times\frac{1}{\mu}\nabla\times E = \lambda E, \\ \nabla\cdot\varepsilon E = 0 \end{cases} \tag{7.10}$$

or

$$\begin{cases} \frac{1}{\mu}\nabla\times\frac{1}{\varepsilon}\nabla\times H = \lambda H, \\ \nabla\cdot\mu H = 0, \end{cases} \tag{7.11}$$

each of which contains only one of the electric and magnetic fields. Here we denote $\lambda = (\omega/c)^2$. This notation will be used from now on. The spectrum of either of these two problems determines the spectrum of M.

We will be mostly concerned with the case when $\mu = 1$, so (7.10) and (7.11) reduce to

$$\begin{cases} \frac{1}{\varepsilon}\nabla\times\nabla\times E = \lambda E, \\ \nabla\cdot\varepsilon E = 0 \end{cases} \tag{7.12}$$

and

$$\begin{cases} \nabla\times\frac{1}{\varepsilon}\nabla\times H = \lambda H, \\ \nabla\cdot H = 0. \end{cases} \tag{7.13}$$

The problem (7.12) can also be rewritten after introducing a new vector field $F = \varepsilon^{1/2}E$ as follows:

$$\begin{cases} -\varepsilon^{-1/2}\Delta\Pi\varepsilon^{-1/2}F = \lambda F, \\ \nabla\cdot\varepsilon^{1/2}F = 0, \end{cases}$$

where Δ is the Laplace operator and Π is the orthogonal projector onto the space of transverse vector fields. This restatement of the problem has proven to be useful, for instance, in localization problems (see [42]). In most cases when we refer to the Maxwell equations, we will mean (7.13).

Note that for $\lambda \neq 0$ the second equation in either of the systems (7.10) or (7.11) is a consequence of the first one. One is thus tempted to eliminate the zero divergency conditions altogether and to study only the "Schrödinger-type" first equations in (7.12) or (7.13), which can be called the eigenvalue problem for the unrestricted Maxwell operator. This introduces a large kernel consisting of longitudal waves (gradients of scalar functions), without changing the spectrum otherwise. Sometimes this trick works well, but in many cases the huge kernels that arise in this approach make analytic and numerical studies harder. This is related to the nonellipticity of the Maxwell operator. In fact, the truth is that the Maxwell operator *is* elliptic, if ellipticity is understood in an appropriate sense. Namely, this operator should not be considered alone, but rather as a part of what is usually called an elliptic complex of operators. Rather than going into the details of the general notion of an elliptic complex (which does not seem to bring any insights about photonic crystals), we will provide a simple explanation of some of the corresponding notions. Let us consider for simplicity the curl operator $\nabla^\times$ instead of the full Maxwell operator M. What are the indications that the curl is not an elliptic operator? If it were, then on a compact manifold its kernel (considered in appropriate spaces) would be finite dimensional, while its range would have finite codimension. Consider the case of the torus $\mathbb{T} = \mathbb{R}^3/\mathbb{Z}^3$, where $\mathbb{Z}^3$ is the integer lattice in $\mathbb{R}^3$. Then the gradient of any function on $\mathbb{T}$ belongs to the kernel of $\nabla^\times$, which shows that the kernel is infinite dimensional. There is a similar situation with the range: every function in the range of $\nabla^\times$ has zero divergence, which means that the range is of infinite codimension. The nonellipticity of curl is also clear from its Fourier domain representation as multiplication (up to a scalar factor) by the matrix

$$\begin{pmatrix} 0 & -\xi_3 & \xi_2 \\ \xi_3 & 0 & -\xi_1 \\ -\xi_2 & \xi_1 & 0 \end{pmatrix}$$

with the determinant identically equal to zero.

The correct point of view at the operator $\nabla^\times$ is to include it into the sequence of operators

$$0 \to C^\infty(\mathbb{T}) \xrightarrow{\nabla} [C^\infty(\mathbb{T})]^3 \xrightarrow{\nabla^\times} [C^\infty(\mathbb{T})]^3 \xrightarrow{\nabla\cdot} C^\infty(\mathbb{T}) \to 0$$

(where the C^∞ spaces can be replaced by appropriate spaces of Sobolev type). This is an example of an elliptic complex. This means that, first, composition of any two consecutive operators is zero. Second, the cohomologies of this complex (i.e., the quotient spaces of the kernel of each next operator modulo the range of the previous one) are finite dimensional. The whole Maxwell operator M can be included into an elliptic complex in a similar way. There is a trick commonly used in geometry that naturally reduces the study of such a complex to a single elliptic matrix operator. A similar technique is known in the study of overdetermined systems of partial differential equations, where it is sometimes called the method of orthogonal extension (see [93]). Let us show how it works in the particular cases of the curl and Maxwell operators. Consider the operator $\nabla^\times$ acting on vector fields E, add one more scalar function f so the operator now acts on pairs (E, f), and define the extended operator as

$$\begin{pmatrix} \nabla^\times & \nabla \\ -\nabla\cdot & 0 \end{pmatrix}.$$

Chapter 7. The Mathematics of Photonic Crystals

One can easily check ellipticity of this extended operator (for instance, by taking the Fourier transform). The subspace of vectors of the form $(\mathbf{E}, 0)$, where $\mathbf{E}$ is divergence free, reduces the extended operator, and on this subspace it coincides with the curl. Analogously, one can include the Maxwell operator M into the larger elliptic operator

$$\mathfrak{M} = \begin{pmatrix} 0 & 0 & -\frac{i}{\varepsilon}\nabla\times & -i\nabla \\ 0 & 0 & i\nabla\cdot\mu & 0 \\ \frac{i}{\mu}\nabla\times & i\nabla & 0 & 0 \\ -i\nabla\cdot\varepsilon & 0 & 0 & 0 \end{pmatrix}$$

acting in the space

$$L^2(\mathbb{R}^3, \varepsilon(x)dx; \mathbb{C}^3) \oplus L^2(\mathbb{R}^3, dx) \oplus L^2(\mathbb{R}^3, \mu(x)dx; \mathbb{C}^3) \oplus L^2(\mathbb{R}^3, dx).$$

Here we denoted by $\nabla\cdot\varepsilon$ the operator acting on a vector field u as $\nabla\cdot\varepsilon u$. One can easily define $\mathfrak{M}$ as a self-adjoint operator. This extension to a larger elliptic problem is often useful in obtaining estimates and in other situations (see, for instance, [22, 93]).

7.2.3 Variational Formulation and Energy

The energy density of the field (E, H) in (7.1) can be defined as

$$\mathcal{E}(x, t) = \frac{1}{2}\left\{\varepsilon(x)|E(x, t)|^2 + \mu(x)|H(x, t)|^2\right\}$$

with the corresponding physical energy

$$E = \int \mathcal{E}(x, t)dx.$$

Each of the problems (7.12) and (7.13) allows a variational formulation of finding stationary points of the ratios

$$\frac{\int |\nabla \times E(x)|^2 \, dx}{\int |E(x)|^2 \varepsilon(x) dx}$$

and

$$\frac{\int |\nabla \times H(x)|^2 \varepsilon^{-1}(x) dx}{\int |H(x)|^2 \, dx},$$

respectively (subject to the natural zero divergence restrictions). This formulation is used, for instance, in the numerical treatment of these problems by finite element methods.

7.2.4 Scaling Properties

The problems (7.12) and (7.13) look similar to the spectral problem for a Schrödinger operator (with $\varepsilon(x)$ playing the role of a "potential," or rather of a metric). Although this analogy is useful, it can be misleading, since the Maxwell operator enjoys many properties different from those of Schrödinger operators. The scaling property is one of them. Consider, for instance, the problem (7.12):

$$\begin{cases} \nabla \times \nabla \times E = \lambda \varepsilon(x) E, \\ \nabla \cdot \varepsilon E = 0. \end{cases}$$

It is straightforward to compute that change of variables $x' = sx$ and simultaneous change of the spectral parameter $\lambda' = \lambda/s^2$ reduces the problem (7.12) to the similar one with the rescaled dielectric function $\varepsilon'(x) = \varepsilon(x/s)$. This means that in rescaling the dielectric function of a medium, we do not need to recompute the spectrum, since its simple rescaling would do. This observation has significant implications. It shows that the Maxwell equations do not have any fundamental length scale besides the requirement that they be macroscopic. For instance, if one finds some spectral phenomenon on the microwave scale, then the similar (rescaled) effect holds in the visible light region of frequencies. This is a significant departure from the Schrödinger case, where the Bohr radius provides a fundamental length scale for potentials. However, one should realize that manufacturing the materials for one length scale (for instance, for the visual light wave length) could be much harder than for another (microwaves).

Another important scaling property deals with the values of the electric permittivity function $\varepsilon(x)$. Assume that it is multiplied by a constant scaling factor s: $\varepsilon'(x) = s\varepsilon(x)$. It is obvious that the spectral problem for the new dielectric function ε' is reduced to the old one by rescaling the eigenvalues according to the formula $\lambda = s\lambda'$. This means that there is no fundamental value of the dielectric constant. In particular, in any homogeneous medium the spectrum always starts at zero, the property that is in striking contrast with the Schrödinger case. Among the important implications are different mechanisms of opening spectral gaps and of creating impurity spectra.

7.2.5 Two-Dimensional Case. TM and TE Polarizations

If a medium has material tensors independent of one of the coordinates, we will call it a "two-dimensional medium." Let us assume that $\mu = 1$ and $\varepsilon(x) = \varepsilon(x_1, x_2)$ is independent on the third coordinate x_3. We will consider the waves propagating in the (x_1, x_2)-plane only. In other words, the electromagnetic field $(\mathbf{E}, \mathbf{H})$ is x_3-independent. It is straightforward to check that on the space of such fields the Maxwell operator

$$M = \begin{pmatrix} 0 & -\frac{i}{\varepsilon}\nabla^\times \\ i\nabla^\times & 0 \end{pmatrix}$$

is reduced by the direct decomposition $S_1 \oplus S_2$, where S_1 consists of the fields $(E_1, E_2, 0, 0, 0, H)$ and S_2 consists of the fields $(0, 0, E, H_1, H_2, 0)$. In other words,

Chapter 7. The Mathematics of Photonic Crystals

S_1 consists of the transverse electric (TE) polarized fields (or H-fields), in which the magnetic field is directed along the x_3 axis and the electric field is normal to this axis. Analogously, S_2 consists of transverse magnetic (TM) polarized fields (E-fields) with the electric field parallel and magnetic field normal to the x_3 axis. One can come to the conclusion that this reduction of the operator is due to the fact that in this case the Maxwell equations are mirror symmetric with respect to any mirror orthogonal to the axis x_3 (see [106]).

On the space S_2 the spectral problem for the Maxwell operator reduces to the scalar problem of Helmholtz type

$$-\Delta E = \lambda \varepsilon E, \qquad (7.14)$$

while on S_1 it reduces to the divergence-type problem

$$-\nabla \cdot \frac{1}{\varepsilon} \nabla H = \lambda H. \qquad (7.15)$$

These two spectral problems also arise when one considers acoustic waves in media with periodically varying parameters. Thus, many results obtained for photonic crystals can be transferred to the case of such waves. The acoustic interpretation also makes the consideration of the three-dimensional analogs of the scalar spectral problems (7.14) and (7.15) meaningful, although they do not reduce the Maxwell operator anymore. We will not, however, concentrate on the acoustic situation.

We would like to mention a rather standard transformation that can be applied to the problem (7.15) in $\mathbb{R}^d$ with smoothly varying $\varepsilon(x)$ to transfer it to a problem resembling (7.14). Here we will assume that the medium is isotropic, and hence the material tensor ε is just a sufficiently smooth periodic function $\varepsilon \geq 1$. We are interested in invertibility of the (suitably defined by quadratic forms) operator

$$L_\lambda u = -\nabla \cdot \frac{1}{\varepsilon} \nabla u - \lambda u$$

in $L^2(\mathbb{R}^d)$. The transformation works as follows:

$$L_\lambda \to H_\lambda = \sqrt{\varepsilon} L_\lambda \sqrt{\varepsilon}. \qquad (7.16)$$

The multiplication by $\sqrt{\varepsilon}$ is an invertible operator in $L^2(\mathbb{R}^d)$ and, if ε is smooth enough, it preserves the domain of the operator L_λ. Thus, the operators L_λ and H_λ are invertible simultaneously. A straightforward calculation shows that

$$H_\lambda = -\Delta + V - \lambda \varepsilon,$$

where

$$V = \frac{3(\nabla \varepsilon)^2}{4\varepsilon^2} - \frac{\Delta \varepsilon}{2\varepsilon}.$$

We conclude that λ is in the spectrum of the operator $-\nabla \cdot \frac{1}{\varepsilon} \nabla$ if and only if the operator H_λ is not invertible, i.e., when λ is in the spectrum of the Schrödinger operator pencil $-\Delta + V - \lambda \varepsilon$.

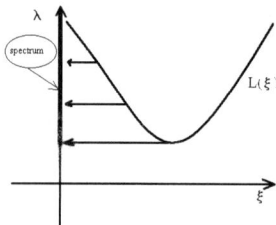

Figure 7.1: The spectrum of the operator of multiplication by $L(\xi)$.

The transform described above is sometimes called Liouville–Green transform and is frequently used in problems of spectral theory (see, for instance, [16, 25, 26, 49, 117]). In particular, nonexistence of bound states for the Schrödinger operator H_λ implies their absence for the operator L_λ. Unfortunately, this transform does not work in the case when the material tensors are piecewise constant, which is the standard situation for photonic crystals.

7.3 Periodic Media and Floquet–Bloch Theory

So far our considerations do not involve any periodicity requirement for the medium. However, as has already been mentioned, the main feature of a (pure) photonic crystal is periodicity of its structure. Let us discuss in very general terms why periodicity is a favorable environment for spectral gaps. In order to do this we need to provide some information about periodic (elliptic) differential operators and Floquet theory, which in the periodic case plays the role of the Fourier transform. Many aspects of this theory are discussed in detail in books and surveys [54, 116, 123, 124, 145, 157, 173, 174, 179, 185]. Some additional references on this subject will be provided later in the text. Many physics books also address this topic, for instance, [5, 36].

Let us start considering a constant coefficient partial differential operator[1] $L(D)$ in $L^2(\mathbb{R}^n)$, where $D = -i\nabla$. In fact, what we will discuss is also applicable to more general convolution operators, where $L(\xi)$ does not have to be a polynomial. The operator is invariant with respect to the (transitive) action of the additive group $\mathbb{R}^n$ on itself via translations. This leads to the natural idea of applying the Fourier transform on this group, which is the standard Fourier transform. After applying the Fourier transform, L becomes the operator of multiplication by the function $L(\xi)$ in $L^2(\mathbb{R}^n)$, where ξ denotes the variable dual to x. It is clear that the spectrum of such an operator coincides with the (closure of the) set of all values of $L(\xi)$. In other words, if we draw the graph of the function $\lambda = L(\xi)$, its projection on the λ-axis produces the spectrum (Figure 7.1).

It is also important to understand when the point spectrum can arise. If there is a nonzero L^2-function $f(\xi)$ and an eigenvalue λ such that $L(\xi)f(\xi) = \lambda f(\xi)$ a.e., one immediately concludes that $L(\xi) = \lambda$ on a set of positive measure. The converse

---

[1] We intentionally avoid here any discussion of exact definition of the operator, its domain, etc.

Chapter 7. The Mathematics of Photonic Crystals

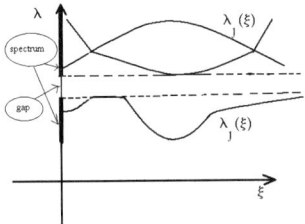

Figure 7.2: The band gap structure of the spectrum in the matrix case.

statement is also correct: positivity of the measure of the set where $L(\xi) = \lambda$ implies existence of an eigenfunction. In the important particular case when $L(\xi)$ is analytic, this would imply that $L(\xi)$ is constant.

Another step toward periodic operators is to consider a system $L(D)$ (i.e., the symbol $L(\xi)$ is a self-adjoint matrix function). It is rather clear then that the spectrum can be found as follows: find the (continuous) eigenvalue branches $\lambda_j(\xi)$ ("dispersion relations" or "band functions") of the matrix function $L(\xi)$ and take their ranges (i.e., project their graphs onto the λ-axis). Each of the branches then provides a band (i.e., a segment) in the spectrum. One can expect that in some cases the bands might have a gap between them (Figure 7.2).

As in the scalar case, existence of the point spectrum is equivalent to existence of flat pieces on the graphs of the band functions, which in the analytic situation implies existence of a constant branch.

Let us now tackle periodic operators. Consider a linear partial differential operator $L(x, D)$, whose coefficients are periodic with respect to a discrete group of translations Γ acting on $\mathbb{R}^d$. Assume, for instance, that Γ is the integer lattice $\mathbb{Z}^d$; this assumption is made for simplicity only and does not restrict generality of our consideration.[2] In analogy with the constant coefficient case, due to invariance of the operator with respect to this group, it is natural to apply the Fourier transform on Γ. "Fourier transform" on $\Gamma = \mathbb{Z}^d$ is in fact the Fourier series, which assigns to a sufficiently fast decaying function $h(n)$ on $\mathbb{Z}^d$ the Fourier series

$$\widehat{h}(k) = \sum_{n \in \mathbb{Z}^d} h(n) e^{ik \cdot n},$$

where $k \in \mathbb{R}^d$ (or $\mathbb{C}^d$). We have to somehow apply this transform to functions defined on $\mathbb{R}^d$. Let $f(x)$ be a function decaying sufficiently fast. We can define its *Floquet transform* (sometimes called *Gelfand transform*) as follows:

$$\mathcal{U}f(x,k) = \sum_{n \in \mathbb{Z}^d} f(x-n) e^{ik \cdot n}. \qquad (7.17)$$

This transform is an analog of the Fourier transform for the periodic case. The parameter k is called *quasi momentum*, and it is an analogue of the dual variable

---

[2]The reader can refer to [5, 106] for a brief introduction into general lattices, Brillouin zones, etc.

in the Fourier transform. Notice that in contrast to the Fourier transform the transformed function still depends on the old variable x. The reason is that the action of the group Γ on $\mathbb{R}^d$ is not transitive, and hence the space of orbits of this action contains more than one point, while in the constant coefficient case we deal with a transitive action of $\mathbb{R}^d$ on itself. One should notice the two following important relations. If we shift x by a period $m \in \mathbb{Z}^d$, then we get the relation

$$(\mathcal{U}f)(x+m,k) = e^{ik \cdot m}(\mathcal{U}f)(x,k). \tag{7.18}$$

This is the *Floquet condition*. It shows that it is sufficient to know the function $\mathcal{U}f(x,k)$ only at one point x on each orbit $x + \mathbb{Z}^d$ in order to recover it completely. For instance, it is sufficient to know it only for $x \in F$, where F is a fundamental domain for the action of $\mathbb{Z}^d$ on $\mathbb{R}^d$. A domain F in $\mathbb{R}^d$ is called a fundamental domain for the action of $\mathbb{Z}^d$ if each orbit has a representative in the closure $\overline{F}$ of F and every point of F is a unique representative in F of its orbit. In other words,

$$\cup_{m \in \mathbb{Z}^d}(\overline{F} + m) = \mathbb{R}^d$$

and $\overline{F} + m$ and $\overline{F}$ can intersect only along their boundaries. One way to find a fundamental domain is to consider all points x that are closer to the origin $0 \in \Gamma$ than to any other point of Γ. An example of a fundamental domain for the action of $\mathbb{Z}^d$ on $\mathbb{R}^d$ by translation is the unit cube

$$W = \left\{ x \in \mathbb{R}^d \mid 0 \leq x_j \leq 1, \, j = 1, \ldots, d \right\}.$$

In physics the fundamental domains are often called *Wigner–Seitz cells*.

The second simple observation is that the function $\mathcal{U}f(x,k)$ is periodic with respect to the quasi momentum k. Indeed,

$$\mathcal{U}f(x, k + 2\pi m) = \mathcal{U}f(x,k), \qquad m \in \mathbb{Z}^d.$$

Notice that the lattice of the periods with respect to k is different from the lattice with respect to which the operator was periodic. Now it is $\Gamma^* = 2\pi \mathbb{Z}^d$, which is the *dual (or reciprocal) lattice* to $\Gamma = \mathbb{Z}^d$. We conclude that k can be considered as an element of the torus $\mathbb{T}^* = \mathbb{R}^d / 2\pi \mathbb{Z}^d$. Another way of saying this is that all information about the function $\mathcal{U}f(x,k)$ is contained in its values for k in the fundamental domain of the dual lattice $\Gamma^* = 2\pi \mathbb{Z}^d$. We can define such a domain B as the set of all vectors k that are closer to the origin than to any other point of Γ^*. In solid-state physics this domain is called the (first) *Brillouin zone*.

As the result, after the Floquet transform one ends up with a function $\mathcal{U}f(x,k)$, which can be considered as a function of k on the torus $\mathbb{T}^*$ (or on the Brillouin zone B) with values in a space of functions of x on the compact Wigner–Seitz cell W. As we will soon see, compactness of the new domain plays the crucial role in the whole Floquet theory.

Now consider the effect of the Floquet transform on a periodic differential operator $L(x,D)$. Due to periodicity, the operator commutes with the transform

$$\mathcal{U}(Lf)(x,k) = L(x, D_x)\mathcal{U}f(x,k),$$

where by the subscript x in D_x we indicate that D differentiates with respect to x rather than k. For each k the operator $L(x, D_x)$ now acts on functions satisfying the corresponding Floquet condition (7.18). In other words, although the differential expression of the operator stays the same, its domain changes with k. If we denote this operator by $L(k)$, we see that the Floquet transform expands the operator L in $L^2(\mathbb{R}^d)$ into the "direct integral" of operators

$$\int_{\mathbb{T}^*}^{\oplus} L(k)dk$$

(see, for instance, discussion of this notion in [157]). This is analogous to the situation of the constant coefficient systems of equations, only instead of matrices $L(\xi)$ we have to deal with operators $L(k)$ in infinite-dimensional spaces. The crucial circumstance is that these operators act on functions defined on a compact manifold (a torus), while the original operator L acted in $\mathbb{R}^d$. Thus, under appropriate ellipticity conditions, these operators have compact resolvents, and hence discrete spectra. Then we can define again the *band functions (dispersion relations)* $\lambda_j(k)$ and obtain a picture analogous to Figure 7.2 with the difference that the number of branches is now infinite. We see that the spectrum is expected to have a band structure, and there is hope of opening spectral gaps.

We will now provide a slightly more detailed discussion of the Floquet transform $\mathcal{U}$ and of its effects on function spaces and differential operators. We will still assume that $\Gamma = \mathbb{Z}^d$, since the case of a general lattice of translations does not at this stage introduce any actual difficulties besides complicating the notations.

Let us introduce an alternative version of the transform $\mathcal{U}$. This version is often useful. We define the transform Φ as follows:

$$\Phi f(x, k) = \sum_{n \in \mathbb{Z}^d} f(x-n) e^{-ik \cdot (x-n)} = e^{-ik \cdot x} \mathcal{U} f(x, k).$$

While the function $\mathcal{U}f(x, k)$ was periodic in k and satisfied the Floquet condition with respect to x, the function $\Phi f(x, k)$ is periodic with respect to x and satisfies a cyclic condition with respect to k:

$$\begin{cases} \Phi f(x+n, k) = \Phi f(x, k), & n \in \Gamma = \mathbb{Z}^d, \\ \Phi f(x, k+\gamma) = e^{-i\gamma \cdot x} \Phi f(x, k), & \gamma \in \Gamma^* = 2\pi\mathbb{Z}^d. \end{cases} \quad (7.19)$$

Now when k changes, the values of $\Phi f(\cdot, k)$ belong to the same space of functions of x on the torus $\mathbb{T} = \mathbb{R}^d / \mathbb{Z}^d$. It is still sufficient, however, to know the values of $\Phi f(x, k)$ for x in the Wigner–Seitz cell W and k in the Brillouin zone B in order to recover the whole function. The transform Φ does not commute with periodic differential operators anymore. A straightforward calculation shows that

$$\Phi(Lf)(x, k) = L(x, D_x + k)\Phi f(x, k) = L(k)\Phi f(\cdot, k). \quad (7.20)$$

So, while we gained a fixed function space, now the differential expression for the operator changes with k.

The main tools needed when one uses Fourier transform are the Plancherel and Paley–Wiener theorems. In the periodic case we need similar statements for the Floquet transforms $\mathcal{U}$ and Φ, since they become crucial in all aspects of spectral theory of periodic operators. Let us first formulate an analogue of the Plancherel theorem. In the theorem below we assume that the natural measures dk on the Brillouin zone B and the dual torus $\mathbb{T}^*$ are normalized.

THEOREM 7.2. *The transforms*

$$\mathcal{U}: L^2(\mathbb{R}^d) \to L^2(\mathbb{T}^*, L^2(W)), \qquad \Phi: L^2(\mathbb{R}^d) \to L^2(B, L^2(\mathbb{T}))$$

are isometric. Their inverse transforms are

$$\Phi^{-1} v(x) = \int_B e^{ix\cdot k} v(x,k) dk$$

and

$$\mathcal{U}^{-1} w(x) = \int_{\mathbb{T}^*} w(x,k) dk,$$

where the function $v(x,k) \in L^2(B, L^2(\mathbb{T}))$ *is considered a periodic function with respect to* $x \in \mathbb{R}^n$ *and* $w(x,k) \in L^2(\mathbb{T}^*, L^2(W))$ *is extended from* W *to all* $x \in \mathbb{R}^n$ *according to the Floquet condition* (7.18).

This theorem, used constantly in solid-state physics since Bloch [28], was introduced into mathematics for spectral analysis of periodic differential operators by Gelfand [85] and further investigated in [149, 191] (see section XIII.16 of [157] and Chapters 2 and 4 of [123] for discussion and further references). The proof is straightforward if one notices that (7.17) is just a Fourier series with coefficients in the Hilbert space $L^2(W)$ and uses the standard Plancherel's theorem for such series. It is easy to prove an analogue of such a theorem for the Sobolev space $H^s(\mathbb{R}^d)$ instead of $L^2(\mathbb{R}^d)$. Namely, it is transformed by Φ isomorphically onto the space $L^2(B, H^s(\mathbb{T}))$. In terms of the transform $\mathcal{U}$ the situation becomes more technical. Let us define for each $k \in \mathbb{T}^*$ the closed subspace H^s_k of the space $H^s(W)$ consisting of restrictions to W of all functions from $H^s_{loc}(\mathbb{R}^d)$ which satisfy the Floquet condition (7.18). It is easy to conclude (see Theorem 2.2.1 in [124]) that

$$\mathcal{E}^s = \bigcup_{k \in \mathbb{T}^*} H^s_k$$

is a Hilbert vector bundle over $\mathbb{T}^*$. Then one can show that the transform $\mathcal{U}$ maps isomorphically the space $H^s(\mathbb{R}^d)$ onto the Hilbert space $L^2(\mathbb{T}^*, \mathcal{E}^s)$ of L^2-sections over $\mathbb{T}^*$ of the bundle $\mathcal{E}^s$.

Let us now move to the Paley–Wiener-type theorems. By this we mean the theorems that describe the images under the Floquet transform of spaces of sufficiently fast decaying functions on $\mathbb{R}^n$. One can notice that while the classical Paley–Wiener theorem deals with spaces of compactly supported functions, such a theorem, although easily provable, has not been useful so far for the Floquet transform.

Chapter 7. The Mathematics of Photonic Crystals

It is a commonplace that Paley–Wiener theorems require extension into the complex domain of the dual variable. The same is true for the Floquet transform. One can see that both transforms $\mathcal{U}$ and Φ can be defined on compactly supported or sufficiently fast decaying functions also for complex quasi momenta k. The bundles $\mathcal{E}^s$ also extend into the complex domain to analytic infinite-dimensional bundles (see Theorem 2.2.1 in [124]). The reader unfamiliar with the technique of infinite-dimensional bundles can think of a closed subspace of a fixed Hilbert space, where the subspace depends analytically upon the parameter k (for instance, there is a projector onto the subspace, which depends analytically on k). One can now obtain analogues of the Paley–Wiener theorem for several spaces of decaying functions. This is done in the Theorem 2.2.2 of [124]. In order to avoid technicalities, we will loosely describe the corresponding results, referring the reader to [124] for details. For instance, the space of functions that belong to $H^s_{loc}(\mathbb{R}^n)$ and decay exponentially in the H^s sense

$$\|f\|_{H^s(W+n)} \leq C e^{-a|n|} \tag{7.21}$$

goes over to the space of sections of the bundle $\mathcal{E}^s$ that are analytic over a specific neighborhood of the real space $\mathbb{R}^d$ in $\mathbb{C}^d$. Availability of the estimate (7.21) for arbitrary $a > 0$ is equivalent to the fact that the function $\mathcal{U}f$ is entire with respect to $\dot{k}$. If the estimate (7.21) is strengthened to require decay of order higher than 1,

$$\|f\|_{H^s(W+n)} \leq C e^{-a|n|^p}, \qquad p > 1,$$

this is reflected in growth estimates on the corresponding entire function. All these theorems are important for periodic partial differential equations and for the spectral theory in particular, as will be mentioned later (one can also refer to Chapter 4 of [124] and to papers [9, 10, 78, 120, 131, 132] for examples of such applications).

Let us now reflect a little bit on the effect that the Floquet transform has on the operators. As we have mentioned already, the periodic operator $L(x, D)$ in $\mathbb{R}^d$ after the Floquet transform becomes a family (in fact, a polynomial with respect to k) of operators $L(k) = L(x, D + k)$. Here each of the operators $L(k)$ acts on the torus $\mathbb{T}$, which is a compact closed manifold. In particular, if L is elliptic, we are dealing with an analytic (polynomial) operator function $L(k)$ whose values are Fredholm operators in appropriate spaces. This enables one to invoke the rich theory of such operator functions (see, for instance, [193] and Chapter 1 of [124] for its discussion and further references).

Ellipticity (or at least hypoellipticity, for instance, parabolicity) of the operator is crucial. It influences not only the technique, but also the results one might expect (see [124]). Here one can see what kind of difficulties can be expected with the Maxwell operator. As we have already discussed before, the Maxwell operator taken alone is not elliptic. The correct idea is to include it into an elliptic complex (or to extend to a larger elliptic operator, which is essentially the same). Consider the example of the homogeneous Maxwell operator $M = (\nabla^\times)^2$ acting from the cokernel of the gradient into the kernel of divergence. Here arises the problem: after the Floquet transform the operator $M(k)$ will act between the cokernel of grad(k) = ($\nabla + ik$) and the kernel of div(k) = ($\nabla + ik$)$\cdot$, where all operators are acting now on periodic functions. It is easy to check by the Fourier series expansion, however, that these spaces (i.e., cokernel and kernel, respectively) do not depend analytically on k. If

$$F(x) = \sum F_\gamma e^{i\gamma \cdot x}$$

is the Fourier series of a periodic vector field F, then

$$(\nabla - ik) \cdot F = \sum i(\gamma + k) \cdot F_\gamma e^{i\gamma \cdot x}.$$

One can see a degeneration of the kernel at the point $k = 0$. Namely, for $k \neq 0$ the condition $(\nabla + ik) \cdot F = 0$ implies that the vectors $(\gamma + k)$ and F_γ are orthogonal, and so F_γ belongs to the two-dimensional orthogonal complement of $(\gamma + k)$. On the other hand, for $k = 0$ the coefficient F_0 can be arbitrary. This means a non-analytic behavior of Ker(div(k)) at $k = 0$. The same thing is true for the cokernel of grad(k). In technical terms this requires one to work with sections of analytic sheaves instead of sections of analytic vector bundles. Although this is possible (see, for instance, [150], where the main result of [123, 124] was extended to the case of elliptic complexes), the technical complications can sometimes be severe.

7.4 Spectra in Periodic Media

In this section we will focus on the spectral properties of periodic (elliptic) differential operators, including the Maxwell operator.

7.4.1 Band Gap Structure

As we have already explained, the spectra of periodic elliptic differential operators exhibit band gap structure. Let us discuss this a little bit more (see [157, 179, 124] for details and references). If we have a self-adjoint periodic operator $L = L(x, D)$ in $L^2(\mathbb{R}^d)$, the Floquet transform expands it into the direct integral of operators $L(k) = L(x, D + k)$ on the torus $\mathbb{T}$.

One can prove the main spectral statement:

$$\sigma(L) = \bigcup_{k \in B} \sigma(L(k)) \qquad (7.22)$$

(see [91, 54, 149, 157, 179, 124]). Due to ellipticity, the spectrum of each $L(k)$ is discrete. If L is bounded from below, the spectrum of $L(k)$ accumulates only at the positive infinity. Let us denote by $\lambda_n(k)$ the nth eigenvalue of $L(k)$ (counted in increasing order with their multiplicity). This continuous function of $k \in B$ is called a *band function* (or one branch of the *dispersion relations*). We conclude that the spectrum $\sigma(L)$ consists of the closed intervals (called the spectral bands)

$$S_n = [\min_k \lambda_n(k), \max_k \lambda_n(k)],$$

Chapter 7. The Mathematics of Photonic Crystals

where $\min_k \lambda_n(k) \to \infty$ when $n \to \infty$. It is well known that for ordinary differential operators of the second order the bands cannot overlap (although they can touch), which explains why it is a generic situation in one dimension that gaps open in the spectrum between adjacent bands (see [157]). In dimensions 2 and higher the bands can and normally do overlap, which makes opening gaps much harder. It is still conceivable that at some selected locations the bands might not overlap and hence open a gap in the spectrum. What we have just described is called the band gap structure of the spectrum for elliptic (or hypoelliptic) periodic differential operators. This is what triggered hopes for creating photonic crystals as dielectric materials of periodic structure.

It is not difficult to derive the band gap structure of the spectrum of the periodic Maxwell operator. This can be done either by including it into an elliptic complex (and following the line of [150]) or by using an orthogonal extension to an elliptic operator, as was discussed above. It looks like this standard derivation of the band gap structure of the spectrum for the periodic Maxwell operator has never been written down, except the two-dimensional version described in [78]. One usually refers to this as "according to the Floquet theory" (with no references provided).

One can make a simple useful remark about the representation (7.22). Namely, not all quasi momenta k are needed in the right-hand side of (7.22). It is sufficient to use any dense subset S of the Brillouin zone B and then take closure of the union of the corresponding spectra:

$$\sigma(L) = \overline{\bigcup_{k \in S} \sigma(L(k))}. \tag{7.23}$$

There are at least two important choices for the subset S. First, as we mentioned in section 7.3, there are values of the quasi momentum that are "bad" for the Maxwell operator (i.e., at which the cokernel of the gradient and the kernel of the divergence lose analyticity). One can just skip these values and then take the closure of the union of the remaining spectra instead. In some cases (like, for instance, in [78]) this works just fine, while it does not eliminate the problem completely in other situations. This trick can also be used in numerics, when some values of quasi momenta cause trouble. Second, it is often useful and commonly used in solid-state physics to represent the spectrum $\sigma(L)$ as the limit of spectra on finite domains. Consider a cube K in $\mathbb{R}^d$ and stretch it: $K_m = mK$, $m = 1, 2, \ldots$. We can naturally define operators L_m in $L^2(K_m)$ using the differential expression $L(x, D)$ with periodic boundary conditions on K_m. If L is elliptic with sufficiently decent coefficients, there is no ambiguity in such a definition. Then one can show that the spectrum $\sigma(L)$ coincides with the closure

$$\overline{\bigcup_m \sigma(L_m)} = \lim_{m \to \infty} \sigma(L_m). \tag{7.24}$$

This is clearly just a particular case of (7.23) when we use the subset of all quasi momenta with components commensurable with a given number. The important relation (7.24) is often proven for specific operators, although it holds for periodic elliptic operators in general and follows from (7.23).

7.4.2 Fermi and Bloch Varieties

We are now going to define two objects of paramount importance for the theory of periodic elliptic (and hypoelliptic) operators. Although they are often used in solid-state physics, their roles are not always completely appreciated. They are analogues of the set of zeros of the symbol of a constant coefficient operator, which is known to determine many properties of such an operator.

Let $L(x, D)$ be a periodic elliptic operator in $\mathbb{R}^d$. We define its *complex Bloch variety* as follows:

$B(L)$ consists of all points $(k, \lambda) \in \mathbb{C}^{d+1}$ such that the equation $L(k)u = \lambda u$ has a nonzero solution $u(x)$ satisfying (7.18).

The *real Bloch variety* $B_R(L)$ is the intersection of $B(L)$ with the real space $\mathbb{R}^{d+1}$. It is clear that the real Bloch variety of the operator L is just the union of graphs of all band functions $\lambda_j(k)$. In other words, the Bloch variety is the graph of the multivalued dispersion relations for the operator L. In particular, the spectrum of the self-adjoint operator L is equal to the projection of $B_R(L)$ onto the λ-axis.

The level sets of the dispersion relations are also of interest. For a given $\lambda \in \mathbb{C}$ we call the *Fermi surface* of the operator L on the level λ the set $F_\lambda(L)$ consisting of all points $k \in \mathbb{C}^d$ such that the equation $L(k)u = \lambda u$ has a nonzero solution $u(x)$ satisfying (7.18). Analogously to the real Bloch variety, we define

$$F_{R,\lambda}(L) = F_\lambda(L) \cap \mathbb{R}^d.$$

It is immediately clear that for a self-adjoint operator L

$$\lambda \in \sigma(L) \iff F_{R,\lambda}(L) \neq \emptyset.$$

One can imagine that when λ changes, the (complex) Fermi surface moves, and the values of λ for which the surface touches the real space constitute the spectrum of the operator.

The following theorem establishes an important property of the Bloch and Fermi varieties.

THEOREM 7.3 ([124]; see also [120, 123]). *The set $B(L)$ coincides with the set of all zeros of an entire function $f(k, \lambda)$ of a finite order in $\mathbb{C}^{d+1}$. (Here an entire function $f(z)$ in C^n is said to be of the finite order p if it satisfies an estimate $|f(z)| \leq C \exp a |z|^p$.) A similar statement holds for the Fermi surface at any level λ.*

One can refer to Theorem 4.4.2, Corollary 3.1.6, and Theorem 3.1.7 of [124] for exact formulations, including the precise order of the entire function (see also further discussion and references in sections 3.5 and 4.7 of [124]). In particular, one concludes that $B(L)$ is an analytic subset of $\mathbb{C}^{d+1}$ in the sense of several complex variables [94], i.e., that it can be locally (and even globally) described by analytic equations (this particular corollary was probably first proven in [191]).

As is explained in [124], a similar statement holds for matrix operators. One can also show that it holds for the Maxwell operator as well (analyticity of $B(L)$ for this case, although without estimates, can be also extracted from [150]).

There is a natural action of the dual lattice Γ^* on the Bloch and Fermi varieties by shifts: $(k,\lambda) \to (k+\gamma,\lambda)$ and $k \to k+\gamma$ correspondingly, where $\gamma \in \Gamma^*$. Considering the case of a constant coefficient operator $L(D)$ one easily finds that

$$B(L) = \{(k,\lambda)|\, L(k+\gamma) - \lambda = 0 \text{ for some } \gamma \in \Gamma^*\}$$

and

$$F_\lambda(L) = \{k|\, L(k+\gamma) - \lambda = 0 \text{ for some } \gamma \in \Gamma^*\}.$$

In other words, one needs to find the set of zeros of the symbol $L(k) - \lambda$ and then take its orbit with respect to Γ^*.

An analytic set X is said to be *reducible* if it can be represented as the union of two smaller analytic subsets: $X = X_1 \cup X_2$. We remind the reader not familiar with this concept that if the function $f(z)$ whose set of zeros is X allows a nontrivial factorization $f = f_1 f_2$, then the sets of zeros of factors reduce X. An analytic set that is not reducible is called *irreducible* [94]. The example of a constant coefficient operator in the previous paragraph shows that one should discuss irreducibility of the Bloch and Fermi varieties only modulo the dual lattice. Irreducibility plays an important role in many problems of the spectral theory of periodic operators: in inverse spectral problems [90, 120], behavior with respect to impurities [131, 132], and others. The irreducibility of $B(L)/\Gamma^*$ was proven for the one-dimensional periodic Schrödinger operator by Kohn [121] and conjectured for the general periodic Schrödinger operator in [6, 120, 145]. It was proven in [120] in two dimensions using an intricate algebrogeometric approach. It is conjectured that $F_\lambda(L)/\Gamma^*$ is also irreducible in this case.

CONJECTURE 7.4. *The varieties $B(L)/\Gamma^*$ and $F_\lambda(L)/\Gamma^*$ are irreducible for any periodic second-order elliptic operator L, including the Maxwell operator.*

This problem looks even harder for the Fermi surface than for the Bloch variety. It was studied in detail for the discrete Schrödinger operator in the book [90] and for the discrete Maxwell operator in [9]. In both cases results on irreducibility of the Fermi surface $F_\lambda(L)/\Gamma^*$ were obtained by methods of algebraic geometry. It was shown in [10] that $F_\lambda(L)/\Gamma^*$ is irreducible for the Schrödinger operator in two dimensions with a separable periodic potential $v_1(x_1) + v_2(x_2)$ and in three dimensions for a separable periodic potential $v_1(x_1) + v_2(x_2, x_3)$.

Another consideration of interest is the following. When λ approaches the spectrum, the Fermi surface approaches the real space, and when λ enters the spectrum, $F_{R,\lambda}$ is not empty. It is natural to expect that when λ goes into the interior of a spectral band, the Fermi surface becomes sufficiently "massive." In fact, one can show that if λ belongs to the interior of a spectral band, then the Fermi surface $F_{R,\lambda}$ as a real analytic set has dimension at least $d - 1$.

It is also natural to assume that one should be able to estimate the distance from the point λ to the spectrum by the distance between the Fermi surface F_λ and the real space. Here is how this argument can go. First, if the Fermi surface is at a certain distance from the real space, this means that the equation $Lu = \lambda u$ has a Floquet–Bloch solution $u = e^{ik \cdot x} v(x)$ with a periodic $v(x)$ and with an estimate on $|\mathrm{Im}\, k|$. In other words, we have an exponential estimate on $u(x)$. Then an argument of the type provided in the section 54 of [91] for the Schrödinger operator should lead to an estimate on the distance $d(\lambda, \sigma(L))$. It would be very interesting to extend this type of an argument to more general operators than Schrödinger (in particular, to the Maxwell operator) and to improve the estimates of [91] to the extent that one can deduce exponential localization estimates obtained in [42] (see section 7.6.3).

Concluding this section, I want to emphasize that analytic properties of the Bloch and Fermi varieties are very important for understanding spectra of corresponding operators: analyticity of these sets imply absolute continuity of the spectrum (section 7.4.3), irreducibility is crucial for inverse spectral problems [120] and for the absence of embedded impurity eigenvalues (section 7.6.4), and the way the Fermi surface approaches the real space is related to embedded eigenvalues (section 7.6.4) and to the exponential localization of impurity modes (section 7.6.3).

7.4.3 Absolute Continuity

As we have already discussed, the spectrum of any periodic elliptic or hypoelliptic operator L has a band gap structure. The natural question is about the type of spectrum that can arise (e.g., absolutely continuous, singular continuous, point). The general expectation is that in principle the spectrum must be absolutely continuous; i.e., no eigenvalues or singular continuous spectrum can arise. In fact, this is not true in general, since one can show existence of periodic elliptic operators of the fourth order that do have point spectrum (see [124, pp. 135–136]). However, there is very little doubt that absolute continuity holds for any second-order periodic elliptic operator, including Maxwell. There is one simple thing one can prove for a periodic elliptic operator of any order: the singular continuous spectrum is empty. The reason is that (as was understood since [28, 85]) the Floquet–Bloch transform represents the operator L as the infinite sum of operators of multiplication by the band functions $\lambda_n(k)$. Another important ingredient is that the band functions are piecewise analytic. Then it is not hard to conclude that each of these multiplication operators either is absolutely continuous or has an eigenvalue. In the latter case, the corresponding band function must be constant on a positive measure set of quasi momenta k and hence constant. This kind of consideration goes back to [184] and is presented in several places, for instance, in [157, 174, 124].

The task of proving absolute continuity of the spectrum now reduces to showing absence of eigenvalues. Although it has been unanimously believed by physicists for a long time, proving this statement happens to be a hard problem. For the Schrödinger case in three dimensions it was proven in the celebrated paper [184] by Thomas and then extended to more general potentials in [157] (see also [14]). Attempts to extend this theorem to more general periodic elliptic operators had failed for about 20 years, except the results of [50] for the Dirac operator and [98] for the magnetic Schrödinger operator with small magnetic potential. Then an avalanche of papers was triggered in 1997 by the paper [24], where absolute continuity was proven in two dimensions for the Schrödinger operator with both magnetic and electric potentials. The same year this result was extended in [180] to any dimension, which required a new technique. The proof of [180] was simplified in [128, 129]. The recent paper [142] contained the absolute continuity result for the two-dimensional Schrödinger operator with periodic metric. The paper [170] contains improved conditions on the potential. One can find more references and a nice survey of known results in [26].

We will now indicate the main thrust of Thomas's proof [184] and of all its extensions. The major step is to use analytic continuation into the domain of complex quasi momenta. The following theorem holds.

THEOREM 7.5 (Theorems 4.1.5 and 4.1.6 in [124]). *Let L be a periodic elliptic operator. Then the following statements are equivalent:*

(a) *The point λ is an eigenvalue of L in $L^2(\mathbb{R}^d)$, i.e., there is a nonzero L^2-solution of the equation $Lu = \lambda u$ in $\mathbb{R}^d$;*

(b) *The Fermi surface F_λ coincides with the whole space $\mathbb{C}^d$;*

(c) *There exists a nonzero solution of the equation $Lu = \lambda u$ in $\mathbb{R}^d$ that decays faster than any exponent:*

$$|u(x)| \leq C \exp(-a|x|) \quad \text{for all } a > 0;$$

(d) *There exists a nonzero solution of the equation $Lu = \lambda u$ in $\mathbb{R}^d$ that decays superexponentially:*

$$|u(x)| \leq C \exp(-|x|^{1+\alpha}) \quad \text{for some } \alpha > 0.$$

In fact, statements (c) and (d) are not needed for the standard proof of absolute continuity, but they are interesting on their own. The exact technical conditions on the operator can be found in [124]. Let us concentrate on the equivalence of (a) and (b), which can be easily explained. As we have already discussed, the operator of multiplication by $\lambda_n(k)$ has an eigenvalue λ if and only if the level set $\lambda_n(k) = \lambda$ has a positive measure. In terms of the Fermi surface $F_{R,\lambda}(L)$ this means that $F_{R,\lambda}(L)$ has a positive measure in $\mathbb{R}^d$. However, as we know already, it is an analytic set. The uniqueness theorems for analytic functions immediately imply that this can happen only when $F_\lambda(L) = \mathbb{C}^d$, and thus the equivalence of (a) and (b) is proven.

Let us interpret this result in a different way. If for each λ we can prove that $F_\lambda \neq \mathbb{C}^d$, then we conclude that there are no eigenvalues and hence that the spectrum is absolutely continuous. Recalling the definition of the Fermi surface, one obtains the following key corollary.

COROLLARY 7.6. *If for any λ there exists a quasi momentum $k \in \mathbb{C}^d$ such that the equation $L(k)u = \lambda u$ has no nontrivial solutions on the torus $\mathbb{T}$, then the spectrum of the operator L is absolutely continuous.*

Now one proves absolute continuity of the spectrum of the Schrödinger operator $-\Delta + v(x)$ with a periodic potential v if one can show the absence of periodic solutions of the equation $(D+k)^2 u + vu = \lambda u$ for an appropriately chosen (depending on λ) quasi momentum k. It is not hard to choose a quasi momentum with a large imaginary part in such a way that the $(D+k)^2$ term dominates the zero-order terms, and hence no nontrivial solutions are allowed (see, for instance, [184, 157, 124] for details). Although the idea stays the same, treatment of more general operators becomes much more complex when one wants to show that $F_\lambda \neq \mathbb{C}^d$.

At this moment we want to address the case of the Maxwell operator, which is of main interest here. Unfortunately, absolute continuity of its spectrum has not been proven yet even for the case of smooth material tensors and for the isotropic medium.[3] Considerations of [128, 129] show that the technology developed in [180] leads for the Maxwell operator with smoothly varying parameters to a model problem that involves a simple covariant Cauchy–Riemann derivative operator on the torus.

For the case of a two-dimensional medium and for the waves propagating in the periodicity plane the result on absolute continuity can be extracted from the known results about operators of the Schrödinger type. Let us recall that, as was discussed in section 7.2.5, in this case the spectral problem for the Maxwell operator splits into the direct sum of two scalar problems:

$$-\Delta u = \lambda \varepsilon(x) u$$

and

$$-\nabla \cdot \frac{1}{\varepsilon} \nabla u = \lambda u.$$

Now the following theorem resolves the problem of absolute continuity in two dimensions (although its first statement holds in any dimension).

THEOREM 7.7. (a) *Under the conditions[4] on the periodic dielectric function $\varepsilon(x)$ that imply the absolute continuity of the spectrum of the Schrödinger operator $(-\Delta - \varepsilon)$, the spectrum of the problem*

$$-\Delta u = \lambda \varepsilon(x) u$$

is absolutely continuous in $\mathbb{R}^d$.

(b) *Let the dielectric tensor $\varepsilon(x)$ be smooth and periodic (not necessarily scalar); then the spectrum of the operator $-\nabla \cdot \varepsilon^{-1} \nabla$ in $L^2(\mathbb{R}^2)$ is absolutely continuous.*

Proof. (a) If $-\Delta u = \lambda \varepsilon(x) u$ has a nonzero L^2-solution, then the Schrödinger operator $(-\Delta - \lambda \varepsilon)$ has a zero eigenvalue, which is impossible according to the known results. (b) This is essentially the result of [142], modulo an application of the transform (7.16).

The equivalence of (a) and (b) in the Theorem 7.5 for Schrödinger operators is essentially due to Thomas [184]. We now want to call the reader's attention to the statements (c) and (d) of this theorem. The proof requires a technique from the several complex variables theory [124]. In principle, these statements suggest a different way of proving absolute continuity of spectra of periodic elliptic operators of the second order. Namely, for such operators existence of a superexponentially

---

[3] When the author was finishing the last revision of this text, the preprint [143] appeared, where the absolute continuity result for the isotropic periodic Maxwell operator was proven.

[4] The best currently known conditions were established in [26, 170]. In dimension $d = 2$ it is that $\varepsilon \in L_{loc}^r(\mathbb{R}^2)$ for some $r > 1$ (or equivalently $\varepsilon \in L^r(W)$, where W is the Wigner–Seitz cell). For $d = 3$ and 4 one requires $\varepsilon \in L_{d/2,\infty}^0(W)$. This means that the function $\rho_\varepsilon(t) = mes\,\{x \in W | |\varepsilon(x)| \geq t\}$ satisfies $\rho_\varepsilon(t) = o(t^{-d/2})$. In dimensions $d \geq 5$ the (nonoptimal) condition is $\varepsilon \in L^{d/2}(W)$. Shen has recently announced the optimal condition for any dimension [171].

decaying solution like in (d) should be an impossible pathology that would violate uniqueness of continuation at infinity (see, for instance, [84, 136, 137, 140], and references therein). If one could prove nonexistence of such solutions, the immediate consequence would be absolute continuity of the spectrum. However, partial differential equation results that guarantee absence of such solutions are probably not currently available for periodic operators.

7.4.4 Spectral Gaps

In this section we will consider the nature, existence, and number of gaps in the spectrum of a periodic operator. This is probably the central issue of the whole photonic crystals theory. Existence of gaps is a prerequisite to most applications of photonic crystals.

Let us discuss briefly one mechanism of opening gaps that exists in the case of a periodic Schrödinger operator $-\Delta + v(x)$. Imagine that we start with a constant potential. Then the spectrum of the operator is continuous and covers a semiaxis $[\alpha, \infty)$. Let us add a localized potential well. This will create a few eigenvalues below the continuous spectrum. The corresponding eigenfunctions (bound states) are localized in a vicinity of the well. Let us now repeat the well periodically with a sufficiently large period. The former bound states can now tunnel to the other wells and hence will not be localized anymore. This will lead to spreading the eigenvalues into narrow bands, which correspondingly will be separated from the rest of the spectrum by gaps. So, the major factor in opening gaps is that by adding a potential one can change the bottom of the spectrum. In the case of photonic crystals, however, this is exactly what is missing. The operators involved in both two- and three-dimensional photonic cases are multiplicative rather than additive perturbations of the corresponding free operators:

$$\frac{1}{\sqrt{\varepsilon}} \left(\nabla^{\times} \right)^2 \frac{1}{\sqrt{\varepsilon}}$$

in three dimensions and

$$-\frac{1}{\sqrt{\varepsilon}} \Delta \frac{1}{\sqrt{\varepsilon}}$$

and

$$-\nabla \cdot \frac{1}{\varepsilon} \nabla$$

in two dimensions (TM and TE polarizations). The outcome is that in all these cases the spectrum starts at zero. Indeed, consider, for instance, the TM case. If ϕ_n is an approximate eigenfunction for $L = -\Delta$ at zero, i.e., if $\|\phi_n\|_{L^2} > c > 0$ and $\|L\phi_n\| \to 0$, then the functions $\psi_n = \sqrt{\varepsilon}\phi_n$ are approximate eigenfunctions for $\left(-\varepsilon^{-1/2} \Delta \varepsilon^{-1/2} \right)$. This shows that the mechanism of opening gaps in the PBG case is different. In particular, while the gaps for the Schrödinger operator can be opened at the bottom of the spectrum, in the photonic case they normally open in the medium-frequency range (see, for instance, [106]).

It is well known that in one dimension (i.e., for the Hill operator) the generic situation is that infinitely many gaps are open (see, for instance, [157]). On the other hand, it is commonly believed that the number of gaps one can open in a periodic medium in dimension higher than 1 is finite. In the case of a periodic Schrödinger operator, this constitutes the Bethe–Sommerfeld conjecture [15], first proven by M. Skriganov (see [52, 96], [110]–[116], [124, 141], [175]–[179], [181, 187] for the discussion of this problem and several different approaches to its proof). The proofs are by no means simple and often employ results from number theory. The following analogue of the Bethe–Sommerfeld conjecture almost certainly holds true.

CONJECTURE 7.8. *In dimensions 2 and higher the spectrum of any photonic crystal (or of its acoustic analogue) has at most a finite number of gaps.*

The main idea of the proof in the case of Schrödinger operators is that the overlap of spectral bands of the free Hamiltonian for sufficiently high energies is so strong that addition of a periodic potential cannot open gaps at these energies. However, in the photonic case one deals with a multiplicative rather than additive perturbation of the free Hamiltonian, which will probably lead to the necessity of involving a different approach to the proof. On the other hand, it looks like it is harder to open gaps in the photonic case, which raises a hope that the proof of finiteness of number of gaps could be simpler than in the solid-state situation.

Let us now address the problem of existence of spectral gaps for the periodic Maxwell operator. While there is a lot of numerical and experimental evidence of it (see the surveys [33, 101, 106, 108, 134, 156, 182, 189]), analytic results on existence of gaps are scarce. We are not aware of any such theorems for the full-vector three-dimensional case, which is the main interest in the PBG theory (the result announced in [75] is erroneous). There are, however, a few cases when existence of gaps was proven for the scalar problems analogous to (7.14) and (7.15) in two and higher dimensions. The authors of [49] studied the Laplace–Beltrami operator

$$L_g f = -\frac{1}{\sqrt{|g|}} \sum_{i,j} \partial_i (g^{ij} \sqrt{|g|} \partial_j f)$$

in $\mathbb{R}^d$ with a conformally flat periodic metric $g_{ij} = a(x)\delta_{ij}$. In the one-dimensional case it reduces to

$$-\left(\frac{1}{\sqrt{a}} \frac{d}{dx}\right)^2,$$

which in turn can be reduced by a simple change of variables to $-d^2/dy^2$. This shows that when $d = 1$ the spectrum of L_g coincides with the positive half-axis and hence has no gaps. This is in contrast with the case of periodic Schrödinger operators, since such an operator in one dimension (the Hill operator) generically possesses infinitely many spectral gaps. Experience with Schrödinger operators also shows that when dimension increases, it becomes increasingly difficult to create spectral gaps. Surprisingly enough, the situation with the periodic Laplace–Beltrami operators is different: while there are no gaps in the spectrum of such an operator in one dimension, it was shown in [49] that in any higher dimension there exist periodic metrics such that the corresponding Laplace–Beltrami operators have gaps in the spectrum. The idea of the proof is that using a procedure similar to the one described in section 7.2.5, one can reduce the operator to a Schrödinger form. If one succeeds in reducing to a Schrödinger operator with a separable potential, then one can use the well-developed theory of spectra of the Hill operators to check existence of gaps. This study was continued in the paper [92], where it was shown that in two dimensions one can achieve any finite number of gaps in the spectrum of a periodic Laplace–Beltrami operator. It is not known whether the number of gaps must always be finite and whether it is not limited in dimensions higher than 2. It is interesting to note that the method used in [92] to show that the number of gaps is not limited in two dimensions is essentially the same one that was applied in [76]–[78] for showing existence of gaps in spectra of some two-dimensional photonic crystals (see description of these results below).

There is not much hope for analytic (rather than purely numerical) prediction of spectral gaps in a general situation. However, when some parameters of the problem approach extremal values (for instance, the dielectric contrast becomes very high, the dielectric regions become very narrow, etc.), one can try to understand the asymptotic situation and therefore to predict the behavior of the spectrum. This is the idea that was employed in [76]–[81] and [8, 100, 126, 127, 172] for studying spectra of the problems (7.14) and (7.15). Due to their specific flavor, we will address these results in the next section.

Suppose that $[a, b]$ is a gap in the spectrum of one of the periodic problems we discuss. This means that a is the maximal value of a band function $\lambda_j(k)$. Analogously, b is the minimal value of another band function. In many cases (some of which will be mentioned later) it is important to know in which way these extrema are attained: are they isolated, nondegenerate, etc.? Unfortunately, there is almost no information about this, except the recent result of [119] on generic simplicity of the endpoints of bands. Probably the only thing known for some periodic operators is the behavior of the band functions at the bottom of the spectrum (which is the upper end of the infinite gap $(-\infty, a]$). The result obtained in [117] concerns a periodic Schrödinger operator $H = -\Delta + V(x)$ in $\mathbb{R}^d$. Let us denote as before

$$H(k) = (D + k)^2 + V(x).$$

Then the band functions $\lambda_j(k)$ provide the eigenvalues of $H(k)$, where $\lambda_1(k)$ is the lowest one.

THEOREM 7.9 (Theorem 2.1 in [117]). *Let ψ_0 be the positive periodic solution of $H\psi_0 = \lambda_1(0)\psi_0$. Then*

$$(\min \psi_0 / \max \psi_0)^2 k^2 \leq \lambda_1(k) - \lambda_1(0) \leq k^2.$$

This theorem implies that the bottom of the spectrum is attained only at the zero quasi momentum $k = 0$, and around that point the lowest band function behaves as

$$\lambda_1(k) = \lambda_1(0) + \gamma(k) + O(k^4),$$

where $\gamma(k)$ is a positive definite quadratic form of k.

The analogous result was recently obtained in [25] for periodic Pauli operators

$$P_\pm = (D - A)^2 \pm B$$

in two dimensions, where $A = (A_1, A_2)$ is a periodic magnetic potential and $B = \partial_1 A_2 - \partial_2 A_1$ is the corresponding magnetic field. Besides, it was shown that the quadratic form $\gamma(k)$ has the form αk^2 with an explicit formula for the coefficient α.

It is interesting to mention that such band edge behavior is closely related to Liouville-type theorems on the structure and dimension of the spaces of polynomially growing solutions of periodic elliptic equations [130].

A similar result about the way the bottom of the spectrum is attained has been obtained recently by Birman and Suslina [21] for the full-vector Maxwell operator in three dimensions (in which case the statement applies to the two first band functions). This, in particular, provides a rigorous justification of the known linear behavior of the band functions $\omega(k)$ at zero frequency (recall that the eigenvalues are related to the frequencies as $\lambda = (\omega/c)^2$). From the physical point of view the situation is rather clear: long waves do not notice the periodic structure of the medium and see it as a homogeneous one. Clearly, some kind of homogenization technique (see [13, 104]) is required in order to find the slope of the dispersion relation close to zero frequency. This was done for several cases in physics papers (see, for instance, [48, 95, 122]) although it looks like a rigorous mathematical analysis is still due.

Any results for the higher gaps of the kind that we described for the bottom of the spectrum would be of great importance. It is very common to see in papers devoted to impurity spectra and localization (see section 7.6.2 below) conditions of the following kind. Let $[a, b]$ be a gap in the spectrum. Then a is the maximum of a band function $\lambda_j(k)$. It is assumed that this function attains its maximum at a single point (or a finite set of points) in the Brillouin zone and that this maximum is nondegenerate (i.e.,

$$\lambda_j(k) = \lambda_j(k_0) + \gamma(k - k_0) + O(|k - k_0|^3),$$

where γ is a positive definite quadratic form). However, it is apparently not known how to verify such a condition, or even how common it is. It is believed that this condition holds generically. The only result in this direction known to the author is that of [119], where the simplicity of the band edge was shown in the generic situation.

7.5 Asymptotic Analysis of High-Contrast PBG Materials

It has been recognized (see [106, 189]) that high dielectric contrast of a photonic crystal favors spectral gaps. Under some circumstances it was also noticed that gaps could benefit from narrowness of optically dense dielectric "walls" separating the air bubbles. It is natural to try to understand what happens in the asymptotic limits when the contrast goes to infinity and the filling fraction of the dielectric (or the air)

Chapter 7. The Mathematics of Photonic Crystals

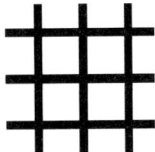

Figure 7.3: Two-dimensional square PBG structure. The dark strips of width δ represent dielectric with $\varepsilon > 1$. The white areas are filled with air ($\varepsilon = 1$).

portion of the medium goes to zero. In this section we will address spectral results known for the PBG materials in such asymptotic situations. One should notice, however, that neither very high contrasts nor very low dielectric filling fractions are currently achievable technologically (for instance, 12 is considered to be a high value for the dielectric contrast). The asymptotic study still makes sense for several reasons. First, it might reveal spectral effects which are hard to recognize otherwise. Second, since often the asymptotic problems are much simpler to study numerically and analytically, they might provide quick ways to estimate the situation. Third, information obtained for the asymptotic models can suggest better algorithms for numerics for the full problem. In particular, one can try to use the spectra and eigenmodes computed for the asymptotic models as seeds for iterative methods for the full problem and/or for creating suitable preconditioners for such methods. One also discovers that asymptotic results can sometimes provide unexpectedly good approximation in the cases when neither the contrast is very high, nor the structure is very thin [8]. One can hope that with further advances in technology the values of parameters closer to the asymptotic limits might become one day technologically feasible. Finally, in the acoustic situation, which is also of interest, one can already achieve such high contrasts. This section is devoted to discussion of the known asymptotic results about PBG materials.

7.5.1 Square Geometry

Probably the first successful asymptotic study of the PBG materials was undertaken in [76]–[78] and in a less detailed form in [172].[5] These papers addressed the square geometry of a two-dimensional PBG medium (Figure 7.3).

The medium has period 1 in both x- and y-directions. The dark areas have thickness $\delta < 1$ and are filled with a dielectric with the dielectric constant $\varepsilon > 1$, while the light areas are filled with air ($\varepsilon = 1$). The dielectric function $\varepsilon(x)$ then takes values ε and 1 in the dielectric and air regions correspondingly. The scaling properties of the Maxwell equations (see section 7.2.4) guarantee that our choice of the period and of the dielectric constant of the "air" regions does not restrict generality of the consideration. The square structure was chosen for its simplicity with the hope that one could understand it and then move on to more complex geometries.

---

[5] We use the word "successful" here since the three-dimensional result announced in [75] was erroneous.

We remind the reader that for the in-plane harmonic waves the Maxwell system reduces to the following two scalar spectral problems (7.15) and (7.14):

$$-\nabla \cdot \frac{1}{\varepsilon(x)} \nabla u = \lambda u$$

(TE polarization) and

$$-\Delta u = \lambda \varepsilon(x) u$$

(TM polarization), where $\lambda = (\omega/c)^2$. The papers [76]–[78] are devoted to the study of these two spectral problems for the square two-dimensional geometry described above in the asymptotic limit when $\varepsilon\delta \to \infty$ and $\varepsilon\delta^2 \to 0$. The TE polarization happens to be the simplest one, and its asymptotic spectral behavior is described by the following result.

THEOREM 7.10 (see [76]). *Let N be an arbitrary positive number and*

$$S_1 = \left\{ \pi^2(n_1^2 + n_2^2) \mid \mathbf{n} = (n_1, n_2) \in \mathbb{Z}^2 \right\}.$$

Denote by σ_{TE} the spectrum of the problem (7.15) for the square geometry described above. Then the Hausdorff distance between $S_1 \cap [0, N]$ and $\sigma_{TE} \cap [0, N]$ tends to zero when $\varepsilon\delta \to \infty$ and $\varepsilon\delta^2 \to 0$. Moreover,

$$d\left(S_1 \cap [0, N], \sigma_{TE} \cap [0, N]\right) \leq C_N \max\left\{ (\varepsilon\delta)^{-1}, \varepsilon\delta^2 \right\},$$

where d denotes the Hausdorff distance.

This theorem says that the spectrum of the TE modes for small values of $(\varepsilon\delta)^{-1}$ and $\varepsilon\delta^2$ concentrates in a small vicinity of the discrete set S_1, and hence large gaps at exactly known locations open up. The reader has probably noticed that the set S_1 to which the spectrum σ_{TE} converges is just the spectrum of the Neumann Laplacian on the unit square (which is the Wigner–Seitz cell of the considered geometry). A more precise description of this result can be found in [76]. An additional observation made in [76] was that the Floquet–Bloch eigenmodes have most of their energy concentrated in the air region.

We would also like to mention that the same result holds for the problem (7.15) for the cubic geometry in three dimensions [76], where one can think of (7.15) as describing acoustic rather than electromagnetic waves.

This study was finished in [78], where the asymptotic behavior of the TM modes (7.14) was investigated. We will present here the main result of [78], omitting some details.

Consider the spectral problem (7.14) for the square structure in two dimensions. Denote by S_2 the following set:

$$S_2 = \left\{ \pi^2(n_1^2 + n_2^2) \mid \mathbf{n} = (n_1, n_2) \in \mathbb{Z}^2 \setminus \{(0,0)\} \right\}.$$

It is clear that S_2 is the spectrum of the Dirichlet Laplacian on the unit square.

THEOREM 7.11 (see [78]). *The spectrum σ_{TM} of the problem (7.14) for the square geometry described above splits into two parts: $\sigma_{TM} = \sigma_1 \cup \sigma_2$.*

Chapter 7. The Mathematics of Photonic Crystals

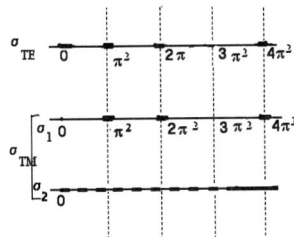

Figure 7.4: The three asymptotic spectra arising in the high-contrast square PBG structure.

If $w = (\varepsilon\delta)^{-1} \to 0$ and $\varepsilon\delta^{4/3} \to 0$, then the following spectral asymptotics hold: Let N be an arbitrary positive number.

(a) The Hausdorff distance between $S_2 \cap [0, N]$ and $\sigma_1 \cap [0, N]$ tends to zero. Moreover,

$$d(S_2 \cap [0, N], \sigma_1 \cap [0, N]) \leq C_N (\varepsilon\delta)^{-1}.$$

(b) There exists a set of disjoint segments

$$\mathcal{D} = \bigcup_{n \geq 0} [D_n^-, D_n^+]$$

not depending on ε and δ such that

$$D_0^- = 0, \quad D_0^+ = 4, \quad D_{n+1}^- > D_n^+, \quad D_n^- \sim 2\pi n, \quad D_n^- \sim 2\pi n + \pi$$

when $n \to \infty$. The spectrum σ_2 allows the representation

$$\sigma_2 \cap [0, N] = \left\{ \bigcup_{n \geq 0} [w_n^- D_n^-, w_n^+ D_n^+] \right\} \cap [0, N],$$

where $w_n^\pm \sim w = (\varepsilon\delta)^{-1}$.

One can find a more precise formulation in [78]. This theorem shows that the two subspectra σ_1 and σ_2 behave differently in our asymptotic limit. The subspectrum σ_1 behaves essentially like σ_{TE}, except for the absence of the band at zero. The bands shrink to the spectrum of the Dirichlet Laplacian on the unit square, therefore becoming almost discrete and opening large gaps at exactly described locations. Another similarity with σ_{TE} is that the eigenmodes are also the air modes, which have most of the energy concentrated in the air bubbles. A completely different behavior is observed in the second subspectrum σ_2. Namely, it splits into narrow bands separated by narrow gaps, both of the asymptotic size $w = (\varepsilon\delta)^{-1}$. Besides, the Floquet–Bloch eigenmodes behave differently: they concentrate in the dielectric regions, quickly dying out in the air. One can attribute this effect to the total internal reflection [103]; i.e., the narrow dielectric regions behave as a waveguide. Figure 7.4 represents these three spectra schematically.

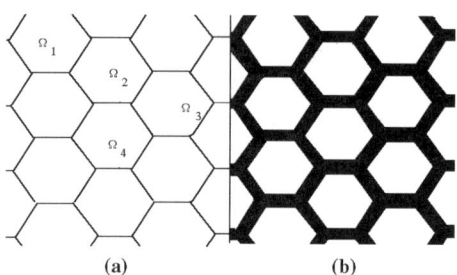

Figure 7.5: (a) The graph Σ and its faces Ω_j. (b) The dielectric PBG material corresponding to the graph Σ. The dark areas of width δ represent dielectric with dielectric constant $\varepsilon > 1$. The white areas are the air bubbles with $\varepsilon = 1$.

One of the by-products of this study is the following statement.

THEOREM 7.12 (see [78]). *For the square geometry of the two-dimensional material, for any given integers N and M the number of gaps in $(\sigma_{TE} \cup \sigma_{TM}) \cap [0, N]$ is at least M for sufficiently small values of $\varepsilon\delta^{4/3}$ and $(\varepsilon\delta)^{-1}$. The spectral bands in $[0, N]$ are of asymptotic size $(\varepsilon\delta)^{-1}$ and are separated by gaps of the same asymptotic size.*

This theorem proves in particular that it is possible to open spectral gaps in PBG materials. It also shows that the TM modes responsible for the subspectrum σ_2 present the main obstacle for the gaps opening, since all other TM and TE waves tend to create an almost discrete spectrum with large gaps.

The proofs of the quoted theorems are rather technical and rely on availability of an exactly solvable model with separable variables in a vicinity of the spectral problem of interest. This approach restricts the consideration to the square case. On the other hand, both the result about splitting the spectrum into subspectra with different asymptotics and the understanding of behavior of the corresponding eigenmodes are of general importance. They will be exploited in our further considerations.

7.5.2 General Two-Dimensional Geometry

The results of the previous section raise several natural questions. The main ones are about the possibility of carrying over a similar analysis for nonsquare geometries, which do not allow separation of variables, and the explanation of the origin of the spectrum σ_2. To some extent, these questions were answered in [79, 80]. The proofs presented in [80] are much simpler than the ones in [76]–[78]. A wide range of PBG geometries is covered. On the other hand, the price paid was a somewhat weaker nature of the results.

Consider a periodic graph Σ on the plane that divides it into compact faces Ω_j. Imagine that all its edges are fattened to the width δ (the dark areas in Figure 7.5) and filled with a dielectric with the dielectric constant $\varepsilon > 1$. The rest of the plane (the white faces Ω_j) is filled with air (Figure 7.5).

We will consider now the asymptotic behavior of the spectrum of TM modes when $\delta \to 0$ and $(\varepsilon\delta)^{-1} \to W < \infty$. We address the TM modes since in the asymptotic limit they are the "worst" modes as far as gaps are concerned (see the previous section). The case of the TE modes will be discussed in [81]. One can notice that in the previous theorems we assumed that $W = 0$. We now allow nonzero (albeit finite) limits of $(\varepsilon\delta)^{-1}$. This is a much more realistic assumption, at least at the current level of technology, since the technologically feasible values of $(\varepsilon\delta)^{-1}$ are of order 1.

Theorem 7.11 shows that the sizes of bands and gaps of the "worst" spectrum σ_2 are of order $(\varepsilon\delta)^{-1}$. It is natural, then, before trying to understand this spectrum, to zoom in on it by introducing a rescaled spectral parameter $D = (\varepsilon\delta)\lambda$. Then the spectral problem (7.14) becomes

$$-\Delta u = (\varepsilon\delta)^{-1} D\varepsilon(x)u. \tag{7.25}$$

THEOREM 7.13 ([80]; see also [79]). *For any positive N the part of the spectrum σ (in terms of the parameter D) of the problem (7.25) that belongs to $[0, N]$ converges to the corresponding part of the spectrum of the following problem:*

$$-\Delta u = D(\delta_\Sigma + W)u. \tag{7.26}$$

Here δ_Σ is the delta function supported by the graph Σ; i.e., for any compactly supported smooth function $\phi(x)$

$$\langle \delta_\Sigma, \phi \rangle = \int_\Sigma \phi(x) dx.$$

There are several comments on this theorem:

(1) All the details, exact definitions of the operators, etc., can be found in [80].

(2) The constant $W = \lim(\varepsilon\delta)^{-1}$ plays the role of a coupling constant. We saw that when $W = 0$ (i.e., in the situation considered in the previous section) the air and dielectric modes decouple.

(3) This theorem allows nonzero values of W, which is much more realistic under the current technological conditions.

(4) Although the statement of the theorem looks similar in spirit to the ones of the previous section, it is in fact weaker for $W = 0$. Indeed, if in the previous section we stated results of convergence of any finite part of the spectrum in terms of the spectral parameter λ, this is now done in terms of the rescaled parameter $D = (\varepsilon\delta)\lambda$. If $D \in [0, N]$, then $\lambda \in [0, (\varepsilon\delta)^{-1}N]$. This shows that when $(\varepsilon\delta) \to \infty$ (i.e., when $W = 0$), we are zooming in on an ever smaller segment of the λ-axis. It is desirable to extend this result to any finite part of the spectrum, in the spirit of Theorem 7.11.

When $W = 0$, the problem (7.26) reduces to

$$-\Delta u = D\delta_\Sigma u. \tag{7.27}$$

In this case the natural domain for consideration of this spectral problem is the graph Σ itself. In order to understand this we need to introduce the notion of the *Dirichlet-to-Neumann (D–N) operator* on the graph Σ. Take a function $\phi(x)$ defined along the edges of the graph Σ. We will further assume that $\phi \in L^2(\Sigma)$, but so far one can think of a sufficiently smooth compactly supported function on the plane restricted to the graph Σ. Take any of the compact faces Ω_j and solve the following Dirichlet problem:

$$\begin{cases} -\Delta u_j(x) = 0, & x \in \Omega_j, \\ u_j|_{\partial \Omega_j} = \phi. \end{cases}$$

Then on each face Ω_j we obtain a harmonic function u_j. Due to the construction, these functions agree across the graph's edges, while their normal derivatives disagree. Let us now define a function ψ on the graph as the sum of all exterior normal derivatives of all the functions u_j:

$$\psi = \sum_j \frac{\partial u_j}{\partial \mathbf{n}_j},$$

where $\mathbf{n}_j$ are the exterior normal vectors to $\partial \Omega_j \subset \Sigma$.

The D–N operator on the graph Σ is the operator

$$\Lambda_\Sigma : \phi \to \psi.$$

It is not hard to define Λ_Σ as a self-adjoint operator on $L^2(\Sigma)$ (see, for instance, [80, 126]).

THEOREM 7.14 (see [80]). *The spectrum of the operator Λ_Σ coincides with the spectrum of the problem* (7.27).

This theorem explains the origin of the "bad" spectrum σ_2 in the previous section: it asymptotically behaves as the spectrum of the D–N operator Λ_Σ rescaled with the small parameter $(\varepsilon\delta)^{-1}$.

The operator Λ_Σ can be thought of as a "pseudodifferential" operator on the graph Σ. Although this is probably possible, we will not try to define the notion of a pseudodifferential operator on graphs. It is known that if the graph is smooth (and in particular has no vertices or loose ends) the operator Λ_Σ is in fact a pseudodifferential operator. D–N operators have been intensively studied recently, in particular due to the needs of inverse problems (see, for instance, [186] and references therein). The only thing different in the photonic situation is that the operator Λ_Σ is a "two-sided" one. This means that in order to define it, we solve Dirichlet problems on both sides of an edge and then take the jump of normal derivatives from both sides, while in standard considerations the Dirichlet problem is solved on only one side, and then the exterior normal derivative at the boundary is taken. For a standard D–N operator it is known that it is pseudodifferential and that its symbol is the square root of the symbol of the Laplace–Beltrami operator on the boundary (see, for instance, [183]). It is easy to conclude then that if the graph Σ is smooth, the operator Λ_Σ is pseudodifferential with the symbol $2|\xi|$ (i.e., the symbol of the operator $2\sqrt{-d^2/ds^2}$, where s is the arc length). This understanding is important for what follows.

The study of [80] was devoted to the case of TM modes in two dimensions only. It is continued in the paper under preparation [81], where the TE modes in two

Chapter 7. The Mathematics of Photonic Crystals

dimensions are treated in a similar asymptotic limit. It is shown, in particular, that the spectrum of the TE modes converges to the spectrum of the following problem:

$$\begin{cases} -\Delta u = \lambda u, & x \in \mathbb{R}^2 - \Sigma, \\ \left[\frac{\partial u}{\partial n}\right] = 0, & x \in \Sigma, \\ \frac{\partial u}{\partial n} = W[u], & x \in \Sigma, \end{cases} \quad (7.28)$$

where $\left[\frac{\partial u}{\partial n}\right]$ and $[u]$ stand for the jumps of $\partial u/\partial n$ and of u, respectively, across Σ and $\partial/\partial n$ is the normal derivative at smooth points of Σ. Some three-dimensional cases were also considered in [81].

7.5.3 Study of the Graph Models

As we saw in the previous section, study of thin high-contrast dielectric structures leads to spectral problems on graphs. In three dimensions, analogous study also leads to similar problems on surface structures. It is interesting to mention that in recent years, due to progress in nanotechnology and microelectronics, problems in thin domains ("fattened" points, graphs, or surfaces) were considered in mesoscopic physics. These are in particular studies of circuits of thin semiconductor strips ("quantum wires"; see a mathematical discussion in [59]), thin superconducting structures [160]–[162]), and others. In all these cases a natural asymptotic consideration was applied, which led to differential problems on graphs. One can also mention related considerations in different branches of mathematics ([30]–[32], [37]–[39], [57, 58, 165, 82, 83, 89, 133], [146]–[148], [163, 166]), chemistry [164], and other areas. The eigenvalue problems that arise in these studies usually look as follows: along each edge of the graph one has the problem

$$-\frac{d^2 u}{ds^2} = \lambda^2 u$$

with "appropriate" boundary conditions at each vertex. These boundary conditions at the vertices are still problematic, since it looks like convergence of spectra on thin domains to spectra on graphs are harder to prove in mesoscopic physics than in the photonic case. The only known results of this kind are probably the ones obtained in [82, 83, 133, 166, 163]. The theorems proved there and some handwaving in other cases show that normally these conditions are probably the following: the function u must be continuous through each vertex, and at each vertex the sum of the outgoing derivatives along each edge must be equal to zero. A further study of this problem is required.

Let us now discuss some spectral properties of the operator Λ_Σ on a periodic graph Σ in the plane. We remind the reader that this is the asymptotic model for the TM waves propagating mostly in the thin dielectric regions along the edges of the graph and that waves of this kind are responsible for the main difficulties in opening spectral gaps in the high-contrast case.

A thorough numerical and analytic study of this operator was done in [8, 126, 127]. We refer the reader to [126] for a description of the numerical algorithm used

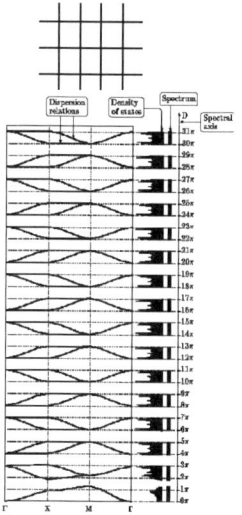

Figure 7.6: Square structure and its spectrum.

for finding spectra of operators Λ_Σ and present here only some of the obtained results.

Figure 7.6 presents the spectrum computed for the square lattice graph formed by the lines $x = n$ and $y = m$ ($n, m \in \mathbb{Z}$) and also explains our graphing system. In this picture the spectral axis is vertical. The first column represents the graphs of several branches of the dispersion relations $D_j(\mathbf{k})$ (we remind the reader that we have a rescaled spectral parameter D instead of the former λ). In order to avoid graphing surfaces, the dispersion relation is commonly graphed only for the values of the quasi momentum $\mathbf{k}$ on the boundary of the irreducible Brillouin zone, which in this case is the triangle with the vertices $\Gamma(0,0)$, $X(\pi, 0)$, and $M(\pi, \pi)$. The second column contains the graph of the density of states over the spectral axis. The third column shows the band gap structure of the spectrum.

Consider now disconnected graphs Σ. For instance, take a circle of a radius less than 0.5 and repeat it periodically with the period group $\mathbb{Z}^2$. One can view the resulting disconnected graph as a model of the structure of thin optically dense dielectric pipes in the air. A similar procedure can be applied to a segment, cross, square, etc., each time yielding a disconnected graph Σ. The numerical study of all of these and of some other disconnected structures produced dispersion relations with band functions that flatten very fast with the growing band number, leading to spectra that consist of very narrow spectral bands and thus are almost discrete for high frequencies. Besides, the spectra appear to be asymptotically periodic. Figure 7.7 represents the results of the calculation for the disconnected structure composed of disjoint circles of radii 0.2.

We present now an analytic result that explains this spectral behavior. It holds in any dimension, not necessarily in two dimensions. Let S be a smooth closed

Chapter 7. The Mathematics of Photonic Crystals 241

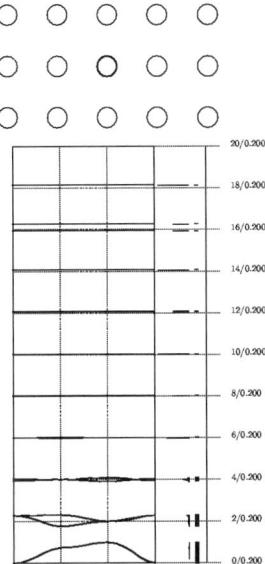

Figure 7.7: The disconnected structure of period 1 consisting of circles of radius 0.2 (top) and its spectrum (bottom).

hypersurface in $\mathbb{R}^d$ and $\Sigma = \cup_{\mathbf{n} \in \mathbb{Z}^d}(S+\mathbf{n})$ be the disjoint union of the integer shifts of S. One can define the D–N operator Λ_Σ on Σ, as was done above for $d = 2$.

THEOREM 7.15 (see [126]). *Let $\{D_n\} \subset \mathbb{R}$ be the (discrete) spectrum of the (positive) Laplace–Beltrami operator Δ_S on the surface S. Then there exists a sequence of positive numbers $\rho_n \to 0$ such that the spectrum of operator Λ_Σ on Σ belongs to the union of intervals*

$$\sigma(N) \subset \bigcup_n \left[2\sqrt{D_n} - \rho_n, 2\sqrt{D_n} + \rho_n\right],$$

and each of these intervals contains a nonempty portion of $\sigma(N)$.

Remark. In fact, if S is smooth, one can guarantee that

$$\rho_n \leq c_p D_n^{-p}$$

for any p. The case when S is a circle can be solved explicitly using Fourier series. It shows that analyticity of S probably implies exponential decay of ρ_n.

Theorem 7.15 explains the "almost discrete" nature of the spectrum and provides its asymptotic location for disconnected smooth structures. For instance, in the two-dimensional case we conclude that the spectrum at higher frequencies must concentrate around values $4\pi n L^{-1}$, where L is the length of S. In particular, for a circle of radius R this leads to $2nR^{-1}$. These numbers are indicated along the spectral axis in Figure 7.7, and one can see perfect agreement with the numerical results. This also provides an explanation of the asymptotic periodicity of the spectrum in two dimensions that was observed in numerics.

A few words are due about the method of proof. First, one can show that for high frequencies the eigenmodes decay very fast in the air regions, so the distinct copies of S essentially decouple. Then one is almost in the situation when the wave is zero on a surface surrounding a copy of S. Now the standard technique shows that we are dealing with a first-order pseudodifferential operator on S which is equal to $2\sqrt{\Delta_S} + R$, where Δ_S is the (positive) Laplace–Beltrami operator on S and R is a smoothing pseudodifferential operator. This in turn leads to the properties of the spectrum claimed in the theorem.

We would like to mention that numerics show a very fast convergence of the asymptotics claimed in the last theorem. So, one can make rather accurate predictions about the spectra using this theorem.

A very restrictive assumption is smoothness of S, since graphs that represent thin dielectric structures will normally have vertices and/or corners. One might expect that if instead of circles we use squares of the same length, the asymptotic nature of the spectrum will stay the same. However, numerical tests show that this is not the case. One could suspect that maybe just the rate of the asymptotic convergence is much slower in the nonsmooth case, but in fact the spectra look systematically shifted from the values predicted according to the formula $4\pi n L^{-1}$. Our current understanding is that this effect is due to the vertices (corners), which require some special boundary conditions. These conditions will be discussed later. So, the treatment of nonsmooth graphs (which are most common) requires additional study.

Connected structures are certainly the most interesting. The paper [126] contained results of computations for different geometries that show how the spectrum reacts to geometry. We will not present all these numerical results here, but rather address an interesting resonance phenomenon observed in [126]. Consider, for instance, the same disconnected circle structure and add dielectric edges connecting the circles along the symmetry axes of the structure. Figure 7.8 represents the computed dispersion relations and spectrum for this model.

One can notice resonance-type behavior: some branches of the dispersion relation become practically flat, and the density of states shows high delta-type peaks at the corresponding locations. As the following result (which at the moment of initial submission of this article was stated as a conjecture) shows, this does not indicate presence of actual eigenvalues, but rather of resonances.

THEOREM 7.16 (see [27]). *For any periodic graph Σ, the spectrum of Λ_Σ is absolutely continuous.*

It is interesting to look at the Floquet–Bloch eigenmodes that correspond to these observed resonances. Figure 7.9 represents the density plot of two such eigenmodes.

What one can see is that the wave is strongly localized at one circle (it is stuck in the loop), in spite of availability of the dielectric edges connecting different circles that allow the wave to propagate. One can also observe that the frequencies at which these resonances occur coincide with a subset of the spectrum computed for the disconnected circle structure. This is not a coincidence. One can show (the corresponding theorem is proven in [126]) that the eigenmodes of the disconnected

Chapter 7. The Mathematics of Photonic Crystals

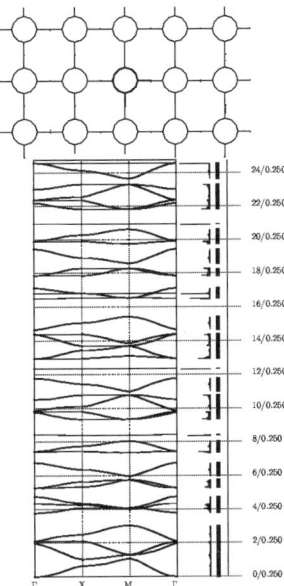

Figure 7.8: The structure of connected circles of radius 0.25 (top) and its spectrum (bottom).

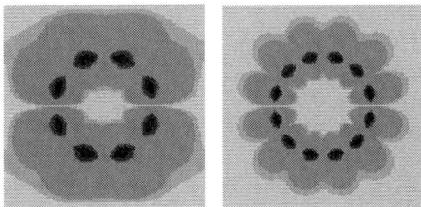

Figure 7.9: Density plots of the first two "localized" eigenmodes for the connected circle structure. Notice that the modes apparently do not propagate along the dielectric edges connecting the circles.

circle structure that are antisymmetric with respect to both symmetry axes of the structure, lead to resonances in the connected structure.

Similar resonant behavior was also observed in [126] for several other geometries, including, for instance, the honeycomb one. There is, however, no complete understanding of this effect. For instance, one can show both analytically and numerically that these resonances do not occur in the square geometry. It is not clear yet what differentiates this geometry from those with resonances. The study of these resonances suggests that it is in principle conceivable to "almost localize" electromagnetic waves in a purely periodic PBG material with no impurities, just by using an appropriate geometry. It is interesting to note that existence of significantly flattened band functions is of practical importance and has recently been used successfully for enhancement of spontaneous emission [29] and lasing [102].

The results of [126] show that there are often infinitely many gaps in the spectrum of the D–N operator Λ_Σ on a periodic graph Σ in the plane. Is the same true for higher dimensions? The following theorem shows that the answer is probably negative.

THEOREM 7.17 (see [126]). *Let the space* $\mathbb{R}^3$ *be tiled with unit cubes and* Σ *be the union of their surfaces. The spectrum of the corresponding D–N operator* Λ_Σ *has only a finite number of gaps. Moreover, there are no gaps in the spectrum for the values of the spectral parameter* $D \geq 40\pi$.

This theorem was proved by separating variables and consequently studying the resulting system of transcendental inequalities. It is interesting to note that the threshold between infinite and finite numbers of gaps lies for the periodic D–N operators between dimensions 2 and 3, while for the periodic Schrödinger operators it is between 1 and 2. The reason is probably that the D–N operator on a graph Σ is to a large extent a one-dimensional, rather than a purely two-dimensional, operator. Similarly, such an operator in three dimensions acts on a surface and hence is to some extent a two-dimensional operator. One should remember, however, that the operator Λ_Σ on a graph Σ still has two-dimensional features; for instance, its spectral bands can overlap.

CONJECTURE 7.18. *For any periodic hypersurface structure* $\Sigma \subset R^d, d \geq 3$, *the number of gaps in the spectrum of* Λ_Σ *is finite.*

Let us now address the most interesting case of nonsmooth graphs Σ. The main feature of Theorem 7.15 is that it reduces a complex pseudodifferential problem to a much simpler (especially in two dimensions) differential one. The question is whether such reduction is possible in the nonsmooth case. It is not clear whether the answer is affirmative in general. However, there are situations when this is possible. First, since the D–N operator is "almost" twice the square root of the negative second derivative with respect to the arc length, it is clear that it is reasonable to consider the eigenvalue problem

$$-\frac{d^2 u}{ds^2} = \left(\frac{D}{2}\right)^2 u \qquad (7.29)$$

along each edge (or maybe

$$(-1)^m \frac{d^{2m} u}{ds^{2m}} = \left(\frac{D}{2}\right)^{2m} u$$

for some integer m). The difficult question arises, however, of what boundary conditions at the vertices and corners one should use. Although the general answer is not known, some special geometries can be treated. The analysis developed in [127], although not completely rigorous, provides an interesting heuristic technique. Due to space limitations, we cannot discuss the details of this method. In order to understand the boundary behavior at a vertex or corner (junction of several edges), one blows it up by applying the Mellin transform in the radial directions from the vertex. Then one needs to study the singularities of analytic continuation of the resulting function. The spectral problem for the D–N operator becomes a functional equation that can be used to study these singularities. We will just present one of the results that can be obtained this way. If one has a symmetric junction of three edges at a vertex and u_j is the restriction of the function to the jth edge, then our analysis leads to the following conditions at the vertex:

$$\begin{cases} u_1(0) = u_2(0) = u_3(0), \\ \sum_{j=1,2,3} \frac{du_j}{ds}(0) = -(\frac{3D}{2}) \cot \frac{\pi}{3} u(0). \end{cases} \quad (7.30)$$

An interesting feature here (besides a funny trigonometric factor) is that the spectral parameter D also enters the boundary conditions. The problem (7.29) with conditions similar to (7.30) leads to simple algebraic equations and hence in many cases can be analyzed analytically. For instance, the dispersion relations for the case of the honeycomb structure with the edge size L can be found explicitly. Namely, one can derive existence of a series of eigenvalues $D = 2n\pi/L$ and of a series of nonflat bands given by

$$D_n(\mathbf{k}) = \frac{2}{L}\left(\pi n + \frac{\pi}{3} \pm \arcsin\sqrt{\frac{1}{4} + \frac{1}{6}\cos k_2 \pm \frac{1}{6}\sqrt{(1+\cos k_1)(1+\cos k_2)}}\right). \quad (7.31)$$

Tests on the disconnected union of three-edge stars, honeycomb structures, and some other geometries lead to an amazing agreement between the differential and pseudodifferential results. Figure 7.10 presents the results of computing the spectrum using the differential model (7.29)–(7.30) and the pseudodifferential operator Λ_Σ for the honeycomb lattice in the plane.

One can see that the pictures differ a little bit for the lowest band functions, but otherwise are practically identical. No rigorous justification of this effect is known. One should note, though, an important difference between the pseudodifferential and differential models. Namely, the almost flat band functions for the pseudodifferential model (left graph) are not exactly flat and do not correspond to actual eigenvalues [27], while one can show that the corresponding bands for the differential model (right) are flat and lead to infinitely degenerate eigenvalues (bound states).

Let us now address the asymptotic problem with $W \neq 0$:

$$-\Delta u = D(\delta_\Sigma + W)u.$$

In this case there is the dielectric-air coupling, and the problem cannot be conveniently reduced to the graph Σ. One of the ways one can handle this is to consider the auxiliary problem with two spectral parameters (c, D):

$$-\Delta u - cu = D\delta_\Sigma u \quad (7.32)$$

and then to intersect its spectrum in the (c, D)-plane with the line $c = WD$. Figure 7.11 represents the results of such calculation for the square structure (i.e.,

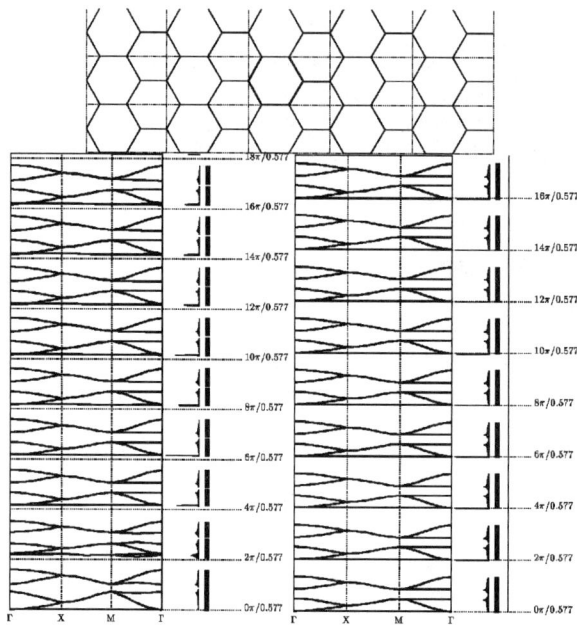

Figure 7.10: The honeycomb structure (top), the spectrum of the corresponding D–N operator (bottom left), and the spectrum of the differential model (bottom right).

formed by the lines $x = n$ and $y = m$, $m, n \in \mathbb{Z}$). The case of the square structure is exactly solvable. Analogous graphs for other geometries can be obtained numerically. The computation of the spectrum is done by fixing c, using the Green's function to rewrite the problem on Σ, and finally numerically finding the spectrum with respect to D. Doing so for many values of c, one can recover the two-dimensional spectrum of the problem.

The D-axis is horizontal and the c-axis is vertical. The shaded areas show the two-dimensional spectrum and the inclined line is $c = D$. One can see that the (c, D)-spectrum shows two distinct patterns. First, almost vertical strips originate at $c = 0$ from the bands of the spectrum of the D–N operator. Another set of narrowing strips goes in the horizontal direction. The horizontal lower edges of these strips indicate that at the corresponding values of c the D-spectrum of the problem (7.32) degenerates and covers the whole real line (and hence, due to an analyticity statement, the whole complex plane). The two different patterns intersect the line $c = D$ over two different subspectra, which correspond to the subspectra σ_1 (horizontal strips) and σ_2 (vertical strips), respectively (these spectra were discussed in section 7.5.1). The next result explains when the spectral degeneration observed at the straight edges of the horizontal strips can occur. This can provide guidance for creating geometry in a way that eliminates or lifts the horizontal pattern higher.

THEOREM 7.19 (see [127]). *The degeneration observed on the picture occurs at a level c if and only if*

Chapter 7. The Mathematics of Photonic Crystals

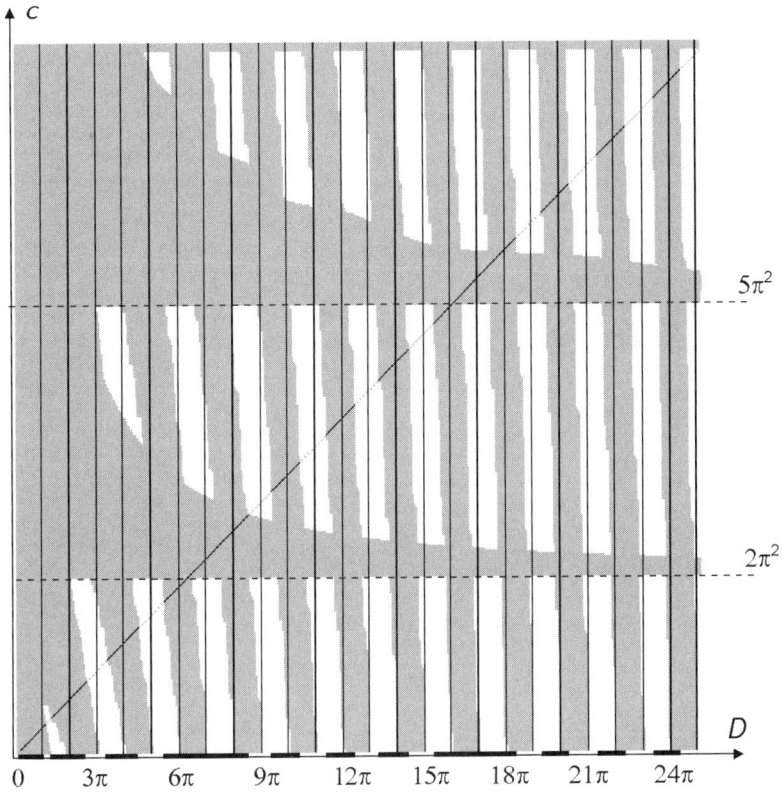

Figure 7.11: Calculation of the spectrum of the problem $-\Delta u - cu = D\delta_\Sigma u$ for the square structure.

(a) c is in the spectrum of the Floquet Laplacian $-\Delta_k = (-i\nabla + k)^2$ on the torus for some real value of the quasi momentum k;

(b) Σ is in the nodal set of an eigenfunction ϕ of $-\Delta_k$ corresponding to the eigenvalue c.

7.5.4 High-Contrast Materials with Dielectric Inclusions

In this section we present the results obtained in [100].

Let $\Omega \subset \mathbb{R}^d$ be an open connected set, which is periodic with respect to $\mathbb{Z}^d$. The complement $M = \Omega^c$ is assumed to have a positive distance from the boundary of the Wigner–Seitz cell $W = [0,1]^d$. Denoting by M_0 the part of M that resides inside W, we see that $M = \bigcup_{\mathbf{n} \in \mathbb{Z}^d}(M_0 + \mathbf{n})$. Let χ_Ω be the characteristic function of the domain Ω. Now consider the spectral problem for the following operator:

$$T_\nu = -\nabla \cdot (1 + \nu\chi_\Omega)\nabla, \qquad \nu \gg 1$$

(the operator must be defined in a standard way through the quadratic form).

In two dimensions this represents the case of TE modes in a high-contrast PBG material formed by an array of optically dense dielectric columns with sections $M_0 + \mathbf{n}, \mathbf{n} \in Z^d$. In three dimensions we may think of this as a model of acoustic waves in a high-contrast periodic medium. We are interested in the asymptotic behavior of the spectrum when the coupling constant ν (and hence the dielectric contrast) tends to infinity. In particular, do the gaps open in the case of high contrast? The positive answer is given by the following theorem.

THEOREM 7.20 (see [100]). (a) Let $\sigma_{n,\nu}$ and $\beta_{n,\nu}$ denote the lower and upper band edges of T_ν, listed in increasing order, so that $\sigma(T_\nu) = \bigcup_n [\sigma_{n,\nu}, \beta_{n,\nu}]$.

Then there exists a sequence of numbers μ_n satisfying $0 = \mu_1 < \mu_2 \leq \cdots$ that interlaces with the eigenvalues δ_n of the Dirichlet Laplacian $(-\Delta_{M_0})$ on M_0,

$$\mu_n \leq \delta_n \leq \mu_{n+1}$$

such that the spectrum of T_ν converges to $\bigcup_n [\mu_n, \delta_n]$ in the sense that for all n

$$\lim_{\nu \to \infty} \sigma_{n,\nu} = \mu_n, \qquad \lim_{\nu \to \infty} \beta_{n,\nu} = \delta_n.$$

(b) If $\delta_{k-1} < \delta_k = \cdots = \delta_m < \delta_{m+1}$ for some $k \leq m$, and additionally there exists an eigenfunction u of $-\Delta_{M_0} u = \delta_k u$ that satisfies

$$\int u \, dx \neq 0,$$

then $\mu_k < \delta_k$ and $\delta_m < \mu_{m+1}$. In particular, a gap opens above δ_m and a band extends below δ_k when $\nu \to \infty$.

Under some mild conditions on regularity on M_0, one can conclude from this theorem that the gap between the first two bands necessarily opens when $\nu \to \infty$. As a by-product one can also extract a statement about absolute continuity of the bottom part of the spectrum.

It was also established in [100] that although the bands in the asymptotic limit are extended and do not shrink into points, the density of states concentrates mostly at the Dirichlet eigenvalues δ_n, so the rest of each band becomes what is often called a pseudogap.

7.6 Defects in a Photonic Crystal

We have dealt so far with purely periodic media only. However, it is well known that practically important modifications of properties of materials can be made by doping them, i.e., by introducing localized or random defects into a purely periodic structure (see, for instance, [5, 106]). In this section we address the mathematics of impurities in PBG materials. The analogous problem for perturbations of periodic Schrödinger operators has been studied intensively in recent decades. We will not describe the corresponding results, referring the reader to the surveys [18]–[20].

7.6.1 Stability of the Essential Spectrum

Consider the dielectric medium described by a periodic electric permittivity $\varepsilon_0(x) \geq 1$, which is assumed to be a bounded measurable function. Then, as we have

Chapter 7. The Mathematics of Photonic Crystals

already discussed, the frequency spectrum of waves propagating in this medium is determined by the spectrum of the Maxwell operator $M_0 = \nabla \times \frac{1}{\varepsilon_0} \nabla \times$ appropriately defined on the subspace of transversal fields.[6] Inserting a localized defect into this periodic structure means adding a compactly supported perturbation $\varepsilon_1(x)$ to $\varepsilon_0(x)$: $\varepsilon = \varepsilon_0 + \varepsilon_1 \geq 1$. Then the operator itself is perturbed: $M = M_0 + M_1$. The first question we want to address is what kind of change of the spectrum this perturbation can bring. A similar question can be posed for the acoustic operator $-\nabla \cdot \frac{1}{\varepsilon} \nabla$. The answer is given by the following result.

THEOREM 7.21. *In both the electromagnetic and acoustic cases introduction of a localized defect does not change the essential spectrum of the operator.*

This theorem was established in the stated form in [72] by using Corollary 4 to Weyl's Theorem XIII.14 in [157]. This required showing that the perturbation is relatively compact in the sense of quadratic forms with respect to a power M_0^n of the unperturbed operator. A similar statement for the acoustic case was also proved in [1]. A general approach that implies this theorem is presented in [16].

Assume now that the unperturbed (acoustic or Maxwell) operator has a gap in the spectrum. Then the last theorem shows that the spectrum that might arise in the gap due to the added defect must consist of isolated eigenvalues of finite multiplicity. The physical meaning of this is rather simple. In the purely periodic medium these values of frequencies are prohibited. If, due to a localized impurity, a wave of a prohibited frequency does arise around the defect, it must decay fast as soon as it enters the unperturbed periodic part of the medium. Thus a bound state (eigenvector) is created. As we will see soon, these impurity modes must decay exponentially (another term used for such waves is "evanescent"). The question is, however, whether these impurity levels actually arise and, if so, in what number. This problem is addressed in the next section, along with a study of the corresponding impurity modes.

7.6.2 Impurity Levels in Spectral Gaps

Probably the first paper where the problem of defect modes was considered in a setting relevant for photonic crystals was [1]. In that paper a divergence-type operator in $\mathbb{R}^d$ was studied:

$$A = -\sum \partial_i a_{ij}(x) \partial_j$$

with a positive definite, Lipschitz continuous, bounded away from zero, and infinity matrix function $a_{ij}(x)$. This operator is perturbed by

$$B = -\sum \partial_i b_{ij}(x) \partial_j$$

with a nonnegative definite matrix b decaying to zero at infinity. Assume that the spectrum of A has a gap and that E belongs to this gap. The perturbed operator

---

[6]We use here the letter M to denote the operator $\nabla \times \frac{1}{\varepsilon} \nabla \times$, while in section 7.1 M was used for a different version of the Maxwell operator.

$A + \kappa B$ is considered for $\kappa > 0$. As is noted in the preceding section, the essential spectra of the operators A and $A + \kappa B$ are the same. Hence, if E belongs to the spectrum of the perturbed operator $A + \kappa \dot{B}$, it must be an eigenvalue of finite multiplicity. One can introduce the counting function

$$N(\kappa, b, E) = \#\{0 < \mu < \kappa | \ E \in \sigma(A + \mu B)\}.$$

The following theorem establishes the possibility of creating an impurity eigenvalue at E if the support of the perturbation is sufficiently large.

THEOREM 7.22 (see [1]). *There exists $R > 0$ such that if $b(x)$ is positive definite for any x in the ball of radius R, then $E \in \sigma(A + \kappa B)$ for some $\kappa > 0$.*

The next statement deals with perturbations of small support. In particular, it shows the impossibility of creating an eigenvalue at E if the support of the perturbation is too small.

THEOREM 7.23 (see [1]). *Let $d \geq 2$. There exists a constant $c_0 > 0$ such that*

$$N(\kappa, b, E) < c_0 R^d$$

for all $R > 0$, $\kappa > 0$, and all $b(x)$ with support in the ball of radius R. In particular

$$N(\kappa, b, E) = 0$$

if the support of $b(x)$ is too small.

The paper [1] also contained a study of the asymptotic behavior of $N(\kappa, b, E)$ for large values of κ under additional conditions on the behavior of $b(x)$ at infinity. One can find some further extensions of these results in [17, 34]. One should also note a difference between dimension $d > 2$, where no impurity spectrum in the gap arises below a threshold value of κ, and $d = 2$, where no such threshold exists.

A series of papers, [72]–[74], attacks the problem of defect modes in a setting coming from the photonic crystal theory. Namely, acoustic

$$A_0 = -\nabla \cdot \frac{1}{\varepsilon_0} \nabla$$

and Maxwell

$$M_0 = \nabla^\times \frac{1}{\varepsilon_0} \nabla^\times$$

operators are considered (the latter one on the subspace of transverse fields). The dielectric function $\varepsilon_0(x)$ is assumed to be a periodic measurable function bounded from above and below by positive constants. Suppose that the spectrum of the operator has a gap. Let us now create a localized defect as follows. Choose a cube with the side l and fill it with a dielectric material with a constant electric permittivity ε. The considerations of the previous section show that only isolated eigenvalues of finite multiplicity can be created inside the gap. The questions are whether such eigenvalues do arise and, if so, then in what quantity. The next theorem guarantees existence of defect eigenvalues if the defect is "strong enough."

Chapter 7. The Mathematics of Photonic Crystals

THEOREM 7.24 (see [72, 73]). *Let (λ_a, λ_b) be a gap in the spectrum of M_0. Also let $\tau \in (\lambda_a, \lambda_b)$ be such that $[\tau(1-\gamma), \tau(1+\gamma)] \subset (\lambda_a, \lambda_b)$ for some $\gamma \in (0, 1)$. If we change the value of $\varepsilon(x)$ to ε in a cube of side l such that*

$$l^2 \varepsilon > \frac{79}{\tau \gamma^2},$$

then the corresponding Maxwell operator has at least one defect eigenvalue inside the segment $[\tau(1-\gamma), \tau(1+\gamma)]$.

The papers [73, 74] also contain important theorems that provide estimates from above of the total number of defect eigenvalues that can arise in the gap. They require, however, some conditions of regularity of the ends of the gaps (see the discussion of this topic in the section 7.4.4). In particular, the following statement on the absence of defect eigenvalues in the case of "weak" defects holds.

THEOREM 7.25 (see [73, 74]). *Let $\varepsilon_0(x)$ be a measurable periodic function such that*

$$0 < \varepsilon_- \leq \varepsilon_0(x) \leq \varepsilon_+ < \infty$$

and (λ_a, λ_b) be a gap in the spectrum of the corresponding Maxwell operator M_0. Let us insert a defect by changing the dielectric function as follows:

$$\varepsilon(x) = \frac{\varepsilon_0(x)}{1 + \theta(x)},$$

where $\theta(x)$ is a measurable function supported inside of a cube of side l and such that

$$-1 < \theta_- \leq \theta(x) \leq \theta_+ < \infty.$$

Then

(a) if the left end λ_a of the gap is regular (in an appropriate sense) and $\theta_- = 0$, then there exists a constant $c > 0$ depending only on λ_a, $\varepsilon_\pm$, and l such that if $\theta_+ < c$, then there are no defect eigenvalues in (λ_a, λ_b);

(b) if the right end λ_a of the gap is regular and $\theta_+ = 0$, then there exists a constant $c > 0$ depending only on λ_b, $\varepsilon_\pm$, and l such that if $|\theta_-| < c$, then there are no defect eigenvalues in (λ_a, λ_b).

The papers [73, 74] also contain an approach to the problem of the mid-gap defect modes based on a version of the Birman–Schwinger method. A Birman–Schwinger-type compact operator depending upon the mid-gap frequency $\lambda \in (\lambda_a, \lambda_b)$ is defined such that its eigenvalues considered as functions of λ completely determine behavior of the defect eigenvalues. This method is probably also suitable for numerical implementation. We refer the reader to the papers [73, 74] for further details.

7.6.3 Exponential Localization

As we mentioned in the section 7.6.1, the impurity modes (eigenfunctions) that arise in spectral gaps due to localized defects are exponentially localized. Although the physics explanation of this effect is rather clear, its rigorous justification and especially determination of the rate of the exponential decay require some work. The main idea is that if A is either acoustic or a Maxwell operator that has a gap (λ_a, λ_b) in its spectrum, then the Green's function $G(\lambda, x, y)$ of $A - \lambda I$ for λ in this gap decays exponentially with $|x - y|$. As soon as one establishes this, the rest is simple. If we have a localized perturbation operator B and an eigenfunction f with the eigenvalue $\lambda \in (\lambda_a, \lambda_b)$, then we get the equation

$$(A - \lambda I) f = -Bf \tag{7.33}$$

or

$$f(x) = -\int G(\lambda, x, y) (Bf)(y) dy.$$

Now the exponential decay of the Green's function together with the local nature of the operator B yield the exponential decay of f. This type of argument was made precise in the papers [72]–[74], where a version of the arguments of [43] was used to get the resolvent estimates.

If an eigenfunction decays as $O(\exp(-|x|/L))$, the constant L is called the radius of localization. It is often important to have some information about this radius. The considerations of [72]–[74] and [43] yield an estimate of the exponential decay of the type $O(\exp(-C \text{dist}(\lambda, \sigma(A)) |x|))$ for a defect eigenfunction corresponding to the eigenvalue λ in a finite gap (λ_a, λ_b) in the spectrum $\sigma(A)$ of the unperturbed operator A. In other words, it estimates the radius of localization from above by the inverse distance to the spectrum of the unperturbed operator. It is known, however, that this estimate is not optimal close to the spectrum. Section 3 of [11] contains a general operator-theoretic approach that improves on the estimates of [43] and enables one to obtain a decay estimate of the form

$$O(\exp(-C\sqrt{|\lambda - \lambda_a| |\lambda - \lambda_b|} |x|)) \tag{7.34}$$

(we remind the reader that λ belongs to the spectral gap (λ_a, λ_b) of the unperturbed operator). Although considerations of [11] were devoted to the magnetic Schrödinger operator only, the approach is rather general and works for acoustic and Maxwell operators as well, as is shown in Appendix 3 of [42]. These estimates actually do not rely on periodicity. There is, however, a different approach, which does employ periodicity of the unperturbed medium. Although it is limited to periodic media only and besides it has not led to the precise estimates (7.34) yet, it might be useful in some circumstances. Here is how it goes. Consider (7.33) and apply the Floquet transform $\mathcal{U}$ to it:

$$A(k)\mathcal{U}f(\cdot, k) = -\mathcal{U}(Bf)(\cdot, k), \tag{7.35}$$

where we denoted by $A(k)$ the operator $A - \lambda I$ restricted to the space of functions on the Wigner–Seitz cell that satisfy the Floquet condition with the quasi momentum k. Due to the local nature of the operator B and Paley–Wiener theorems for the

Floquet transform, the vector function $\mathcal{U}(Bf)(\cdot, k)$ is analytic with respect to k in a neighborhood of the real space. Taking into account that we are at some distance from the spectrum of the operator A (since λ is in a spectral gap), it is possible to show that the analytic operator function $A(k)$ is invertible in a neighborhood of the real space. This statement is equivalent to the following: distance to the spectrum of A can be estimated using the distance of the complex Fermi surface to the real space (see section 7.4.2). Then solving (7.35), $\mathcal{U}f(k) = -A(k)^{-1}\mathcal{U}(Bf)(k)$, we derive analyticity of $\mathcal{U}f(k)$ in a neighborhood of the real space, which in turn, due to a Paley–Wiener theorem for the Floquet transform, implies that f decays exponentially. Implementation of this program for the Schrödinger case using results of [91] leads to an estimate weaker than (7.34). It would be interesting to extend this to the Maxwell case and to achieve (7.34).

7.6.4 Embedded Impurity Levels

In the discussions of the previous two sections we considered a background self-adjoint operator A (an elliptic periodic differential operator) with a gap in the spectrum and then added a local perturbation operator B. Then we discussed the behavior of the impurity spectrum in the gap. However, the natural question arises of whether the impurity eigenvalues can arise inside the spectrum of the operator A rather than in its gaps. Such eigenvalues (if they exist) are called the *embedded* ones. If this does occur, then we have a peculiar situation. Consider, for instance, the Schrödinger operator with a periodic potential $A = -\Delta + v(x)$ and add to it a localized potential $w(x)$. If there is an impurity eigenvalue λ of $-\Delta + v(x) + w(x)$ that resides inside the spectrum of A, then the corresponding bound state u of the electron is very strange: the electron has sufficient energy to propagate (since $\lambda \in \sigma(A)$), but for some reason it stays attached to the defect. There is large literature devoted to discussion of embedded eigenvalues (see, for instance, the book [55]). It is known that if the impurity potential does not decay sufficiently fast, then embedded eigenvalues can occur. There are plenty of results saying that if the perturbation decays fast enough, then there are no embedded eigenvalues. However, no such results appear to cover the case of a periodic background operator (even for Schrödinger operators). The only exception is the one-dimensional result of [158, 159] that states that for sufficiently fast decaying perturbations of the Hill operator no embedded eigenvalues arise. Probably the only known multidimensional result of this kind is proved in [131, 132] (papers [88, 118] contain theorems about discreteness of the set of embedded eigenvalues). Let us introduce some notation first. We denote by $H_0 = -\Delta + q(x)$ the unperturbed Schrödinger operator with a periodic potential q, whose spectrum has the band structure

$$\sigma(H_0) = \bigcup_{i \geq 1} [a_i, b_i].$$

We now add a decaying perturbation potential $v(x)$ to get the operator $H = -\Delta + q(x) + v(x)$.

THEOREM 7.26 (see [131, 132]). *If a real periodic potential $q(x)$ belongs to $L^\infty(\mathbb{R}^d)$ ($d \leq 3$), the operator H_0 satisfies Conjecture 7.4 about the Fermi surface, and the impurity potential $v(x)$ is measurable and satisfies the estimate*

$$|v(x)| \leq Ce^{-|x|^r} \text{ for some } r > 4/3 \qquad (7.36)$$

almost everywhere in $\mathbb{R}^d$, then the spectrum of H contains no embedded eigenvalues. In other words,

$$\{\lambda_j\} \cap \bigcup_{i \geq 1} (a_i, b_i) = \emptyset,$$

where $\{\lambda_j\}$ is the impurity point spectrum of H.

As was mentioned in section 7.4.2, in the case when the potential $q(x)$ is separable, Conjecture 7.4 holds true. Hence, in this case the theorem claims that no embedded impurity spectra can arise.

It would be very interesting to extend this result to arbitrary periodic potentials and to other periodic elliptic operators of interest, including the ones arising in PBG studies. However, this is probably a difficult task, since the considerations of [131, 132] show that validity of Conjecture 7.4 is crucial. In fact, we believe that the following conjecture holds true.

CONJECTURE 7.27. *If for a periodic self-adjoint elliptic operator A and a point $\lambda \in \sigma(A)$ there is an irreducible component of the complex Fermi surface $F_\lambda(A)$ that does not intersect the real space, then there exists a local perturbation operator B such that λ is an eigenvalue for $A + B$.*

This conjecture is supported by the following example. For a fourth-order self-adjoint ordinary differential equation with periodic coefficients the Fermi surface F_λ is discrete and contains four points. When λ belongs to the spectrum, two of these points (irreducible components) can be complex. In this particular case we do have irreducible components "hidden" in the complex domain. One can construct an example of such an equation and of a local perturbation that leads to an embedded eigenvalue [151].

7.6.5 Linear Defects and Waveguides

Besides localized impurities linear defects are of great importance for applications. By a linear defect we mean a strip (column) of a dielectric, whose dielectric properties differ from the ones dictated by the underlying periodic structure. For instance, imagine a row of a homogeneous dielectric material inserted into a periodic structure. It is conceivable that such a row might support a propagating mode, whose frequency falls into the frequency gap of the background periodic material. In this case such a mode must be evanescent when it leaves the defect. In other words, one creates a perfect optical waveguide without standard drawbacks of the fiber-optic cables, like leakage through sharp bends. This explains attention paid to this topic in physics literature (see, for instance, [60, 106, 138, 139] and references therein). Although a rather extensive study was done numerically and experimentally, no rigorous mathematical analysis of the problem is available. Some statements are easy to prove. For instance, similarly to the case of the localized defects one can show that the propagating modes with frequencies in the gap must be evanescent in the periodic part of the crystal. This can be done either by using the estimates of the exponential decay of the Green's function or by employing the Paley–Wiener theorems for the Floquet transform (see section 7.3). It would be interesting to have a study similar to the one done in [72]–[74] that would guarantee existence or nonexistence of the propagating modes depending on the properties of the linear defect. When this is done, a study is due of transmission through a bend in a linear defect (a numerical study of this problem was done in [139]).

7.6.6 Anderson Localization

An important and extensively studied part of the photonic crystal research is Anderson localization of classical (for instance, electromagnetic or acoustic) waves in a periodic medium perturbed by random impurities. While the study of a similar phenomenon for the Schrödinger operator has attracted a lot of attention from mathematicians, the case of classical waves has been considered in only a handful of articles. We, however, cannot address this problem here due to space limitations. A large survey article could probably be written on this topic alone. The reader can consult with the physics surveys [4, 107]–[109] and with the recent publications [42], [68]–[71], and [73] that rigorously established important results on existence of Anderson localization of acoustic and electromagnetic waves.

7.7 Some Numerical Methods and Optimization

The numerical approaches commonly used in the photonic crystals theory amount to the plane wave (Fourier expansion) methods, transfer matrix methods, finite-difference time-domain methods, and some others. The surveys [101, 152, 189] describe most of these techniques rather well. Links to websites containing descriptions of algorithms and codes can be found in [153, 190]. So, in this section we will briefly discuss only a few recent developments in this area.

7.7.1 Finite Elements and Vector Elements

The finite element method has been successfully used in many applied areas, including electromagnetics. We address the reader to the book [105] for a survey of electromagnetics applications. A finite element approach to computing dispersion relations and spectra of two-dimensional PBG materials was developed independently in the papers [7, 50]. Although the algorithms developed there are not identical, they are very close. The method is applicable to both TE and TM modes described by (7.15) and (7.14), respectively. First, a mesh is generated that has the same periodicity as the problem. In [7] the mesh generator Easymesh 1.4 created by Bojan Niceno, University of Trieste, was used. This generator, which produces high-quality triangular two-dimensional meshes, adjusts the mesh to the prescribed geometry of the air-dielectric interfaces. A square mesh was utilized in [50]. Consider the TE polarization (the TM polarization is handled similarly)

$$-\nabla \cdot \frac{1}{\varepsilon(x)} \nabla \psi = \lambda \psi.$$

The algorithm handles arbitrary lattices of periods, but as before we will concentrate on the case when the structure is one-periodic with respect to each variable. For the Floquet waves with the quasi momentum k the problem reduces to

$$-(\nabla + ik) \cdot \frac{1}{\varepsilon(x)} (\nabla + ik) u = \lambda u \qquad (7.37)$$

on periodic functions u. One can rewrite (7.37) as follows:

$$\int_\mathbb{T} \frac{1}{\varepsilon(x)} (\nabla + ik) u \cdot \overline{(\nabla + ik) v} dx = \lambda \int_\mathbb{T} u \bar{v} dx.$$

Here $\mathbb{T}$ is the two-dimensional torus $\mathbb{T} = \mathbb{R}^2/\mathbb{Z}^2$, $\mathbb{Z}^2$ is the two-dimensional integer lattice, u is the eigenmode, and v is an arbitrary periodic function from $H^1(\mathbb{T})$. Using the mesh, a basis of functions $\phi_j(x)$ is chosen (in [7] the basis functions are linear, and in [50], bilinear on each element). Representing $u = \sum \xi_j \phi_j$ and then choosing $v = \phi_l$, we get a generalized eigenvalue problem

$$A(k)\xi = \lambda B \xi \qquad (7.38)$$

on the corresponding subspace of linear combinations of the basis functions in $L_2(\mathbb{T})$. Here

$$A_{jl} = \int_T \frac{1}{\varepsilon(x)} (\nabla + ik) \phi_j \cdot \overline{(\nabla + ik) \phi_l} dx$$

and

$$B_{jl} = \int_T \phi_j \overline{\phi_l} dx.$$

Now the task is to solve numerically the generalized eigenvalue problem (7.38) to find the band functions $\lambda_j(k)$. Since the matrices A and B are very sparse, in order to cut the memory requirements and to increase the speed of calculations, one wants to use eigenvalue solvers that employ this sparsity pattern efficiently. In both papers [7, 50] versions of the subspace iteration method were used. The algorithm described in [7] uses the SICOR (simultaneous coordinate overrelaxation) method [169], while [50] is based on the (similar in spirit) subspace preconditioning method developed in [35]. The advantage of [50] is usage of clever preconditioners of two types. First, moving along a path in the Brillouin zone, the algorithm uses the results obtained for the previous value of the quasi momentum as a seed for the current one. Second, each iteration step involves solving the problem for a homogeneous medium. These preconditioners significantly speed up the convergence.

Testing of both algorithms shows good numerical convergence and good agreement with previously known numerical results and with explicitly solvable models. The algorithms are fast. Due to economical use of memory and employing sparseness, they can handle significantly larger meshes in comparison with the number of modes that one can use with the plane wave methods. The finite element method is also known to capably handle nonsmooth interfaces and singular solutions, the factors that significantly slow down convergence of Fourier series. For instance, the analysis of the high-contrast PBG structures presented in this survey shows existence of modes (the dielectric modes that led to the "bad" spectrum σ_2) would be hard to catch with the plane wave methods.

The full-vector three-dimensional case can also be handled by the finite element method, but in this case the method is known to lead to spurious spectra [105]. Using the so-called vector (or edge) elements one completely (or almost completely) eliminates this problem [105] (see [144] for mathematical theory of vector elements). This project was realized in [51].

7.7.2 Using Soluble Models

Analysis of the two-dimensional square structure done in [76]–[78] led in [64, 155] to development of an unusual method of computing spectral characteristics of PBG materials. Namely, in the case of the square structure one can find exactly solvable models in a vicinity of both problems (7.14) and (7.15). If one now finds explicit eigenfunctions and spectra for these approximate models, one hopes that they represent a good basis of functions to use for the accurate model. For instance, one can use Galerkin-type methods, or any other variation on the theme. This was done a little bit differently in the cited papers, but the general ideas are the same. The results presented in [64, 155] agree very well with each other, and also with the computations presented in [7]. The drawback of this approach is that it relies on existence of an analytically solvable model sufficiently close to the one that we want to solve, which is probably a rather exceptional situation.

7.7.3 Optimization

The question of optimizing a PBG structure comes naturally to mind. How should one change geometric and physical parameters of a medium in order to widen an existing gap or to try to open a new gap between a couple of bands? Until recently, noone had tried to consider this as an optimization problem in the technical sense. This was done for the first time for the TM modes in two dimensions in [44]. The results are rather promising. The Helmholtz equation $\Delta u + \lambda^2 \varepsilon u = 0$ in two dimensions is considered, where the electric permittivity ε is a measurable periodic function satisfying fixed bounds $0 < c_1 \leq \varepsilon(x) \leq c_2 < \infty$. The idea is to start with a dielectric function ε_0 in this class for which existence of a gap between the bands $\lambda_j(k)$ and $\lambda_{j+1}(k)$ is known, i.e., $\lambda_j(k) < \alpha < \lambda_{j+1}(k)$ for all k in the Brillouin zone B. Then one considers the goal of maximizing the function

$$G(\varepsilon) = \inf_{k \in B} (\min\{\alpha - \lambda_j(k), \lambda_{j+1}(k) - \alpha\})$$

over the set of dielectric functions satisfying

$$c_1 \leq \varepsilon(x) \leq c_2.$$

The problem is with nonsmoothness of the goal function. This forces us to create a clever generalized gradient ascent algorithm, where the generalized gradient is understood in the sense described in [40]. Due to the multivaluedness of the generalized gradient, choosing the directions on each step involves solving an auxiliary linear programming problem. Although convergence of the algorithm was not rigorously established, the results of the performed numerical experiments are very encouraging [44]. The TE case was recently treated in a similar manner in [45]. This direction of study definitely deserves further development. One also notices that the optimization procedure involves multiple computations of spectra of PBG materials. This explains the need to have efficient methods of computing the PBG spectra like those described in section 7.7.1.

7.8 Conclusions

I would divide the mathematical problems of the photonic crystals theory into two broad categories. The first one consists of problems whose answers are known with a high level of certainty, while justification of these answers is hard to achieve. I can mention here the problems of absence of bounded states (localized waves) in a purely periodic photonic crystal, finiteness of the number of gaps, absence of embedded impurity eigenvalues, and some others. Although neither physicists nor mathematicians doubt what the correct answers to these questions are, our inability to provide rigorous proofs shows that sufficient understanding of these phenomena has probably not been achieved yet. Another category consists of problems whose resolution could have an immediate impact on applications. Among these I would mention developing tools of analytic prediction of existence and size of gaps depending on the geometric and physical parameters of the medium, understanding the behavior of the impurity spectra, creating significantly flattened bands, and studying properties of PBG waveguides, nonlinear effects, tunable crystals, Anderson localization, and many other phenomena. Some of the outstanding problems are mentioned in the text. Many more can be easily found in the available physics literature.

I hope that the reader is persuaded by now that the field of photonic crystals research is an applied mathematician's dream: it is of high practical importance; its mathematical models are practically exact; it involves great mathematics ranging from algebraic geometry to several complex variables, to functional analysis, to numerics—you name it; most mathematical problems are largely unexplored.

Acknowledgments

I want to express my gratitude to many people. First, to Professor Alex Figotin, who attracted me to this beautiful area of research and with whom I spent countless hours discussing PBG materials. His influence was crucial for my research and probably for the research of some other people working on mathematical problems of PBG materials. I also want to thank my colleagues H. Ammari, G. Bao, W. Axmann, M. Birman, E. Bonnetier, M. Boroditsky, R. Carlson, J. M. Combes, B. DeFacio, D. Dobson, P. Exner, L. Friedlander, Yu. Godin, J. C. Guillot, E. Harrel, J. W. Haus, R. Hempel, Yu. Karpeshina, A. Klein, H. Knörrer, F. Klopp, L. Kunyansky, S. Levendorskiî, S. Molchanov, S. Novikov, V. Palamodov, V. Papanicolaou, Y. Pinchover, I. Ponomarev, G. Rosenblum, J. Rubinstein, M. Schatzman, D. Sievenpiper, A. Sobolev, M. Solomyak, T. Suslina, A. Tip, B. Vainberg, S. Venakides, and H. Zeng for information and discussions. Thanks also go to W. Axmann, M. Birman, O. Kuchment, M. Mogilevsky, T. Suslina, and to the reviewers for their comments about the manuscript.

This research was partly sponsored by the NSF through grant DMS 9610444 and by the Department of the Army, Army Research Office, through a DEPSCoR grant. The author thanks the NSF and the ARO for this support. The content of this paper does not necessarily reflect the position or the policy of the federal government, and no official endorsement should be inferred.

References

[1] S. Alama, M. Avellaneda, P. A. Deift, and R. Hempel, *On the existence of eigenvalues of a divergence form operator $A+\lambda B$ in a gap of $\sigma(A)$*, Asymptotic Anal., 8 (1994), pp. 311–314.

[2] S. Alama, P. A. Deift, and R. Hempel, *Eigenvalue branches of the Schrödinger operator $H-\lambda W$ in a gap of $\sigma(H)$*, Comm. Math. Phys., 121 (1989), pp. 291–321.

[3] H. Ammari, N. Béreux, and E. Bonnetier, *Analysis of the radiation properties of a planar antenna on a photonic crystal structure*, submitted to SIAM J. Appl. Math.

[4] P. Anderson, *The question of classical localization. A theory of white paint?*, Philosophical Magazine B, 52 (1985), pp. 505–509.

[5] N. W. Ashcroft and N. D. Mermin, *Solid State Physics*, Holt, Rinehart and Winston, New York, London, 1976.

[6] J. E. Avron and B. Simon, *Analytic properties of band functions*, Ann. of Phys., 110 (1978), pp. 85–101.

[7] W. Axmann and P. Kuchment, *An efficient finite element method for computing spectra of photonic and acoustic band-gap materials I. Scalar case*, J. Comput. Phys., 150 (1999), pp. 468–481.

[8] W. Axmann, P. Kuchment, and L. Kunyansky, *Asymptotic methods for thin high contrast 2D PBG materials*, J. Lightwave Techn., 17 (1999), pp. 1996–2007.

[9] D. Bättig and J. C. Guillot, *The Fermi Surface for the Discretized Maxwell Equations*, Journées "Equations aux Dérives Partielles" (Saint Jean de Monts, 1991), Exp. No. XI, 6, Ecole Polytech., Palaiseau, 1991.

[10] D. Bättig, H. Knörrer, and E. Trubowitz, *A directional compactification of the complex Fermi surface*, Compositio Math., 79 (1991), pp. 205–229.

[11] J. M. Barbaroux, J. M. Combes, and P. D. Hislop, *Localization near band edges for random Schrödinger operators*, Helv. Phys. Acta, 70 (1997), pp. 16–43.

[12] M. M. Beaky, J. B. Burk, H. O. Everitt, M. A. Haider, and S. Venakides, *Two-dimensional photonic crystal Fabry-Perot resonators with lossy dielectrics*, IEEE Trans. Microwave Theory Tech., 47 (1999), pp. 2085–2091.

[13] A. Bensoussan, J. L. Lions, and G. Papanicolaou, *Asymptotic Analysis of Periodic Structures*, North-Holland, Amsterdam, 1980.

[14] A. Berthier, *On the point spectrum of Schrödinger operators*, Ann. Scie. École Norm. Sup. (4), 15 (1982), pp. 1–15.

[15] G. Bethe and A. Sommerfeld, *Elektronentheorie der Metalle*, Springer-Verlag, Berlin, New York, 1967.

[16] M. Sh. Birman, *On the spectrum of singular boundary-value problems*, Mat. Sb., 55 (1961), pp. 125–174. English transl. in Eleven Papers on Analysis, Amer. Math. Soc. Transl. Ser. 2, 53, AMS, Providence, RI, 1966, pp. 23–60.

[17] M. Sh. Birman, *Discrete spectrum of the periodic elliptic operator with a differential perturbation*, Journees "Équations aux Derivées Partielles" (Saint-Jean-de-Monts, 1994), Exp. No. XIV, Ecole Polytech., Palaseau, France, 1994.

[18] M. Sh. Birman, *The discrete spectrum of the periodic Schrödinger operator perturbed by a decreasing potential*, Algebra i Analiz, 8 (1996), pp. 3–20. English transl. in St. Petersburg Math. J., 8 (1997), pp. 1–14.

[19] M. Sh. Birman, *The discrete spectrum in gaps of the perturbed periodic Schrödinger operator. I. Regular perturbations*, in Boundary Value Problems, Schrödinger Operators, Deformation Quantization, Math. Top. 8, Akademie-Verlag, Berlin, 1995, pp. 334–352.

[20] M. Sh. Birman, *The discrete spectrum in gaps of the perturbed periodic Schrödinger operator. II. Nonregular perturbations*, Algebra i Analiz, 9 (1997), pp. 62–89.

[21] M. Sh. Birman, private communication, December 1998.

[22] M. Sh. Birman and M. Solomyak, L^2-*theory of the Maxwell operator in arbitrary domains*, Russian Math. Surveys, 42 (1987), pp. 75–96.

Chapter 7. The Mathematics of Photonic Crystals

[23] M. Sh. Birman and M. Solomyak, *Self-adjoint Maxwell operator in arbitrary domains*, Algebra i Analiz, 1 (1989), pp. 96–110; translation in Leningrad Math. J., 1 (1990), pp. 99–115.

[24] M. Sh. Birman and T. A. Suslina, *The two-dimensional periodic magnetic Hamiltonian is absolutely continuous*, Algebra i Analiz, 9 (1997), pp. 32–48 (in Russian); translation in St. Petersburg Math. J., 9 (1998), pp. 21–32.

[25] M. Sh. Birman and T. A. Suslina, *Two-dimensional periodic Pauli operator. The effective masses at the lower edge of the spectrum*, in Mathematical Results in Quantum Mechanics (QMath7, Prague, June 22–26, 1998), J. Dittrich, P. Exner et al., eds., Oper. Theory Adv. Appl. 108, Birkhäuser, Basel, 1999, pp. 13–31.

[26] M. Sh. Birman and T. A. Suslina, *Periodic magnetic Hamiltonian with a variable metric. The problem of absolute continuity*, Algebra i Analiz, 11 (1999); English translation in St. Petersburg Math J., 11 (2000), no. 2, pp. 203–232.

[27] M. Sh. Birman and T. A. Suslina, private communication, 1999.

[28] F. Bloch, *Über die Quantenmechanik der Electronen in Kristallgittern*, Z. Phys., 52 (1928), pp. 555–600.

[29] M. Boroditsky, R. Vrijen, T. F. Krauss, R. Coccioli, R. Bhat, and E. Yablonovitch, *Spontaneous emission extraction and Purcell enhancement from thin-film 2-D photonic crystals*, J. Lightwave Techn., 17 (1999), pp. 2096–2112.

[30] L. Borcea and G. Papanicolaou, *Network approximation for transport properties of high contrast materials*, SIAM J. Appl. Math., 58 (1998), pp. 501–539.

[31] L. Borcea, J. G. Berryman, and G. Papanicolaou, *Network asymptotics for high contrast impedance tomography*, in Inverse Problems in Geophysical Applications (Yosemite, CA, 1995), H. W. Engl, A. K. Louis, and W. Rindell, eds., SIAM, Philadelphia, PA, 1997, pp. 287–303.

[32] L. Borcea, J. G. Berryman, and G. Papanicolaou, *High-contrast impedance tomography*, Inverse Problems, 12 (1996), pp. 835–858.

[33] C. M. Bowden, J. P. Dowling, and H. O. Everitt, eds., *Development and applications of materials exhibiting photonic band gaps*, Journal Opt. Soc. Amer. B, 10 (1993), pp. 280–413.

[34] S. I. Boyarchenko and S. Z. Levendorskiĭ, *An asymptotic formula for the number of eigenvalue branches of a divergence form operator $A + \lambda B$ in a spectral gap of A*, Comm. Partial Differential Equations, 22 (1997), pp. 1771–1786.

[35] J. H. Bramble, A. V. Knyazev, and J. E. Pasciak, *A subspace preconditioning algorithm for eigenvector/eigenvalue computation*, Adv. Comput. Math., 6 (1996), pp. 159–189.

[36] L. Brillouin, *Wave Propagation in Periodic Structures: Electric Filters and Crystal Lattices*, 2nd ed., Dover, NY, 1953.

[37] R. Carlson, *Hill's equation for a homogeneous tree*, Electron. J. Differential Equations, 23 (1997), pp. 1–30.

[38] R. Carlson, *Adjoint and self-adjoint operators on graphs*, Electron. J. Differential Equations, 6 (1998), pp. 1–10.

[39] R. Carlson, *Inverse eigenvalue problems on directed graphs*, Trans. Amer. Math. Soc., 351 (1999), pp. 4069–4088.

[40] F. Clarke, *Optimization and Nonsmooth Analysis*, SIAM, Philadelphia, 1990.

[41] J. M. Combes, *Spectral problems in the theory of photonic crystals*, in Mathematical Results in Quantum Mechanics (QMath7, Prague, June 22–26, 1998), J. Dittrich, P. Exner et al., eds., Operator Theory Adv. Appl. 108, Birkhäuser, Basel, 1999, pp. 33–46.

[42] J. M. Combes, P. D. Hislop, and A. Tip, *Band edge localization and the density of states for acoustic and electromagnetic waves*, Ann. Inst. H. Poincaré Phys. Théor., 70 (1999), pp. 381–428.

[43] J. M. Combes and L. Thomas, *Asymptotic behavior of eigenfunctions for multiparticle Schrödinger operators*, Comm. Math. Phys., 34 (1973), pp. 251–270.

[44] S. J. Cox and D. C. Dobson, *Maximizing band gaps in two-dimensional photonic crystals*, SIAM J. Appl. Math., 59 (1999), pp. 2108–2120.

[45] S. J. Cox and D. C. Dobson, *Band structure optimization of two-dimensional photonic crystals in H-polarization*, J. Comp. Phys., 158 (2000), pp. 214–224.

[46] B. Dahlberg and E. Trubowitz, *A remark on two dimensional potentials*, Comment. Math. Helv., 57 (1982), pp. 130–134.

[47] L. Danilov, *Spectrum of the Dirac operator in $\mathbf{R}^n$*, Teor. Math. Fiz., 85 (1990), pp. 41–53.

[48] S. Datta, C. T. Chan, K. M. Ho, and C. M. Soukoulis, *Effective dielectric constant of periodic composite structures*, Phys. Rev. B, 48 (1993), pp. 14936–14943.

[49] E. B. Davies and E. Harrell, *Conformally flat Riemannian metrics, Schrödinger operators, and semiclassical approximation*, J. Differential Equations, 66 (1987), pp. 165–188.

[50] D. C. Dobson, *An efficient method for band structure calculations in 2D photonic crystals*, J. Comput. Phys., 149 (1999), pp. 363–376.

[51] D. C. Dobson, J. Gopalakrishnan, and J. E. Pasciak, *An efficient method for band structure calculations in 3D photonic crystals*, J. Comput. Phys., 161 (2000), pp. 668–679.

[52] H. J. S. Dorren and A. Tip, *Maxwell equations for non-smooth media; fractal and pointlike objects*, J. Math. Phys., 32 (1991), pp. 3060–3070.

[53] B. A. Dubrovin, A. T. Fomenko, and S. P. Novikov, *Modern Geometry-Methods and Applications, Part* I: *The Geometry of Surfaces, Transformation Groups, and Fields*, Springer-Verlag, New York, 1991.

[54] M. S. P. Eastham, *The Spectral Theory of Periodic Differential Equations*, Scottish Acad. Press, Edinburgh, London, 1973.

[55] M. S. P. Eastham and H. Kalf, *Schrödinger-Type Operators with Continuous Spectra*, Pitman, Boston, 1982.

[56] D. E. Edmunds and W. Evans, *Spectral Theory and Differential Operators*, Oxford Science Publications, Clarendon Press, Oxford, UK, 1990.

[57] W. D. Evans and D. J. Harris, *Fractals, trees and the Neumann Laplacian*, Math. Ann., 296 (1993), pp. 493–527.

[58] W. D. Evans and Y. Saito, *Neumann Laplacians on domains and operators on associated trees*, to appear in Quart. J. Math. Oxford.

[59] P. Exner and P. Seba, *Electrons in semiconductor microstructures: A challenge to operator theorists*, in Proceedings of the Workshop on Schrödinger Operators, Standard and Nonstandard (Dubna 1988), World Scientific, Singapore, 1989, pp. 79–100.

[60] S. Fan, J. N. Winn, A. Devenyi, J. C. Chen, R. D. Meade, and J. D. Joannopoulos, *Guided and defect modes in periodic dielectric waveguides*, J. Opt. Soc. Amer. B, 12 (1995), pp. 1267–1272.

[61] A. Figotin, *Existence of gaps in the spectrum of periodic dielectric structures on a lattice*, J. Statist. Phys., 73 (1993), pp. 571–585.

[62] A. Figotin, *Photonic pseudogaps in periodic dielectric structures*, J. Statist. Phys., 74 (1994), pp. 443–446.

[63] A. Figotin, *High contrast photonic crystals*, in Diffuse Waves in Complex Media, J.-P. Fouque, ed., Kluwer, Norwell, MA, 1999, pp. 109–136.

[64] A. Figotin and Yu. Godin, *The computation of spectra of some 2D photonic crystals*, J. Comp. Phys., 136 (1997), pp. 585–598.

[65] A. Figotin, Yu. Godin, and I. Vitebsky, *Tunable photonic crystals*, Phys. Rev. B, 57 (1998), pp. 2841–2848.

[66] A. Figotin and V. Gorentsveig, *Localized electromagnetic waves in a layered periodic dielectric medium with a defect*, Phys. Rev. B, 58 (1998), pp. 180–188.

[67] A. Figotin and I. Khalfin, *Bound states of a one-band model for 3D periodic medium*, J. Comput. Phys., 138 (1997), pp. 153–170.

[68] A. Figotin and A. Klein, *Localization phenomenon in gaps of the spectrum of random lattice operators*, J. Statist. Phys., 75 (1994), pp. 997–1021.

[69] A. Figotin and A. Klein, *Localization of electromagnetic and acoustic waves in random media. Lattice model*, J. Statist. Phys., 76 (1994), pp. 985–1003.

[70] A. Figotin and A. Klein, *Localization of classical waves I: Acoustic waves*, Comm. Math. Phys., 180 (1996), pp. 439–482.

[71] A. Figotin and A. Klein, *Localization of classical waves II: Electromagnetic waves*, Comm. Math. Phys., 184 (1997), pp. 411–441.

[72] A. Figotin and A. Klein, *Localized classical waves created by defects*, J. Statist. Phys., 86 (1997), pp. 165–177.

[73] A. Figotin and A. Klein, *Localization of light in lossless inhomogeneous dielectrics*, J. Opt. Soc. Amer. A, 15 (1998), pp. 1423–1435.

[74] A. Figotin and A. Klein, *Midgap defect modes in dielectric and acoustic media*, SIAM J. Appl. Math., 58 (1998), pp. 1748–1773.

[75] A. Figotin and P. Kuchment, *Band-gap structure of the spectrum of periodic Maxwell operators*, J. Statist. Phys., 74 (1994), pp. 447–458.

[76] A. Figotin and P. Kuchment, *Band-gap structure of the spectrum of periodic and acoustic media. I. Scalar model*, SIAM J. Appl. Math., 56 (1996), pp. 68–88.

[77] A. Figotin and P. Kuchment, *Band-gap structure of the spectrum of periodic and acoustic media. II. 2D photonic crystals*, Report 1995-1, Math. Dept., University of North Carolina at Charlotte, 1995.

[78] A. Figotin and P. Kuchment, *Band-gap structure of the spectrum of periodic and acoustic media. II. 2D photonic crystals*, SIAM J. Appl. Math., 56 (1996), pp. 1561–1620. (An abridged version of [77].)

[79] A. Figotin and P. Kuchment, *2D photonic crystals with cubic structure: Asymptotic analysis*, in Wave Propagation in Complex Media, G. Papanicolaou, ed., IMA Vol. Math. Appl. 96, 1997, pp. 23–30.

[80] A. Figotin and P. Kuchment, *Spectral properties of classical waves in high contrast periodic media*, SIAM J. Appl. Math., 58 (1998), pp. 683–702.

[81] A. Figotin and P. Kuchment, *Asymptotic Models of High Contrast Periodic Photonic and Acoustic Media* (tentative title), *Parts* I *and* II, in preparation.

[82] M. Freidlin, *Markov Processes and Differential Equations: Asymptotic Problems*, Lectures Math. ETH Zürich, Birkhäuser-Verlag, Basel, 1996.

[83] M. Freidlin and A. Wentzell, *Diffusion processes on graphs and the averaging principle*, Ann. Probab., 21 (1993), pp. 2215–2245.

[84] R. Froese, I. Herbst, M. Hoffmann-Ostenhof, and T. Hoffmann-Ostenhof, L^2-*lower bounds to solutions of one-body Schrödinger equations*, Proc. Roy. Soc. Edinburgh Sect. A, 95 (1983), pp. 25–38.

[85] I. M. Gelfand, *Expansion in eigenfunctions of an equation with periodic coefficients*, Dokl. Akad. Nauk. SSSR, 73 (1950), pp. 1117–1120.

[86] A. Georgieva, T. Kriecherbauer, and S. Venakides, *Wave propagation and resonance in a one-dimensional nonlinear discrete periodic medium*, SIAM J. Appl. Math., 60 (1999), pp. 272–294.

[87] A. Georgieva, T. Kriecherbauer, and S. Venakides, *1:2 resonance mediated second harmonic generation in a 1-D nonlinear discrete periodic medium*, submitted.

[88] C. Gerard and F. Nier, *The Mourre theory for analytically fibered operators*, J. Funct. Anal., 152 (1998), pp. 202–219.

[89] N. Gerasimenko and B. Pavlov, *Scattering problems on non-compact graphs*, Theoret. Math. Phys., 75 (1988), pp. 230–240.

[90] D. Gieseker, H. Knörrer, and E. Trubowitz, *The Geometry of Algebraic Fermi Curves*, Academic Press, Boston, 1992.

[91] I. M. Glazman, *Direct Methods of Qualitative Spectral Analysis of Singular Differential Operators*, translated from the Russian by the Israel Program for Scientific Translations, Jerusalem, 1965; Daniel Davey & Co., New York, 1966.

[92] E. L. Green, *Spectral theory of Laplace-Beltrami operators with periodic metrics*, J. Differential Equations, 133 (1997), pp. 15–29.

[93] I. Gudovich and S. Krein, *Boundary value problems for overdetermined systems of partial differential equations*, Differencial'nye Uravnenija i Primenen.—Trudy Sem. Processy Optimal. Upravlenija. I Sekcija Vyp., 9 (1974), pp. 1–145 (in Russian).

[94] R. C. Gunning and H. Rossi, *Analytic Functions of Several Complex Variables*, Prentice-Hall, Englewood Cliffs, NJ, 1965.

[95] P. Halevi, A. A. Krokhin, and J. Arriaga, *Photonic crystal optics and homogenization of 2D periodic composites*, Phys. Rev. Lett., 82 (1999), pp. 719–722.

[96] B. Helffer and A. Mohamed, *Asymptotic of the density of states for the Schrödinger operator with periodic electric potential*, Duke Math. J., 92 (1998), pp. 1–60.

[97] R. Hempel, *Second order perturbations of divergence type operators with a spectral gap*, in Operator Calculus and Spectral Theory (Lambrecht, 1991), Oper. Theory Adv. Appl. 57, Birkhäuser, Basel, 1992, pp. 117–126.

[98] R. Hempel and I. Herbst, *Strong magnetic fields, Dirichlet boundaries, and spectral gaps*, Comm. Math. Phys., 164 (1995), pp. 237–259.

[99] R. Hempel and I. Herbst, *Bands and gaps for periodic magnetic Hamiltonians*, in Partial Differential Operators and Mathematical Physics, Oper. Theory Adv. Appl. 78, Birkhäuser Basel, 1995, pp. 175–184.

[100] R. Hempel and K. Lienau, *Spectral properties of periodic media in the large coupling limit*, Comm. Partial Differential Equations, 25 (2000), pp. 1445–1470.

[101] P. M. Hui and N. F. Johnson, *Photonic band-gap materials*, in Solid State Physics, Vol. 49, H. Ehrenreich and F. Spaepen, eds., Academic Press, New York, 1995, pp. 151–203.

[102] K. Inoue, M. Sasada, J. Kuwamata, K. Sakoda, and J. W. Haus, *A two-dimensional photonic crystal laser*, Japan J. Appl. Phys. 2, 38 (1999), pp. L157–L159.

[103] J. D. Jackson, *Classical Electrodynamics*, Wiley, New York, 1962.

[104] V. V. Jikov, S. M. Kozlov, and O. A. Oleinik, *Homogenization of Differential Operators and Integral Functionals*, Springer-Verlag, Berlin, 1994.

[105] J. Jin, *The Finite Element Method in Electromagnetics*, Wiley, New York, 1993.

[106] J. D. Joannopoulos, R. D. Meade, and J. N. Winn, *Photonic Crystals, Molding the Flow of Light*, Princeton University Press, Princeton, NJ, 1995.

[107] S. John, *Strong localization of photons in certain disordered dielectric superlattices*, Phys. Rev. Lett., 58 (1987), pp. 2486–2489.

[108] S. John, *The localization of waves in disordered media*, in Scattering and Localization of Classical Waves in Random Media, Ping Sheng, ed., World Scientific, Singapore, 1990, pp. 1–66.

[109] S. John, *Localization of light*, Phys. Today, May 1991.

[110] Yu. Karpeshina, *Geometrical background for the perturbation theory of the polyharmonic operator with periodic potential*, in Topological Phases in Quantum Theory (Dubna, 1988), World Scientific, Teaneck, NJ, 1989, pp. 251–276.

[111] Yu. Karpeshina, *Analytic perturbation theory for a periodic potential*, Izv. Akad. Nauk SSSR Ser. Math., 52 (1989), no. 1, pp. 45–65; English translation in Math. USSR-Izvestiya, 34 (1990).

[112] Yu. Karpeshina, *Perturbation theory for Schrödinger operator with a periodic potential*, in Schrödinger Operators. Standard and Non-Standard, World Scientific, Teaneck, NJ, 1990, pp. 131–145.

[113] Yu. Karpeshina, *Perturbation theory for the Schrödinger operator with a periodic potential*, in Proc. Steklov Inst. Math., 3 (1991), pp. 109–145.

[114] Yu. Karpeshina, *Perturbation formula for the Schrödinger operator with a nonsmooth periodic potential*, Math. USSR Sb., 71 (1992), pp. 101–124.

[115] Yu. Karpeshina, *On the density of states for a periodic Schrödinger operator*, preprint, Ark. Mat. 38 (2000), no. 1, pp. 111–137.

[116] Yu. Karpeshina, *Perturbation Theory for the Schrödinger Operator with a Periodic Potential*, Lecture Notes in Math. 1663, Springer-Verlag, New York, 1997.

[117] W. Kirsch and B. Simon, *Comparison theorems for the gap of Schrödinger operators*, J. Funct. Anal., 75 (1987), pp. 396–410.

[118] F. Klopp, *Resonances for perturbations of a semi-classical periodic Schrödinger operator*, Ark. Mat., 32 (1994), pp. 323–371.

[119] F. Klopp and J. Ralston, *Endpoints of the Spectrum of Periodic Operators Are Generically Simple*, preprint, 1999.

[120] H. Knörrer and E. Trubowitz, *A directional compactification of the complex Bloch variety*, Comm. Math. Helv., 65 (1990), pp. 114–149.

[121] W. Kohn, *Analytic properties of Bloch waves and Wannier functions*, Phys. Rev., 115 (1959), pp. 809–821.

[122] A. A. Krokhin, J. Arriaga, and P. Halevi, *Speed of light in a 2D photonic crystal in the low-frequency limit*, Phys. A, 241 (1997), pp. 52–57.

[123] P. Kuchment, *Floquet theory for partial differential equations*, Russian Math. Surveys, 37 (1982), pp. 1–60.

[124] P. Kuchment, *Floquet Theory for Partial Differential Equations*, Birkhäuser, Basel, 1993.

[125] P. Kuchment, *To the Floquet theory of periodic difference equations*, in Geometrical and Algebraical Aspects in Several Complex Variables, Cetraro, Italy, June 1989, EditEl, Naples, 1991, pp. 203–209.

[126] P. Kuchment and L. Kunyansky, *Spectral properties of high contrast band-gap materials and operators on graphs*, Experiment. Math., 8 (1999), pp. 1–28.

[127] P. Kuchment and L. Kunyansky, *Differential operators on graphs and photonic crystals*, to be submitted to Adv. Comp. Math.

[128] P. Kuchment and S. Levendorskiî, *On the absolute continuity of spectra of periodic elliptic operators*, in Mathematical Results in Quantum Mechanics (QMath7, Prague, June 22–26, 1998), J. Dittrich, P. Exner, et al., eds., Operator Theory Adv. Appl. 108, Birkhäuser, Basel, 1999, pp. 291–297.

[129] P. Kuchment and S. Levendorskiî, *On the structure of spectra of periodic elliptic operators*, preprint 00-388, in http://www.ma.utexas.edu/mp_arc, submitted.

[130] P. Kuchment and Y. Pinchover, *Integral representations and Liouville theorems for solutions of periodic elliptic equations*, to appear in J. Funct. Anal.

[131] P. Kuchment and B. Vainberg, *On embedded eigenvalues of perturbed periodic Schrödinger operators*, in Spectral and Scattering Theory (Newark, DE, 1997), Plenum, New York, 1998, pp. 67–75.

[132] P. Kuchment and B. Vainberg, *Absence of embedded eigenvalues for perturbed Schrödinger operators with periodic potentials*, Comm. PDE, 25 (2000), pp. 1809–1826.

[133] P. Kuchment and H. Zeng, *Convergence of Spectra of Mesoscopic Systems Collapsing onto a Graph*, preprint 00-308 in http://www.ma.utexas.edu/mp_arc, to appear in J. Math. Anal. Appl.

[134] G. Kurizki and J. W. Haus, eds., *Photonic band structures*, J. Mod. Opt., 41 (1994), no. 2, a special issue.

[135] L. D. Landau and E. M. Lifshitz, *The Classical Theory of Fields*, 4th ed., Pergamon Press, Oxford, New York, 1987.

[136] E. M. Landis, *Some problems in the qualitative theory of second-order elliptic equations*, Russian Math. Surveys, 18 (1963), pp. 1–62.

[137] E. M. Landis, *On the behavior of solutions of higher order elliptic equations in unbounded domains*, Trans. Moscow Math. Soc., 31 (1976), pp. 30–54.

[138] A. R. McGurn, *Green's-function theory for row and periodic defect arrays in photonic band structures*, Phys. Rev. B, 53 (1996), pp. 7059–7064.

[139] A. Mekis, J. C. Chen, I. Kurland, S. Fan, P. Villeneuve, and J. D. Joannopoulos, *High transmission through sharp bends in photonic crystal waveguides*, Phys. Rev. Lett., 77 (1996), pp. 3787–3790.

[140] V. Meshkov, *On the possible rate of decay at infinity of solutions of second order partial differential equations*, Mat. Sb., 182 (1991), pp. 364–383; English translation in Math. USSR Sb., 72 (1992), pp. 343–351.

[141] A. Mohamed, *Asymptotic of the density of states for the Schrödinger operator with periodic electromagnetic potential*, J. Math. Phys., 38 (1997), pp. 4023–4051.

[142] A. Morame, *Absence of singular spectrum for a perturbation of a two-dimensional Laplace-Beltrami operator with periodic electro-magnetic potential*, J. Phys. A, 31 (1998), pp. 7593–7601.

[143] A. Morame, *The Absolute Continuity of the Spectrum of Maxwell Operator in Periodic Media*, Preprint #99-308 in the Texas Math Physics archive http://www.ma.utexas.edu/mp_arc

[144] J.-C. Nédélec, *Mixed finite elements in $\mathbb{R}^3$*, Numer. Math., 35 (1980), pp. 315–341.

[145] S. Novikov, *Two-dimensional Schrödinger operators in the periodic fields*, in Current Problems in Mathematics 23, VINITI, Moscow, 1983, pp. 3–32.

[146] S. Novikov, *Schrödinger operators on graphs and topology*, Russian Math Surveys, 52 (1997), pp. 177–178.

[147] S. Novikov, *Discrete Schrödinger operators and topology*, Asian Math. J., 2 (1999), pp. 841–853.

[148] S. Novikov, *Schrödinger operators on graphs and symplectic geometry*, in The Arnoldfest: Proceedings of a Conference in Honour of V. I. Arnold for His Sixtieth Birthday, E. Bierstone, B. Khesin, A. Khovanskii, and J. E. Marsden, eds., AMS, Providence, RI, 1999.

[149] F. Odeh and J. B. Keller, *Partial differential equations with periodic coefficients and Bloch waves in crystals*, J. Math. Phys., 5 (1964), pp. 1499–1504.

[150] V. Palamodov, *Harmonic synthesis of solutions of elliptic equations with periodic coefficients*, Ann. Inst. Fourier, 43 (1993), pp. 751–768.

[151] V. Papanicolaou, private communication, April 1999.

[152] J. B. Pendry, *Calculating photonic band structure*, J. Phys.: Condens. Matter, 8 (1996), pp. 1085–1108.

[153] Photonic and Acoustic Band-Gap Bibliography, http://home.earthlink.net/~jpdowling/pbgbib.html

[154] *Photonic Crystals and Photonic Microstructures*, Special issue of IEEE Proceedings—Optoelectronics, 145 (1998), no. 6.

[155] I. Ponomarev, *Separation of variables in the computation of spectra in 2D photonic crystals*, SIAM J. Appl. Math., 61 (2000), pp. 1202–1218.

[156] J. Rarity and C. Weisbuch, eds., *Microcavities and Photonic Bandgaps: Physics and Applications*, Proceedings of the NATO Advanced Study Institute: Quantum Optics in Wavelength-Scale Structures, Cargese, Corsica, August 26–September 2, 1995, NATO Adv. Sci. Inst., Kluwer, Dordrecht, the Netherlands, 1996.

[157] M. Reed and B. Simon, *Methods of Modern Mathematical Physics, Vol. IV: Analysis of Operators*, Academic Press, New York, 1978.

[158] F. S. Rofe-Beketov, *A test for the finiteness of the number of discrete levels introduced into the gaps of a continuous spectrum by perturbations of a periodic potential*, Soviet Math. Dokl., 5 (1964), pp. 689–692.

[159] F. S. Rofe-Beketov, *Spectrum perturbations, the Knezer-type constants and the effective mass of zones-type potentials*, in Constructive Theory of Functions 84, Bulgarian Academy of Sciences, Sofia, 1984, pp. 757–766.

[160] J. Rubinstein and M. Schatzman, *Spectral and variational problems on multiconnected strips*, C. R. Acad. Sci. Paris Sér. I Math., 325 (1997), pp. 377–382.

[161] J. Rubinstein and M. Schatzman, *Asymptotics for thin superconducting rings*, J. Math. Pures Appl. (9), 77 (1998), pp. 801–820.

[162] J. Rubinstein and M. Schatzman, *On multiply connected mesoscopic superconducting structures*, Sémin. Théor. Spectr. Géom. 15, Univ. Grenoble I, Saint-Martin-d'Hères, France, 1998, pp. 207–220.

[163] J. Rubinstein and M. Schatzman, *Variational Problems on Multiply Connected Thin Strips* I: *Basic Estimates and Convergence of the Laplacian Spectrum*, preprint, 1999.

[164] K. Ruedenberg and C. W. Scherr, *Free-electron network model for conjugated systems. I. Theory*, J. Chem. Phys., 21 (1953), pp. 1565–1581.

[165] Y. Saito, *Convergence of the Neumann Laplacians on Shrinking Domains*, preprint, 1999.

[166] M. Schatzman, *On the eigenvalues of the Laplace operator on a thin set with Neumann boundary conditions*, Appl. Anal., 61 (1996), pp. 293–306.

[167] A. Scherer, T. Doll, E. Yablonovitch, H. O. Everett, and J. A. Higgins, eds., *Special section on electromagnetic crystal structures, design, synthesis, and applications (optical)*, J. Lightwave Techn., 17 (1999), no. 11.

[168] A. Scherer, T. Doll, E. Yablonovitch, H. O. Everett, and J. A. Higgins, eds., *Special section on electromagnetic crystal structures, design, synthesis, and applications (microwave)*, IEEE Transactions on Microwave Theory and Techniques, 47 (1999), no. 11.

[169] H. R. Schwarz, *Finite Element Methods*, Academic Press, London, 1988.

[170] Z. Shen, *On absolute continuity of the periodic Schrödinger operators*, Preprint ESI 597, 1998, http://www.esi.ac.at, to appear in Internat. Math. Res. Notes.

[171] Z. Shen, *The Periodic Schrödinger Operator with Potentials in the Morrey-Companato Class*, Preprint #99-15, Math. Dept., Univ. of Kentucky, Lexington, KY, 1999 and #99-455 in the Texas Math Physics archive, 1999, http://www.ma.utexas.edu/mp_arc.

[172] T. J. Shepherd and P. J. Roberts, *Soluble two-dimensional photonic-crystal model*, Phys. Rev. E, 55 (1997), pp. 6024–6038.

[173] M. Shubin, *Spectral theory and the index of elliptic operators with almost periodic coefficients*, Uspekhi Mat. Nauk, 34 (1979), pp. 95–135; English translation in Russian Math. Surveys, 34 (1979), pp. 109–157.

[174] J. Sjostrand, *Microlocal analysis for the periodic magnetic Schrödinger equation and related questions*, in Microlocal Analysis and Applications, Lecture Notes in Math. 1495, Springer-Verlag, Berlin, 1991, pp. 237–332.

[175] M. M. Skriganov, *Proof of the Bethe-Sommerfeld conjecture in dimension two*, Soviet Math. Dokl., 20 (1979), pp. 956–959.

[176] M. M. Skriganov, *On the Bethe-Sommerfeld conjecture*, Soviet Math. Dokl., 20 (1979), pp. 89–90.

[177] M. M. Skriganov, *Proof of the Bethe-Sommerfeld Conjecture in Dimension Three*, preprint, LOMI, P-6-84, Leningrad, 1984 (in Russian).

[178] M. M. Skriganov, *The spectrum band structure of the three dimensional Schrödinger operator with periodic potential*, Invent. Math., 80 (1985), pp. 107–121.

[179] M. M. Skriganov, *Geometric and arithmetic methods in the spectral theory of multidimensional periodic operators*, Proc. Steklov Inst. Math., 171 (1985), pp. 1–117; English translation in Proc. Steklov Inst. Math., 1987, no. 2.

[180] A. Sobolev, *Absolute continuity of the periodic magnetic Schrödinger operator*, Invent. Math., 137 (1999), pp. 85–112.

[181] A. Sobolev, *A Lecture at the Spectral Theory Workshop*, International E. Schrödinger Institute, Matrei, Austria, July 1999.

[182] C. M. Soukoulis, ed., *Photonic band gap materials*, Proceedings of the NATO ASI on Photonic Band Gap Materials, Elounda, Crete, Greece, June 18–30, 1995, NATO Adv. Sci. Inst. Series, Kluwer, Dordrecht, the Netherlands, 1996.

[183] J. Sylvester and G. Uhlmann, *Inverse boundary value problems at the boundary-continuous dependence*, Comm. Pure Appl. Math., XLI (1988), pp. 197–219.

[184] L. E. Thomas, *Time dependent approach to scattering from impurities in a crystal*, Comm. Math. Phys., 33 (1973), pp. 335–343.

[185] E. C. Titchmarsh, *Eigenfunction Expansions Associated with Second-Order Differential Equations, Part* II, Clarendon Press, Oxford, UK, 1958.

[186] G. Uhlmann, *Inverse boundary value problems and applications*, Astérisque, 207 (1992), pp. 153–211.

[187] O. A. Veliev, *Asymptotic formulas for the eigenvalues of the multidimensional Schrödinger operator and Bethe-Sommerfeld conjecture*, Funktsional. Anal. i Prilozhen., 21 (1987), pp. 1–15; English translation in Funct. Anal. Appl., 21 (1987), pp. 87–99.

[188] S. Venakides, M. Haider, and V. Papanicolaou, *Boundary integral calculations of 2-d electromagnetic scattering by photonic crystal Fabry–Perot structures*, SIAM J. Appl. Math., 60 (2000), pp. 1686–1706.

[189] P. R. Villaneuve and M. Piché, *Photonic band gaps in periodic dielectric structures*, Prog. Quant. Electr., 18 (1994), pp. 153–200.

[190] Yurii A. Vlasov, *The ultimate collection of photonic band gap research links*, http://www.neci.nj.nec.com/homepages/vlasov/photonic.html

[191] C. Wilcox, *Theory of Bloch waves*, J. Anal. Math., 33 (1978), pp. 146–167.

[192] E. Yablonovitch, *Inhibited spontaneous emission in solid-state physics and electronics*, Phys. Rev. Lett., 58 (1987), pp. 2059–2062.

[193] M. Zaidenberg, S. Krein, P. Kuchment, and A. Pankov, *Banach bundles and linear operators*, Russian Math. Surveys, 30 (1975), pp. 115–175.

Chapter 8

Mathematical Analysis and Numerical Approximation of Optical Waveguides

Anne-Sophie Bonnet-Ben Dhia and
Patrick Joly

The importance of optical telecommunications has dramatically increased since the 1960s, based on both the use of *optical fibers* (Figure 8.1) for long-distance transmissions [24] and the use of miniaturized *integrated optical circuits* instead of the classical electrical ones [17]. Among the various advantages of these optical systems over their electrical counterpart, let us mention their low losses, their high immunity from electromagnetic interferences, and their small size and light weight.

This is why optical waveguides have recently been the object of many theoretical and numerical studies: these studies are not restricted to the optical fiber, which can be seen as the most typical example of optical waveguide, but they also deal with various stratified optical microguides (such as *the rib waveguide*), which are fundamental elements of the optical circuits.

Let us first point out that an optical waveguide is what is usually called an *open* waveguide, as opposed to the *closed* electromagnetic waveguides: this means that, while the latter have a bounded cross section limited by a metallic wall, optical waveguides are supposed to have an unbounded cross section, which gives rise to specific difficulties from both the theoretical and the numerical points of view.

In open waveguides, as in closed ones, the propagation may be analyzed thanks to a modal analysis. The *modes* of a waveguide are waves of the form

$$\Re(u(x_1, x_2)e^{i(\omega t - \beta x_3)}), \quad \omega \in \mathbb{R}, \quad \beta \in \mathbb{R}, \tag{8.1}$$

where x_3 denotes the coordinate in the direction of invariance of the guide, while x_1 and x_2 denote the transverse coordinates. A mode is said to be *guided* if it has a finite transverse energy and *radiating* otherwise.

An important difference between closed and open waveguides is that, while the propagation in closed waveguides is completely described with the help of a countable infinity of guided modes (see, for instance, [11]), the theory of open waveguides includes two types of modes, the guided modes and the radiating ones. For a given β, the former generally constitute a finite set and the latter a continuum. Moreover, in an open waveguide, the existence of guided modes, which is straightforward in a closed waveguide, depends on the optical characteristics of the waveguide and on the frequency.

The equations that characterize the guided modes are derived by plugging the expression (8.1) into the second-order *Maxwell's equations*. This way, the problem can be written (at least formally) as *a parameterized eigenvalue problem* of the form

$$A(\beta)u = \omega^2 u,$$

where $A(\beta)$ is a symmetric operator depending on the parameter β. In other words, the determination of the guided modes is equivalent to the determination of the eigenvalues of the operator $A(\beta)$. From the mathematical point of view, the main difference between open and closed waveguides is that $A(\beta)$ has a compact resolvent for the case of closed waveguides, but not for the case of open ones.

Notice that the practical problem to solve is generally the inverse one, which consists of the determination of β as a function of ω. However, this inverse problem has unusual mathematical properties, and its theory is still not satisfactory, which is why it will not be considered in this book. The two approaches ($\omega \to \beta$ and $\beta \to \omega$) could probably be related by going back to the time-dependent problem, using techniques of deformation of paths in the complex plane. The *leaky modes* [28] ($\beta \in \mathbb{C}\backslash\mathbb{R}$) appear naturally in this context, as the poles of a meromorphic function [41].

There are two basic examples of optical waveguides that can be completely studied by analytical calculations. The first one is the three-layers planar waveguide (also called the *slab* waveguide) composed of three layers of dielectric materials, the central one having the smallest light velocity. The second one is the circular step-index fiber which is made up of a glass core, whose cross section is circular, surrounded by a cladding with another dielectric material. Again, the light velocity is supposed to be lower in the core than in the cladding. For these two examples, an explicit expression of *the dispersion relation* of the guided modes, which links the pulsation ω to the propagation constant β, may be derived and analyzed (see, for example, [28], [29], [38]).

However, for general open waveguides, analytical solutions are not available. Thus several theoretical and numerical methods have been developed to estimate and/or approximate the guided modes.

- The theoretical methods provide mathematical proofs of existence, or nonexistence, of guided modes for arbitrary waveguides, characterized by their velocity profiles. Generally, this profile must satisfy some "guiding condition" to ensure the existence of guided modes; this condition expresses the fact that the velocity is minimal in the region where the mode is confined. Theoretical studies also provide estimates for the number of guided modes as a function of the parameter β.

The tools used are those of the modern analysis of partial differential equations, especially those concerning *the spectral theory of linear self-adjoint operators*. We assume here that the reader is familiar with basic concepts of abstract and applied functional analysis (in particular, Sobolev spaces). However, for convenience, we have given in the appendix a brief summary of the results concerning self-adjoint operators and their spectral theory.

- The numerical methods provide approximations of the dispersion curves $\omega_j(\beta)$ and of the associated fields $u_j(\beta)$. Most of them couple analytical representations of the field and finite element approximations and lead to a nonlinear finite-dimensional eigenvalue problem of the form

$$A(\beta,\omega)u = \omega^2 u.$$

Contrary to closed waveguides, the analysis of open waveguides cannot be carried out for a completely arbitrary velocity profile. Indeed, both theoretical and numerical techniques are based on a fundamental assumption, which says that the cross section of the waveguide differs only by a compactly supported perturbation from a simple reference medium. In practice, *simple* means that the reference medium can be completely analyzed by explicit calculations. This assumption is satisfied by optical fibers: the reference medium is then simply a homogeneous one. It also holds for optical integrated waveguides, the reference medium in this case being a stratified medium. In the following, we will speak of an optical fiber when considering a waveguide with a homogeneous cladding, and of an optical integrated waveguide when considering a waveguide with a stratified cladding. The results as well as the methods depend critically on the nature of the cladding. Optical fibers and optical integrated waveguides will therefore be studied in two different sections.

The outline of the chapter is the following. The general mathematical framework is given in the first section. The theoretical study of the optical fiber is the object of the second section (it is an abridged version of [20], which completes and extends the results of [2] and [4]). The third section is devoted to the theoretical study of integrated optical waveguides. The results are new, generalizing to the vector model those of [12] and [13] established for the scalar weakly guiding approximation. In the last section, we give a detailed presentation of a numerical method for the computation of the guided modes in an optical fiber (which is again an abridged version of [22]). The general features of the method are pointed out and the extension to other waveguides is discussed.

8.1 Mathematical Formulation of the Problem

8.1.1 Position of the Problem

We consider a three-dimensional dielectric linear isotropic medium occupying the whole space $\mathbb{R}^3$. We denote by (x, x_3), with $x = (x_1, x_2) \in \mathbb{R}^2$, the generic point of $\mathbb{R}^3$. We assume that the propagation medium has a cylindrical structure in the sense that it is invariant under any translation in the x_3-direction. This means that the dielectric permittivity ε and the magnetic permeability μ are functions of the only transverse variable x. See Figure 8.2. We assume that they are measurable, strictly positive, and bounded functions. We introduce

$$\begin{cases} \varepsilon_- = \inf_{x \in \mathbb{R}^2} \varepsilon(x) > 0, & \varepsilon^+ = \sup_{x \in \mathbb{R}^2} \varepsilon(x) < +\infty, \\ \mu_- = \inf_{x \in \mathbb{R}^2} \mu(x) > 0, & \mu^+ = \sup_{x \in \mathbb{R}^2} \mu(x) < +\infty, \end{cases} \quad (8.2)$$

and the local propagation velocity $c(x)$:

$$c(x)^2 = (\varepsilon(x)\mu(x))^{-1}. \quad (8.3)$$

We shall assume piecewise regularity, which is not restrictive with respect to the applications (these assumptions can be weakened; see [42]):

$$\begin{cases} \mathbb{R}^2 = \overline{\Omega_0} \cup \overline{\Omega_1} \cup \cdots \cup \overline{\Omega_N}, & \Omega_j \cap \Omega_l = \emptyset \text{ for } j \neq l, \\ \text{for all } 0 \leq j \leq N, & \varepsilon = \varepsilon_j, \ \mu = \mu_j \ \text{in } \Omega_j, (\varepsilon_j, \mu_j) \in W^{2,\infty}(\mathbb{R}^2)^2. \end{cases} \quad (8.4)$$

As we said in the introduction, the propagation medium is supposed to be a local perturbation (local with respect to the transverse variables) of another cylindrical reference medium characterized by $\varepsilon_{ref}(x), \mu_{ref}(x)$.

$$\exists R > 0 \ / \ |x| \geq R \ \Rightarrow \ \varepsilon(x) = \varepsilon_{ref}(x), \ \mu(x) = \mu_{ref}(x). \quad (8.5)$$

The two examples (Figure 8.1 and Figure 8.3) we treat differ by the nature of the reference medium.

(i) For the *optical fiber*, the reference medium is homogeneous:

$$\varepsilon_{ref}(x) = \varepsilon_\infty > 0, \quad \mu_{ref}(x) = \mu_\infty > 0. \quad (8.6)$$

(ii) For the *integrated optical guide*, the reference medium is stratified:

$$\varepsilon_{ref}(x) = \varepsilon_\infty(x_2), \quad \mu_{ref}(x) = \mu_\infty(x_2), \quad (8.7)$$

where we assume the existence of $a > 0$ such that

$$\begin{array}{l} x_2 \geq a \ \Rightarrow \ \varepsilon_\infty(x_2) = \varepsilon_1 > 0, \ \mu_\infty(x_2) = \mu_1 > 0, \\ x_2 \leq -a \ \Rightarrow \ \varepsilon_\infty(x_2) = \varepsilon_2 > 0, \ \mu_\infty(x_2) = \mu_2 > 0. \end{array} \quad (8.8)$$

We suppose, for example, that

$$\varepsilon_1 \mu_1 \geq \varepsilon_2 \mu_2. \quad (8.9)$$

Such a reference medium will be called a *planar waveguide*.

The electric field $\mathbb{E}$ and the magnetic field $\mathbb{H}$ in such media are governed by Maxwell's equations:

$$\varepsilon \frac{\partial \mathbb{E}}{\partial t} - \text{rot}\,\mathbb{H} = 0, \quad \mu \frac{\partial \mathbb{H}}{\partial t} + \text{rot}\,\mathbb{E} = 0. \quad (8.10)$$

Chapter 8. Optical Waveguides   277

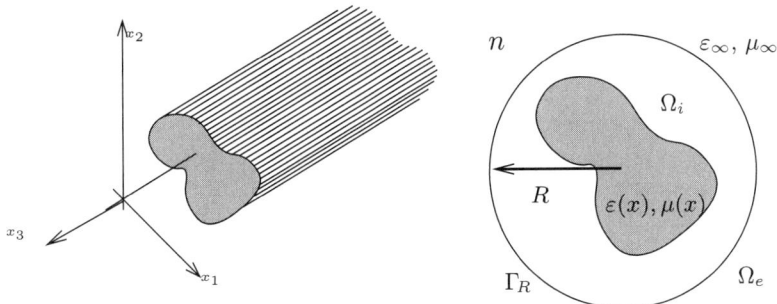

Figure 8.1: The optical fiber.   Figure 8.2: The cross section.

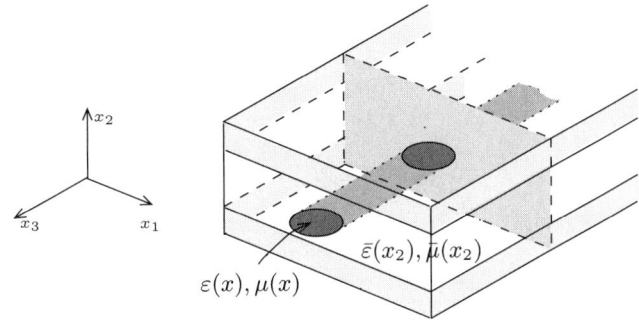

Figure 8.3: Integrated optical waveguide.

In this chapter, we shall be concerned with very particular solutions of (8.10) called *guided modes* or *guided waves*. More generally, we define a *mode* as a solution of the form (V^t stands for the transpose of V):

$$\begin{cases} \mathbb{E}(x, x_3, t) = \Re e \left\{ (E_1(x), E_2(x), E_3(x))^t \exp i(\omega t - \beta x_3) \right\}, \\ \mathbb{H}(x, x_3, t) = \Re e \left\{ (H_1(x), H_2(x), H_3(x))^t \exp i(\omega t - \beta x_3) \right\}, \end{cases} \quad (8.11)$$

where

- $\omega > 0$ is the pulsation of the wave,
- $\beta > 0$ is the wavenumber in the x_3-direction.

(8.11) can be seen as a harmonic plane wave propagating without any distortion in the x_3-direction with a velocity $V = \omega/\beta$ (the phase velocity). A *guided mode* is, by definition, a mode that satisfies the additional condition

$$\int_{\mathbb{R}^2} (\varepsilon \mid E \mid^2 + \mu \mid H \mid^2) \, dx < +\infty, \quad (8.12)$$

which expresses that the transverse electromagnetic energy is finite. In practice, this means that the energy of the mode remains confined in some bounded region of the cross section. That is why, for instance, guided waves differ from the usual plane waves in a homogeneous medium. Of course, the fact that the coefficients $\varepsilon(x)$ and $\mu(x)$ vary at least locally plays a fundamental role: when these coefficients are constant or if they depend only on one space variable (i.e., for the reference media), it is well known that guided waves do not exist. Moreover, even when they exist, guided modes do not exist for any value of ω and β; ω and β must be linked by some relation $\omega = f(\beta)$, called the *dispersion relation*. Therefore, the two main questions we would like to address are as follows:

(i) What conditions on $\varepsilon(x)$ and $\mu(x)$ can ensure the existence of guided waves?

(ii) What are the properties of the dispersion relation of the modes?

We derive the equations of the problem by substituting (8.11) into (8.10). We eliminate H and introduce the new unknown $u = (u_1, u_2, u_3)$, where

$$u_1 = E_1, \quad u_2 = E_2, \quad u_3 = iE_3, \qquad (8.13)$$

which leads to a system of equations with real coefficients (the E-formulation):

$$\varepsilon^{-1} \mathrm{rot}_\beta^* (\mu^{-1} \mathrm{rot}_\beta u) = \omega^2 u, \qquad (8.14)$$

where the differential operator rot_β is defined for $u \in \mathcal{D}'(\mathbb{R}^2)$ by

$$\mathrm{rot}_\beta u = \left(\frac{\partial u_3}{\partial x_2} - \beta u_2,\ \beta u_1 - \frac{\partial u_3}{\partial x_1},\ \frac{\partial u_2}{\partial x_1} - \frac{\partial u_1}{\partial x_2} \right)^t \qquad (8.15)$$

and rot_β^* is the adjoint of rot_β: $\mathrm{rot}_\beta^* = \mathrm{rot}_{-\beta}$. Condition (8.12) is equivalent to

$$\int_{\mathbb{R}^2} (\varepsilon \mid u \mid^2 + \mu^{-1} \mid \mathrm{rot}_\beta u \mid^2)\, dx < +\infty. \qquad (8.16)$$

Remark 8.1. The choice of eliminating H is arbitrary. We could have eliminated E to end up with the H-formulation, which looks like (8.14) after permutation of the rôles of ε and μ. Playing with the two formulations can be useful (cf. Theorem 8.7, for instance) for symmetrizing some results with respect to ε and μ.

The problem (8.14), (8.16) appears as a family of eigenvalue problems for second-order differential operators parametrized by β in which ω^2 plays the role of the eigenvalue and u the role of the corresponding eigenvector. Note that even if the original problem was posed in $\mathbb{R}^3$, (8.14), (8.16) is a two-dimensional problem posed in the cross section of the propagation medium (namely, $\mathbb{R}^2$). Except for the case of the circular step-index optical fiber (see [29] for instance), one cannot generally expect to solve such a problem explicitly. That is why we need to use functional analysis methods to get existence results as well as qualitative properties and/or quantitative estimates for the solutions and numerical methods for their effective computations. This requires an appropriate mathematical framework.

8.1.2 Mathematical Setting

We first introduce our functional framework. Consider the Hilbert space

$$H_\varepsilon = L^2(\mathbb{R}^2)^3, \qquad (8.17)$$

Chapter 8. Optical Waveguides

which is equipped with the scalar product

$$(u,v)_\varepsilon = \int_{\mathbb{R}^2} \varepsilon u.v \, dx. \tag{8.18}$$

The associated norm is denoted by $|u|_\varepsilon$. In what follows, for any three-dimensional vector field $u(x) = (u_1(x), u_2(x), u_3(x))$, we shall denote by $\boldsymbol{u} = (u_1, u_2)$ the transverse field, so that we can write u or $(\boldsymbol{u}, u_3)$ indifferently. Define also

$$V_\varepsilon = \{u \in H_\varepsilon \ / \ \mathrm{rot}_\beta u \in H_\varepsilon\}. \tag{8.19}$$

It is immediate to verify that

$$V_\varepsilon = \{u = (\boldsymbol{u}, u_3) \in H(\mathrm{rot}; \mathbb{R}^2) \times H^1(\mathbb{R}^2)\}, \tag{8.20}$$

where

$$H(\mathrm{rot}; \mathbb{R}^2) = \left\{\boldsymbol{u} \in L^2(\mathbb{R}^2)^2 \ / \ \mathrm{rot}\boldsymbol{u} = \frac{\partial u_2}{\partial x_1} - \frac{\partial u_1}{\partial x_2} \in L^2(\mathbb{R}^2)\right\}.$$

The space V_ε is a Hilbert space for the norm

$$\|u\|_{V_\varepsilon}^2 = \int_{\mathbb{R}^2} (|u|^2 + |\nabla u_3|^2 + |\mathrm{rot}\boldsymbol{u}|^2) \, dx. \tag{8.21}$$

Finally, we denote by $\tilde{A}(\beta)$ the unbounded operator in H_ε defined by

$$\begin{cases} D(\tilde{A}(\beta)) = \{u \in V_\varepsilon \ / \ \mathrm{rot}_\beta^*(\mu^{-1}\mathrm{rot}_\beta u) \in H_\varepsilon\}, \\ \tilde{A}(\beta)u = \varepsilon^{-1}\mathrm{rot}_\beta^*(\mu^{-1}\mathrm{rot}_\beta u) \quad \text{for all } u \in D(\tilde{A}(\beta)). \end{cases} \tag{8.22}$$

Since we are only interested in nonzero eigenvalues ($\omega > 0$), a first step of the formulation is to get rid of the infinite-dimensional kernel of $\tilde{A}(\beta)$. This is a classical difficulty with Maxwell's equations. It is linked to nondivergence-free vector fields. Let us introduce the differential operators

$$\begin{aligned} \mathrm{div}_\beta u &= \frac{\partial u_1}{\partial x_1} + \frac{\partial u_2}{\partial x_2} - \beta u_3 = \mathrm{div}\,\boldsymbol{u} - \beta u_3, \\ \nabla_\beta \varphi &= \left(\frac{\partial \varphi}{\partial x_1}, \frac{\partial \varphi}{\partial x_2}, \beta\varphi\right)^t. \end{aligned} \tag{8.23}$$

We have

$$\mathrm{div}_\beta(\mathrm{rot}_\beta^*) = 0, \quad \mathrm{rot}_\beta(\nabla_\beta) = 0. \tag{8.24}$$

Therefore, if u solves (8.14) with $\omega \neq 0$, it satisfies

$$\mathrm{div}_\beta(\varepsilon u) = 0. \tag{8.25}$$

On the other hand, for any φ in $H^1(\mathbb{R}^2)$, $v = \nabla_\beta \varphi$ belongs to $D(\tilde{A}(\beta))$ and $\tilde{A}(\beta)v = 0$. We exploit this observation in the next two lemmas.

LEMMA 8.1. *The following orthogonal decomposition holds (with respect to $(.,.)_\varepsilon$):*

for all $v \in H_\varepsilon$, $v = u + \nabla_\beta \varphi$, $u \in H_\varepsilon(\beta)$, $\varphi \in H^1(\mathbb{R}^2)$, where $H_\varepsilon(\beta)$ is the closed subspace of H_ε defined by

$$H_\varepsilon(\beta) = \{u \in H_\varepsilon \ / \ \mathrm{div}_\beta(\varepsilon u) = 0\}. \tag{8.26}$$

Proof. Introduce the unique solution φ in $H^1(\mathbb{R}^2)$ of $\mathrm{div}_\beta(\varepsilon \nabla_\beta \varphi) = \mathrm{div}_\beta(\varepsilon v)$ and set $u = v - \nabla_\beta \varphi$. This gives the decomposition. The orthogonality stems from Green's formula. □

LEMMA 8.2.

(a) Kernel $\tilde{A}(\beta) = \{\nabla_\beta \varphi, \ \varphi \in H^1(\mathbb{R}^2)\}$,
(b) Range $\tilde{A}(\beta) \subset H_\varepsilon(\beta)$.

Proof. The inclusion (b) is a consequence of (8.24), which also implies that $\{\nabla_\beta \varphi, \varphi \in H^1(\mathbb{R}^2)\} \subset$ Kernel $\tilde{A}(\beta)$. Reciprocally, if u belongs to Kernel $\tilde{A}(\beta)$, then by Green's formula, we have

$$\int_{\mathbb{R}^2} \mu^{-1} |\mathrm{rot}_\beta u|^2 \, dx = 0 \implies \mathrm{rot}_\beta u = 0.$$

Coming back to the definition of rot_β (see (8.15)), we get $u = \nabla_\beta \varphi$, with $\varphi = \frac{1}{\beta} u_3$, which completes the proof. □

Thus, we only need to work with the restriction of the operator $\tilde{A}(\beta)$ to the Hilbert space $H_\varepsilon(\beta)$, considered as an unbounded operator $A(\beta)$ in $H_\varepsilon(\beta)$:

$$\begin{cases} D(A(\beta)) = \{u \in V_\varepsilon \cap H_\varepsilon(\beta) \ / \ \mathrm{rot}_\beta^*(\mu^{-1}\mathrm{rot}_\beta u) \in L^2(\mathbb{R}^2)^3\}, \\ A(\beta)u = \varepsilon^{-1}\mathrm{rot}_\beta^*(\mu^{-1}\mathrm{rot}_\beta u), \end{cases} \tag{8.27}$$

and the formulation of the problem is

$$\begin{cases} \text{Find } u \in D(A(\beta)), \ \omega^2 > 0 \ / \\ A(\beta)u = \omega^2 u, \ u \neq 0. \end{cases} \tag{8.28}$$

8.1.3 Self-Adjointness

Introduce the bilinear form on $V_\varepsilon(\beta) \times V_\varepsilon(\beta)$:

$$a(\beta; u, v) = \int_{\mathbb{R}^2} \mu^{-1} \mathrm{rot}_\beta u \cdot \mathrm{rot}_\beta v \, dx, \tag{8.29}$$

where

$$V_\varepsilon(\beta) = V_\varepsilon \cap H_\varepsilon(\beta) = \{u \in V_\varepsilon \ / \ \mathrm{div}_\beta(\varepsilon u) = 0\}. \tag{8.30}$$

By Green's formula, one has

$$(A(\beta)u, v)_\varepsilon = a(\beta; u, v) \quad \text{for all } (u, v) \in D(A(\beta)) \times V_\varepsilon(\beta). \tag{8.31}$$

The fact that $a(\beta; ., .)$ is symmetric and positive implies that $A(\beta)$ is symmetric and positive. Moreover, we have the following lemma.

LEMMA 8.3. *The bilinear form $a(\beta; ., .)$ is coercive in the space $V_\varepsilon(\beta)$. More precisely, if $c_- = \inf_{x \in \mathbb{R}^2} c(x)$, then*

$$a(\beta; u, u) \geq c_-^2 \int_{\mathbb{R}^2} \varepsilon \left(|\, rot\boldsymbol{u}\,|^2 + |\, \nabla u_3 \,|^2 \right) dx + c_-^2 \beta^2 \int_{\mathbb{R}^2} \varepsilon \left(|\, \boldsymbol{u}\,|^2 + 2\,|\, u_3\,|^2 \right) dx.$$

Chapter 8. Optical Waveguides

Proof. By definition, we have

$$a(\beta; u, u) = \int_{\mathbb{R}^2} \mu^{-1} |\operatorname{rot}_\beta u|^2 \, dx = \int_{\mathbb{R}^2} \varepsilon c^2 |\operatorname{rot}_\beta u|^2 \, dx$$

$$\geq c_-^2 \int_{\mathbb{R}^2} \varepsilon |\operatorname{rot}_\beta u|^2 \, dx$$

$$= c_-^2 \int_{\mathbb{R}^2} \varepsilon(|\nabla u_3 - \beta \boldsymbol{u}|^2 + |\operatorname{rot} \boldsymbol{u}|^2) \, dx$$

$$= c_-^2 \int_{\mathbb{R}^2} \varepsilon(|\nabla u_3|^2 + |\operatorname{rot} \boldsymbol{u}|^2 + \beta^2 |\boldsymbol{u}|^2) \, dx$$

$$- 2\beta c_-^2 \int_{\mathbb{R}^2} \varepsilon \boldsymbol{u} . \nabla u_3 \, dx.$$

By Green's formula, since $\operatorname{div}(\varepsilon \boldsymbol{u}) = \varepsilon \beta u_3$, we have

$$- \int_{\mathbb{R}^2} \varepsilon \boldsymbol{u} \cdot \nabla u_3 \, dx = \int_{\mathbb{R}^2} \operatorname{div}(\varepsilon \boldsymbol{u}) u_3 \, dx = \beta \int_{\mathbb{R}^2} \varepsilon |u_3|^2 \, dx,$$

from which the result follows immediately. □

From classical characterizations of self-adjoint operators [35] we deduce the main result of this section.

THEOREM 8.1. *For any $\beta > 0$, the operator $A(\beta)$ is self-adjoint, bounded from below. Moreover, if $\sigma(A(\beta))$ denotes the spectrum of $A(\beta)$, the inclusion below holds:*

$$\sigma(A(\beta)) \subset [c_-^2 \beta^2, +\infty).$$

We thus have to determine the *point spectrum* of self-adjoint operators. The difficulty comes from the *unboundedness* of the cross section. It is a difficulty for the numerical approximation since, in practice, we shall have to bound the computational domain artificially. It is also a mathematical difficulty since the operators we study *do not have a compact resolvent*. They have a nonempty essential spectrum (a purely continuous one, as we shall see), and the existence of eigenvalues is not obvious. The next step in the analysis is the determination of this essential spectrum. From now on, the results differ if one considers the optical fiber or the integrated optics waveguide.

8.2 Mathematical Analysis of Optical Fibers

In this section, we make the assumptions (8.6).

8.2.1 The Essential Spectrum

We now determine the essential spectrum of $A(\beta)$. The idea is to consider our case as a perturbation of the homogeneous medium corresponding to $\varepsilon = \varepsilon_\infty$ and $\mu = \mu_\infty$. Indeed, the essential spectrum is invariant under compact perturbations (see [37], [35]). For the homogeneous medium, the corresponding operator, say $A_\infty(\beta)$, has a purely continuous spectrum that one determines very easily using the Fourier transform. One gets $\sigma(A_\infty(\beta)) = [c_\infty^2 \beta^2, +\infty)$, where $c_\infty^2 = (\varepsilon_\infty \mu_\infty)^{-1}$. The idea is to prove that $\sigma_{ess}(A(\beta)) = \sigma(A_\infty(\beta))$.

LEMMA 8.4. $\sigma_{ess}(A(\beta)) \supset [c_\infty^2 \beta^2, +\infty)$.

Proof. We are going to prove that any real number λ greater than $c_\infty^2 \beta^2$ belongs to $\sigma(A_\infty(\beta))$ by constructing explicitly an associated *singular sequence* (see the appendix for the link between the essential spectrum and the singular sequences). The idea of the construction starts from plane waves propagating in a homogeneous medium that we truncate appropriately in order to avoid the perturbation (i.e., the region where the coefficients vary). Let $k = (k_1, k_2) \in \mathbb{R}^2$ and set

$$u_k(x) = (-k_2 \cos(k.x), k_1 \cos(k.x), 0).$$

One easily verifies that

$$\text{div}_\beta u_k = 0, \quad \text{rot}^*_\beta(\text{rot}_\beta u_k) = (\beta^2 + |k|^2) u_k .$$

We choose k such that $\lambda = c_\infty^2(\beta^2 + |k|^2)$. Let ψ be a cut-off function satisfying

$$\begin{cases} \psi \in C_0^\infty(\mathbb{R}^2), \quad \text{supp } \psi \subset \{x \in \mathbb{R}^2 / R \leq |x| \leq 4R\}, \quad 0 \leq \psi \leq 1, \\ \psi(x) \equiv 1 \quad \text{for } 2R \leq |x| \leq 3R, \end{cases}$$

where R is given by (8.5). We construct the sequence $u^n = (\boldsymbol{u}^n, u_3^n)$ such that $\text{div}_\beta(\varepsilon u^n) = 0$, by taking

$$\begin{cases} \boldsymbol{u}^n(x) = C^n \psi\left(\frac{x}{n}\right) \boldsymbol{u}_k(x), \\ u_3^n(x) = \frac{C^n}{n\beta} \nabla \psi\left(\frac{x}{n}\right) \cdot \boldsymbol{u}_k(x) . \end{cases}$$

As ε and μ are constant outside $|x| < R$, it is clear that the sequence u^n belongs to $D(A(\beta))$ for $n \geq 1$. We choose the constant $C^n > 0$ such as

$$|u^n|_\varepsilon^2 = \int_{\mathbb{R}^2} \varepsilon |u^n|^2 \, dx = 1.$$

A simple calculation shows that $C^n \sim C/n, C > 0$. It follows that $\| u^n \|_{L^\infty} \searrow 0$ and thus, as $|u^n|_\varepsilon = 1$, that u^n converges to 0 weakly in H_ε. To conclude that $c_\infty^2(\beta^2 + |k|^2)$ belongs to the essential spectrum of $A(\beta)$, it suffices to check by an explicit computation that (see [20])

$$|A(\beta)u^n - c_\infty^2(\beta^2 + |k|^2)u^n|_\varepsilon = O\left(\frac{1}{n}\right). \qquad \square$$

For the second step of the proof, we need a compactness result which can be proved easily (see [20]) using the results of Weber [43].

PROPOSITION 8.1. *Introduce the Hilbert space*

$$H(rot, \text{div}_\varepsilon, \mathbb{R}^2) = \{\boldsymbol{u} \in L^2(\mathbb{R}^2)^2 \ / \ rot\boldsymbol{u} \in L^2(\mathbb{R}^2) \ , \ \text{div}(\varepsilon\boldsymbol{u}) \in L^2(\mathbb{R}^2)\},$$

equipped with the norm

$$\| \boldsymbol{u} \|_\varepsilon^2 = \int_{\mathbb{R}^2} (|\boldsymbol{u}|^2 + |rot\boldsymbol{u}|^2 + |\text{div}(\varepsilon\boldsymbol{u})|^2) \, dx. \tag{8.32}$$

Chapter 8. Optical Waveguides

Then, if $B_R = \{x/ \mid x \mid \leq R\}$, the mapping $\boldsymbol{u} \to \boldsymbol{u}\mid_{B_R}$ is compact from $H(\text{rot}, \text{div}_\varepsilon, \mathbb{R}^2)$ into $L^2(B_R)^2$.

This second step also uses a particular decomposition of $a(\beta; u, u)$.

LEMMA 8.5. *The following identity holds:*

$$\text{for all } u \in V_\varepsilon(\beta), \quad a(\beta; u, u) = c_\infty^2 \beta^2 \mid u \mid_\varepsilon^2 + p(\beta; u, u) + c(\beta; u, u), \quad (8.33)$$

where

$$\begin{cases} p(\beta; u, u) = \displaystyle\int_{\mathbb{R}^2} \varepsilon(c^2 \mid \text{rot}\boldsymbol{u} \mid^2 + c^2 \mid \nabla u_3 \mid^2 + \beta^2 c_\infty^2 \mid u_3 \mid^2) \, dx, \\ c(\beta; u, u) = \beta^2 \displaystyle\int_{\mathbb{R}^2} \varepsilon(c^2 - c_\infty^2) \mid \boldsymbol{u} \mid^2 \, dx - 2\beta \displaystyle\int_{\mathbb{R}^2} \varepsilon(c^2 - c_\infty^2) \nabla u_3 \cdot \boldsymbol{u} \, dx, \end{cases}$$

with the properties:
(a) $p(\beta; u, u) \geq 0$ for all $u \in V_\varepsilon(\beta)$,
(b) $u^n \rightharpoonup u$ in $V_\varepsilon(\beta)$ weakly $\implies \lim_{n\to+\infty} c(\beta; u^n, u^n) = c(\beta; u, u)$.

Proof. We start from the identity (see the proof of Lemma 8.3)

$$a(\beta; u, u) = \int_{\mathbb{R}^2} \varepsilon c^2 \{\mid \text{rot}\boldsymbol{u}\mid^2 + \mid \nabla u_3\mid^2 + \beta^2 \mid \boldsymbol{u}\mid^2\} \, dx \\ - 2\beta \int_{\mathbb{R}^2} \varepsilon c^2 \nabla u_3 \cdot \boldsymbol{u} \, dx. \quad (8.34)$$

We transform the last term of (8.34) as follows:

$$-\int_{\mathbb{R}^2} \varepsilon c^2 \nabla u_3 . \boldsymbol{u} \, dx = -\int_{\mathbb{R}^2} \varepsilon(c^2 - c_\infty^2)\nabla u_3 \cdot \boldsymbol{u} \, dx - c_\infty^2 \int_{\mathbb{R}^2} \varepsilon \nabla u_3 . \boldsymbol{u} \, dx$$

$$= -\int_{\mathbb{R}^2} \varepsilon(c^2 - c_\infty^2)\nabla u_3 \cdot \boldsymbol{u} \, dx + \beta c_\infty^2 \int_{\mathbb{R}^2} \varepsilon \mid u_3 \mid^2 \, dx.$$

(We have used integration by parts and the fact that $\text{div}(\varepsilon\boldsymbol{u}) = \beta\varepsilon u_3$.) The decomposition (8.33) follows easily.

Next, as property (a) is immediate, we only have to check property (b). Assume that $u^n \rightharpoonup u$ weakly in $V_\beta(\varepsilon)$; then $(\boldsymbol{u}^n, u_3^n)$ is bounded in $H(\text{rot}; \mathbb{R}^2) \times H^1(\mathbb{R}^2)$ and $\text{div}(\varepsilon\boldsymbol{u}^n)$ is bounded in $L^2(\mathbb{R}^2)$. Thanks to Proposition 8.1,

$$(\boldsymbol{u}^n, u_3^n) \longrightarrow (\boldsymbol{u}, u_3) \text{ strongly in } L^2(B_R)^3.$$

Therefore, as $(c^2 - c_\infty^2)$ has compact support included in B_R, it follows that

$$\lim_{n\to+\infty} \int_{\mathbb{R}^2} \varepsilon(c^2 - c_\infty^2) \mid \boldsymbol{u}^n \mid^2 \, dx = \int_{\mathbb{R}^2} \varepsilon(c^2 - c_\infty^2) \mid \boldsymbol{u}\mid^2 \, dx,$$

$$\lim_{n\to+\infty} \int_{\mathbb{R}^2} \varepsilon(c^2 - c_\infty^2)\nabla u_3^n . \boldsymbol{u}^n \, dx = \int_{\mathbb{R}^2} \varepsilon(c^2 - c_\infty^2)\nabla u_3 \cdot \boldsymbol{u} \, dx,$$

which completes the proof. $\square$

LEMMA 8.6. $\inf \sigma_{ess}(A(\beta)) \geq c_\infty^2 \beta^2$.

Proof. If $\sigma \in \sigma_{ess}(A(\beta))$, there exists a sequence u^n in $D(A(\beta))$ such that

$$\begin{cases} \mid u^n \mid_\varepsilon = 1, \; u^n \rightharpoonup 0 & \text{weakly in } H_\varepsilon(\beta), \\ A(\beta)u^n - \sigma u^n \to 0 & \text{strongly in } H_\varepsilon(\beta). \end{cases} \quad (8.35)$$

Therefore, $a(\beta; u^n, u^n)$ tends to σ and from Lemma 8.3 we deduce that u^n is bounded in $V_\varepsilon(\beta)$. Thus, u^n converges weakly to 0 in $V_\varepsilon(\beta)$. By (8.33), we have

$$a(\beta; u^n, u^n) = c_\infty^2 \beta^2 + p(\beta; u^n, u^n) + c(\beta; u^n, u^n)$$
$$\geq c_\infty^2 \beta^2 + c(\beta; u^n, u^n).$$

Let n go to $+\infty$; as $\lim_{n\to+\infty} c(\beta; u^n, u^n) = 0$ (property (b)), we get $\sigma \geq c_\infty^2 \beta^2$. □

We now regroup Lemmas 8.4 and 8.5 in the main result of this section.

THEOREM 8.2. $\sigma_{ess}(A(\beta)) = [c_\infty^2 \beta^2, +\infty)$.
Consequently, the *discrete spectrum* satisfies

$$\sigma_d(A(\beta)) \subset (c_-^2 \beta^2, c_\infty^2 \beta^2). \tag{8.36}$$

An immediate consequence is the following *nonexistence* result.

LEMMA 8.7. *If $c(x)^2 \geq c_\infty^2$ almost everywhere, the discrete spectrum of $A(\beta)$ is empty.*

The inclusion (8.36) also means that any eigenvalue in the discrete spectrum of $A(\beta)$ is necessarily strictly smaller than the lower bound $c_\infty^2 \beta^2$ of its essential spectrum. This implies that all these eigenvalues can be characterized with the help of the Min-Max Principle (see section 8.2.3).

Remark 8.2. If $\omega^2 \in \sigma_d(A(\beta))$ and u is a guided mode associated to ω^2, it satisfies $-\Delta u + (\beta^2 - \omega^2/c_\infty^2)u = 0$ outside B_R. By (8.36), $\beta^2 - \omega^2/c_\infty^2 > 0$, so that we deduce that u is exponentially decaying outside B_R:

$$|u(x)| \leq C\, e^{-(\beta^2 - \omega^2/c_\infty^2)^{\frac{1}{2}}|x|}, \quad |x| > R \tag{8.37}$$

(use, for instance, the Green's function of the operator $-\Delta + (\beta^2 - \omega^2/c_\infty^2)$).

8.2.2 Absence of Eigenvalues in the Essential Spectrum

We prove in this section that there is no eigenvalue of $A(\beta)$ embedded in its essential spectrum, except possibly the lower bound $c_\infty^2 \beta^2$. Note that this means physically that guided modes necessarily propagate more slowly than the waves in the reference medium.

THEOREM 8.3. *The operator $A(\beta)$ has no eigenvalue in $(c_\infty^2 \beta^2, +\infty)$.*

Proof. In this proof we shall use the partition of $\mathbb{R}^2$ given by (8.4). Without loss of generality, we assume that $\Omega_0 = \{x/|x| > R\}$ and that the open sets Ω_j are connected and numbered in such a way that the open sets $\mathcal{O}_k$ defined by

$$\overline{\mathcal{O}_k} = \bigcup_{j=0}^{k} \overline{\Omega_j}$$

are connected. Note that $\mathcal{O}_N = \mathbb{R}^2$.

Suppose by contradiction that u is an eigenfunction associated to some eigenvalue ω^2 of $A(\beta)$, with $\omega^2 > c_\infty^2 \beta^2$. We are going to show by induction that u vanishes in $\mathcal{O}_k$ for all $k \in \{0, \ldots, N\}$. We first prove that $u = 0$ in $\mathcal{O}_0$.

Since $\text{div}_\beta u = 0$ in $|x| \geq R$, it is clear that (use the identity $\text{rot}_\beta^*(\text{rot}_\beta) = \nabla_\beta(\text{div}_\beta) - \Delta + \beta^2$)

$$\Delta u + \left(\frac{\omega^2}{c_\infty^2} - \beta^2\right) u = 0 \quad \text{for } |x| \geq R.$$

As $u \in L^2(\mathbb{R}^2)$, Rellich's theorem (see [36]) implies that $u = 0$ for $|x| \geq R$.

Chapter 8. Optical Waveguides   285

Now, we assume that u vanishes in $\mathcal{O}_{k-1}, k \geq 1$. Let us introduce $(\varepsilon_k, \mu_k) \in W^{2,\infty}(\mathbb{R}^2)^2$, which are bounded from below by strictly positive constants and coincide with (ε, μ) in Ω_k. We claim that u is the solution of

$$\mathrm{rot}^*_\beta(\mu_k^{-1}\mathrm{rot}_\beta u) = \varepsilon_k \omega^2 u \quad \text{in} \quad \mathcal{O}_k.$$

Indeed, for $v \in \mathcal{D}(\mathcal{O}_k)$, it is not difficult to see that

$$\begin{cases} \displaystyle\int_{\mathcal{O}_k} \mu_k^{-1}\mathrm{rot}_\beta u \cdot \mathrm{rot}_\beta v \, dx = \int_{\Omega_k} \mu^{-1}\mathrm{rot}_\beta u \cdot \mathrm{rot}_\beta v \, dx, \\ \displaystyle\int_{\mathcal{O}_k} \varepsilon_k u \cdot v \, dx = \int_{\Omega_k} \varepsilon u \cdot v \, dx \, . \end{cases}$$

Using the formulas $((e_1, e_2, e_3)$ denotes the canonical basis of $\mathbb{R}^3$):

$$\begin{cases} \mathrm{rot}^*_\beta(\mu_k^{-1}\mathrm{rot}_\beta u) = \mu_k^{-1}\mathrm{rot}^*_\beta(\mathrm{rot}_\beta u) - \mu_k^{-2}\mathrm{rot}\,\boldsymbol{u}\,\nabla\mu_k \times e_3 \\ \qquad\qquad\qquad +\mu_k^{-2}\nabla\mu_k \cdot (\nabla u_3 - \beta \boldsymbol{u})e_3, \end{cases}$$

$$\begin{cases} \nabla_\beta(\mathrm{div}_\beta\,u) = \nabla_\beta(\varepsilon_k^{-1}\,\mathrm{div}_\beta(\varepsilon_k u)) - \displaystyle\sum_{j=1}^{2} \varepsilon_k^{-1}\nabla\varepsilon_k \cdot \dfrac{\partial \boldsymbol{u}}{\partial x_j} \\ \qquad\qquad - \displaystyle\sum_{j=1}^{2} \dfrac{\partial}{\partial x_j}(\varepsilon_k^{-1}\nabla\varepsilon_k) \cdot \boldsymbol{u}\,e_j - \beta(\varepsilon_k^{-1}\nabla\varepsilon_k) \cdot \boldsymbol{u}\,e_3, \end{cases}$$

we see, as $\mathrm{div}_\beta(\varepsilon_k u) = 0$ and $\mathrm{rot}^*_\beta(\mathrm{rot}_\beta) = \nabla_\beta(\mathrm{div}_\beta) - \Delta + \beta^2$, that in $\mathcal{O}_k$

$$\begin{cases} \Delta u = (\beta^2 - \omega^2\varepsilon_k\mu_k)u - \mu_k^{-1}\mathrm{rot}\,\boldsymbol{u}\nabla\mu_k \times e_3 + \mu_k^{-1}\nabla\mu_k \cdot (\nabla u_3 - \beta\boldsymbol{u})e_3 \\ \qquad - \displaystyle\sum_{j=1}^{2}\left\{\varepsilon_k^{-1}\nabla\varepsilon_k \cdot \dfrac{\partial \boldsymbol{u}}{\partial x_j} + \dfrac{\partial}{\partial x_j}(\varepsilon_k^{-1}\nabla\varepsilon_k) \cdot \boldsymbol{u}\right\}e_j - \beta(\varepsilon_k^{-1}\nabla\varepsilon_k) \cdot \boldsymbol{u}\,e_3 \, . \end{cases}$$

Then, by elliptic regularity, we deduce that $u \in H^2_{loc}(\mathcal{O}_k)^3$ and that

$$|\Delta u(x)| \leq C(|\nabla u(x)| + |u(x)|) \quad \text{a.e. } x \text{ in } \mathcal{O}_k,$$

where the constant C depends on $\|\varepsilon_k\|_{W^{2,\infty}}$ and $\|\mu_k\|_{W^{1,\infty}}$. We finally apply a unique continuation theorem [42] to conclude that u vanishes in $\mathcal{O}_k$. □

Consequently, if we except $c_\infty^2 \beta^2$, all eigenvalues of $A(\beta)$ belong to the discrete spectrum and are located below the essential spectrum. As a consequence, they can be characterized by the Min-Max Principle.

8.2.3 Existence of Eigenvalues in the Discrete Spectrum

The Min-Max Principle. Let us introduce, for $m \geq 1$,

- $\mathcal{V}^\varepsilon_m(\beta)$: the set of m-dimensional subspaces of $V_\varepsilon(\beta)$,

- $\mathcal{H}^\varepsilon_m(\beta)$: the set of m-dimensional subspaces of $H_\varepsilon(\beta)$, and define, for any subset F of $H_\varepsilon(\beta)$:

$$F^\perp = \{u \in H_\varepsilon(\beta) \ / \ (u,v)_\varepsilon = 0 \quad \text{for all } v \in F\}.$$

For any $m \geq 1$, we introduce the real numbers given by the two equivalent formulas (see the appendix):

$$\begin{aligned}
s_m(\beta) &= \inf_{E \in \mathcal{V}_m^\varepsilon(\beta)} \sup_{u \in E \setminus \{0\}} \frac{a(\beta\,;u,u)}{|u|_\varepsilon^2} \\
&= \sup_{F \in \mathcal{H}_{m-1}^\varepsilon(\beta)} \inf_{u \in F^\perp \cap V_\varepsilon(\beta)} \frac{a(\beta\,;u,u)}{|u|_\varepsilon^2}.
\end{aligned} \tag{8.38}$$

Using the H formulation and the equivalence between both formulations, we can derive two other expressions of the min-max $s_m(\beta)$ by simply inverting the roles of ε and μ.

As functions of β, these are continuous functions. More precisely, we have the following theorem.

THEOREM 8.4. *For all $m \geq 1$, the functions $\beta \to s_m(\beta)$, $\beta > 0$ are continuous and locally Lipschitz.*

The proof of this result is purely technical. It is essentially a consequence of the smooth dependence of $a(\beta;u,u)$ with respect to β. We refer the reader to [20] for further details.

Remark 8.3. Applying Kato's perturbation theory [25], one can even prove that each function $\beta \to s_m(\beta)$ is piecewise analytic.

The Min-Max Principle applied to the operator $A(\beta)$ can be stated as follows.

THEOREM 8.5. *The sequence $s_m(\beta)$ is nondecreasing and converges to $c_\infty^2 \beta^2$. Moreover, for each $m \geq 1$, the following alternative holds:*

(a) *if $s_m(\beta) < c_\infty^2 \beta^2$, then $A(\beta)$ admits at least m eigenvalues strictly smaller than $c_\infty^2 \beta^2$ and $\{s_1(\beta), s_2(\beta), \ldots, s_m(\beta)\}$ are the m first eigenvalues of $A(\beta)$.*

(b) *if $s_m(\beta) = c_\infty^2 \beta^2$, then $s_j(\beta) = c_\infty^2 \beta^2$ for any $j \geq m$ and $A(\beta)$ has at most $(m-1)$ eigenvalues strictly smaller than $c_\infty^2 \beta^2$.*

We deduce a rule for proving existence or nonexistence of eigenvalues:

(i) If one can construct a subspace of *test functions* E of $V_\varepsilon(\beta)$ such that

$$\dim E = m \quad \text{for all } u \in E, \quad a(\beta\,;u,u) - c_\infty^2 \beta^2 \,|\,u\,|_\varepsilon^2 < 0,$$

then $A(\beta)$ has at least m eigenvalues strictly smaller than $c_\infty^2 \beta^2$.

(ii) If one can construct a subspace F of dimension $m-1$ of $H_\varepsilon(\beta)$ such that

$$\text{for all } u \in F^\perp \cap V_\varepsilon(\beta), \quad a(\beta\,;u,u) - c_\infty^2 \beta^2 \,|\,u\,|_\varepsilon^2 \geq 0,$$

then $A(\beta)$ has at most $m-1$ eigenvalues strictly smaller than $c_\infty^2 \beta^2$.

From now on, we shall denote by $N(\beta)$ the (possibly infinite) number of eigenvalues of $A(\beta)$ in $(c_-^2 \beta^2, c_\infty^2 \beta^2)$. $N(\beta)$ is also characterized by

$$\begin{cases} m \leq N(\beta) \iff s_m(\beta) < c_\infty^2 \beta^2, \\ m > N(\beta) \iff s_m(\beta) = c_\infty^2 \beta^2. \end{cases}$$

Chapter 8. Optical Waveguides

The Main Existence Result

THEOREM 8.6. *Assume that $c_- < c_\infty$. Then for any $m \geq 1$, there exists $\beta_m \geq 0$ such that for any $\beta > \beta_m$, $A(\beta)$ admits at least m eigenvalues strictly smaller than $c_\infty^2 \beta^2$.*

Proof. Since ε and μ are piecewise $W^{2,\infty}$, by definition of c_-, for any $\gamma > 0$ small enough, there exists a disk $D_\gamma \subset B_R$ such that $\varepsilon \in W^{2,\infty}(D_\gamma)$ and

$$\text{a.e. } x \in D_\gamma, \quad c(x)^2 < c_-^2 + \gamma. \tag{8.39}$$

The idea is to localize the test functions in this disk. For $m \geq 1$, let $\{\boldsymbol{u}_k^\gamma, 1 \leq k \leq m\}$ be m linearly independent two-dimensional vector fields satisfying

$$\boldsymbol{u}_k^\gamma \in C_0^\infty(\mathbb{R}^2), \quad \text{supp } \boldsymbol{u}_k^\gamma \subset D_\gamma$$

and define

$$(u_3)_k^\gamma = \frac{1}{\varepsilon \beta} \text{div}(\varepsilon \boldsymbol{u}_k^\gamma) = \frac{1}{\beta} \text{div}(\boldsymbol{u}_k^\gamma) + \frac{1}{\beta} \frac{\nabla \varepsilon}{\varepsilon} \cdot \boldsymbol{u}_k^\gamma.$$

As $\varepsilon \in W^{2,\infty}(D_\gamma)$, it is clear that, for all $k \leq m$,

$$(u_3)_k^\gamma \in H^1(\mathbb{R}^2), \text{ supp } (u_3)_k^\gamma \subset D_\gamma \text{ and } u_k^\gamma = (\boldsymbol{u}_k^\gamma, (u_3)_k^\gamma) \in V_\varepsilon(\beta). \tag{8.40}$$

Now define the m-dimensional subspace of $V_\varepsilon(\beta)$:

$$E_m^\gamma = \text{span } [u_1^\gamma, u_2^\gamma, \ldots, u_m^\gamma]. \tag{8.41}$$

Let $u = (\boldsymbol{u}, u_3) \in E_m^\gamma$. The same manipulation as in the proof of Lemma 8.3 (simply use (8.39) instead of $c^2 \geq c_-^2$) leads to (since $\beta u_3 = \varepsilon^{-1} \text{div}(\varepsilon \boldsymbol{u})$)

$$a(\beta; u, u) \leq (c_-^2 + \gamma) \left\{ \beta^2 \mid \boldsymbol{u} \mid_\varepsilon^2 + \int_{\mathbb{R}^2} (\varepsilon \mid \text{rot}\boldsymbol{u} \mid^2 + \varepsilon^{-1} \mid \text{div } \varepsilon \boldsymbol{u} \mid^2) \, dx \right.$$
$$\left. + \frac{1}{\beta^2} \int_{\mathbb{R}^2} \varepsilon \mid \nabla(\varepsilon^{-1} \text{div } \varepsilon \boldsymbol{u}) \mid^2 \, dx \right\}.$$

Denote by $M_1(m, \gamma)$ and $M_2(m, \gamma)$ the two positive constants defined by

$$M_1(m, \gamma) = \sup_{u \in E_m^\gamma \setminus 0} \frac{\int_{\mathbb{R}^2} (\varepsilon \mid \text{rot}\boldsymbol{u} \mid^2 + \varepsilon^{-1} \mid \text{div } \varepsilon \boldsymbol{u} \mid^2) \, dx}{\mid \boldsymbol{u} \mid_\varepsilon^2},$$
$$M_2(m, \gamma) = \sup_{u \in E_m^\gamma \setminus 0} \frac{\int_{\mathbb{R}^2} \varepsilon \mid \nabla(\varepsilon^{-1} \text{div}(\varepsilon \boldsymbol{u})) \mid^2 \, dx}{\mid \boldsymbol{u} \mid_\varepsilon^2} \tag{8.42}$$

(note that because of the generalized divergence-free condition, $u \neq 0 \iff \boldsymbol{u} \neq 0$). $M_1(m, \gamma)$ and $M_2(m, \gamma)$ are finite since E_m^γ has finite dimension and does not depend on β. We deduce that, since $\mid \boldsymbol{u} \mid_\varepsilon^2 \leq \mid u \mid_\varepsilon^2$,

$$\text{for all } u \in E_m^\gamma, \quad a(\beta; u, u) \leq (c_-^2 + \gamma) \left\{ \beta^2 + M_1(m, \gamma) + \frac{M_2(m, \gamma)}{\beta^2} \right\} \mid u \mid_\varepsilon^2.$$

Since $c_- < c_\infty$, we can choose $\gamma > 0$ such that $c_-^2 + \gamma < c_\infty^2$. Then it is clear that for β large enough

$$\text{for all } u \in E_m^\gamma, \quad u \neq 0, \qquad a(\beta; u, u) - c_\infty^2 \beta^2 \mid u \mid_\varepsilon^2 < 0.$$

This concludes the proof. □

In fact, the previous proof shows that

$$\frac{s_m(\beta)}{\beta^2} \leq (c_-^2 + \gamma) \left\{ 1 + \frac{M_1(m,\gamma)}{\beta^2} + \frac{M_2(m,\gamma)}{\beta^4} \right\}$$

for arbitrarily small γ. Joined with $s_m(\beta) \geq c_-^2 \beta^2$, this result proves that

$$\lim_{\beta \longrightarrow +\infty} \frac{s_m(\beta)}{\beta^2} = c_-^2, \tag{8.43}$$

i.e., that the high-frequency limit of the phase velocity of the guided modes is the minimum propagation velocity c_-.

Remark 8.4. As a consequence of (8.43), it is not difficult to prove (see [4]) that, when $\beta \longrightarrow +\infty$, the guided mode is concentrated in the region where c reaches its minimum. This is, in some sense, a finer result than (8.37).

The Notion of Threshold

Theorem 8.6 points out a priori the existence of critical values of β which are thresholds for the apparition of eigenvalues in the discrete spectrum of $A_\varepsilon(\beta)$. This leads us to introduce the notion of upper thresholds

$$\begin{aligned}\beta_m^* &= \inf\{\beta_m > 0 \ / \ \text{for all } \beta > \beta_m, A_\varepsilon(\beta) \text{ admits at least } m \\ &\qquad \text{eigenvalues strictly smaller than } c_\infty^2 \beta^2 \} \\ &= \inf\{\beta_m \ / \ s_m(\beta) < c_\infty^2 \beta^2, \quad \text{for all } \beta > \beta_m\}.\end{aligned} \tag{8.44}$$

In opposition, we also introduce the notion of lower threshold β_m^0 defined by

$$\begin{aligned}\beta_m^0 &= \sup\{\beta_m \ / \ \text{for all } \beta \leq \beta_m, A_\varepsilon(\beta) \text{ admits at most } m-1 \\ &\qquad \text{eigenvalues strictly smaller than } c_\infty^2 \beta^2 \} \\ &= \sup\{\beta_m \ / \ \beta < \beta_m \implies s_m(\beta) = c_\infty^2 \beta^2 \}.\end{aligned} \tag{8.45}$$

It is immediate to verify that both sequences β_m^0 and β_m^* are nondecreasing and that, since the functions $s_m(\beta)$ are continuous (cf. Theorem 8.4),

$$\text{for all } m \geq 1, \quad \beta_m^0 \leq \beta_m^*. \tag{8.46}$$

Also, a lot of properties of guided waves can be expressed in function of the thresholds β_m^0 and β_m^*. That is why our next section will be entirely devoted to the study of these two sequences.

Comparison Results

We give here useful comparison results between different propagation media. We consider two propagation media characterized by $(\varepsilon_1(x), \mu_1(x))$ and $(\varepsilon_2(x), \mu_2(x))$.

Chapter 8. Optical Waveguides

THEOREM 8.7. *Assume that*

$$\varepsilon_1(x) \leq \varepsilon_2(x) \text{ and } \mu_1(x) \leq \mu_2(x) \quad a.e. \ x \in \mathbb{R}^2; \tag{8.47}$$

then, with obvious notation,

$$\text{for all } m \geq 1, \quad \text{for all } \beta > 0, \quad s_m^1(\beta) \geq s_m^2(\beta). \tag{8.48}$$

Consequently, if moreover $\varepsilon_{1,\infty} = \varepsilon_{2,\infty}$ *and* $\mu_{1,\infty} = \mu_{2,\infty}$, *then*

$$N_1(\beta) \leq N_2(\beta) \quad \text{for all } \beta, \quad \beta_m^{1,*} \geq \beta_m^{2,*} \quad \text{and} \quad \beta_m^{1,0} \geq \beta_m^{2,0}. \tag{8.49}$$

Proof.

(a) The result is true if $\varepsilon_1(x) = \varepsilon_2(x) = \varepsilon(x)$. Let us set $a^j(\beta\,;u,u) = \int_{\mathbb{R}^2} \mu_j^{-1} \mid \text{rot}_\beta u \mid^2 dx$, $j = 1, 2$. We have

$$a^1(\beta\,;u,u) \geq a^2(\beta\,;u,u) \quad \text{for all } u \in V_\varepsilon(\beta).$$

Then, by the Min-Max Principle

$$s_m^1(\beta) \geq s_m^2(\beta).$$

(b) The result is true if $\mu_1(x) = \mu_2(x) = \mu(x)$. It suffices to invert the roles of ε and μ and to use the H formulation.

(c) Now consider the intermediate medium defined by

$$\tilde{\varepsilon}(x) = \varepsilon_1(x) \text{ and } \tilde{\mu}(x) = \mu_2(x).$$

Let us denote by $\tilde{s}_m(\beta)$ the corresponding min-max. By (a) $s_m^1(\beta) \geq \tilde{s}_m(\beta)$ and by (b) $\tilde{s}_m(\beta) \geq s_m^2(\beta)$. Therefore, $s_m^1(\beta) \geq s_m^2(\beta)$. The other inequalities (8.49) derive directly from this one. □

8.2.4 About the Thresholds

A (Quasi) Characterization

We first derive a generalized eigenvalue problem of which the thresholds are the solution. The idea is very simple, at least formally, and consists in passing to the limit in the eigenvalue problem for guided modes when ω^2 tends to $c_\infty^2 \beta^2$. The only, but essential, difficulty lies in the functional framework needed for the justification of the limit procedure. We need to take into account a change of behavior at infinity, which can be expected from (8.37). Let us introduce the weight function

$$\rho(x) = (1+\mid x \mid^2)^{-1}(\text{Log}(2+\mid x \mid^2))^{-2} > 0. \tag{8.50}$$

We first define the space of two-dimensional vector fields,

$$H_\rho(\text{rot}, \mathbb{R}^2) = \{\, \boldsymbol{u} \in L^2_{loc}(\mathbb{R}^2)^2 \,,\, \rho^{\frac{1}{2}}\boldsymbol{u} \in L^2(\mathbb{R}^2)^2 \,,\, \text{rot}\boldsymbol{u} \in L^2(\mathbb{R}^2) \,\},$$

and then the space of three-dimensional vector fields,

$$V_{\rho,\varepsilon}(\beta) = \{u \in H_\rho(\text{rot}, \mathbb{R}^2) \times H^1(\mathbb{R}^2) \ / \ \text{div}_\beta(\varepsilon u) = 0\}. \tag{8.51}$$

$V_{\rho,\varepsilon}(\beta)$ is a Hilbert space for the norm:

$$|||u|||_\rho^2 = \int_{\mathbb{R}^2} \left(\rho \mid \boldsymbol{u} \mid^2 + \mid \text{rot}\boldsymbol{u} \mid^2\right) dx + \int_{\mathbb{R}^2} \left(\mid u_3 \mid^2 + \mid \nabla u_3 \mid^2\right) dx.$$

PROPOSITION 8.2.
(a) *In the space $V_{\rho,\varepsilon}(\beta)$, the mapping*

$$u \longrightarrow \left(\int_{B_R} \mid \boldsymbol{u} \mid^2 dx + \int_{\mathbb{R}^2} \mid \left(\text{rot}\boldsymbol{u} \mid^2 + \mid u_3 \mid^2 + \mid \nabla u_3 \mid^2\right) dx\right)^{\frac{1}{2}}$$

is a norm equivalent to the norm $|||.|||_\rho$.

(b) *The mapping $u \to \boldsymbol{u}\mid_{B_R}$ is compact from $V_{\rho,\varepsilon}(\beta)$ into $L^2(B_R)^2$.*

(c) *The subspace $V_{\rho,\varepsilon}^{comp}(\beta)$ of compactly supported functions in $V_{\rho,\varepsilon}(\beta)$ is dense in $V_{\rho,\varepsilon}(\beta)$.*

Proof. We refer to [20]. (b) directly follows from Proposition 8.1. (a) derives from a generalized Hardy's inequality that justifies the choice of the weight ρ. (c) is obtained by adapting a truncation process. □

Note in particular that if $u \in V_{\rho,\varepsilon}(\beta)$, the behaviors at infinity which are allowed are different for $\boldsymbol{u}$ and u_3. Also note that the bilinear forms $p(\beta; u, v)$ and $c(\beta; u, v)$, introduced in Lemma 8.5, are defined and continuous in $V_{\rho,\varepsilon}(\beta)$. We can thus introduce the *threshold problem*:

$$\begin{cases} \text{Find } \beta > 0 \ / \ \exists u \neq 0 \in V_{\rho,\varepsilon}(\beta) \\ \text{for all } v \in V_{\rho,\varepsilon}(\beta), \quad p(\beta; u, v) + c(\beta; u, v) = 0. \end{cases} \tag{8.52}$$

THEOREM 8.8. *If β is a threshold such that $0 < \beta < +\infty$, then it is the solution of the threshold equation (8.52).*

Proof. (1) By definition of the upper and lower thresholds, any of them can be characterized as the limit β of a decreasing sequence $(\beta_p)_{p \geq 1}$ such that, for each p, there exists u_p in $D(A(\beta_p))$ and ω_p^2 satisfying

$$\begin{cases} A(\beta_p) u_p = \omega_p^2 u_p, \quad u_p \neq 0, \\ \lim_{p \to +\infty} \omega_p^2 = c_\infty^2 \beta^2, \quad \omega_p^2 \in (c_-^2 \beta_p^2, c_\infty^2 \beta_p^2). \end{cases} \tag{8.53}$$

In particular, as $\omega_p^2 < c_\infty^2 \beta_p^2$, we have

$$p(\beta_p; u_p, u_p) + c(\beta_p; u_p, u_p) = (\omega_p^2 - c_\infty^2 \beta_p^2) \mid u_p \mid_\varepsilon^2 < 0. \tag{8.54}$$

This implies in particular that $\boldsymbol{u}_p$ does not vanish in B_R so that we can impose

$$\int_{B_R} \varepsilon \mid \boldsymbol{u}_p \mid^2 dx = 1. \tag{8.55}$$

Chapter 8. Optical Waveguides

Since $u_p = (\boldsymbol{u}_p, u_{3,p})$ belongs to $V_\varepsilon(\beta_p)$, the vector field $\tilde{u}_p = (\boldsymbol{u}_p, \tilde{u}_{3,p} = \beta_p u_{3,p})$ belongs to the fixed space $V_\varepsilon(1)$. Moreover, $p(\beta_p\ ; u_p, u_p) = \tilde{p}(\beta_p\ ; \tilde{u}_p, \tilde{u}_p)$ and $c(\beta_p\ ; u_p, u_p) = \tilde{c}(\beta_p\ ; \tilde{u}_p, \tilde{u}_p)$, where

$$\tilde{p}(\beta\ ; \tilde{u}, \tilde{u}) = \int_{\mathbb{R}^2} \varepsilon c^2 \left(|\operatorname{rot} \boldsymbol{u}|^2 + \frac{1}{\beta^2} |\nabla \tilde{u}_3|^2 \right) dx + c_\infty^2 \int_{\mathbb{R}^2} \varepsilon |\tilde{u}_{3,p}|^2\ dx,$$

$$\tilde{c}(\beta\ ; \tilde{u}, \tilde{u}) = \beta^2 \int_{B_R} \varepsilon (c^2 - c_\infty^2) |\boldsymbol{u}|^2\ dx - 2 \int_{B_R} \varepsilon (c^2 - c_\infty^2) \nabla \tilde{u}_3 \cdot \boldsymbol{u}\ dx.$$

Using Young's inequality,

$$2\varepsilon (c^2 - c_\infty^2) \nabla \tilde{u}_3 \cdot \boldsymbol{u} \leq \eta |\nabla \tilde{u}_3|^2 + \frac{\varepsilon^2 (c^2 - c_\infty^2)^2 |\boldsymbol{u}|^2}{\eta} \tag{8.56}$$

with $\eta = \dfrac{\varepsilon c^2}{2\beta_p^2}$, which leads to

$$\begin{aligned} c(\beta_p\ ; u_p, u_p) \geq{} & \beta_p^2 \int_{\mathbb{R}^2} \varepsilon (c^2 - c_\infty^2)[1 - 2(c^2 - c_\infty^2)/c^2] |\boldsymbol{u}_p|^2\ dx \\ & - \frac{1}{2\beta_p^2} \int_{\mathbb{R}^2} \varepsilon c^2 |\nabla \tilde{u}_{3,p}|^2\ dx\,. \end{aligned} \tag{8.57}$$

Substituting (8.57) into (8.54) we obtain (with $C = \| (c^2 - c_\infty^2)(2(c_\infty^2/c^2) - 1) \|_{L^\infty}$):

$$\begin{aligned} & \int_{\mathbb{R}^2} |\operatorname{rot} \boldsymbol{u}_p|^2 \varepsilon c^2\ dx + \frac{1}{2\beta_p^2} \int_{\mathbb{R}^2} |\nabla \tilde{u}_{3,p}|^2 \varepsilon c^2\ dx \\ & + c_\infty^2 \int_{\mathbb{R}^2} \varepsilon |\tilde{u}_{3,p}|^2\ dx \leq C\,\beta_p^2. \end{aligned} \tag{8.58}$$

Since $\operatorname{div}(\varepsilon \boldsymbol{u}_p) = \varepsilon \tilde{u}_{3,p}$, we deduce from (8.58) and Proposition 8.2 (b) that

$$\tilde{u}_p \text{ is bounded in } V_{\rho,\varepsilon}(1). \tag{8.59}$$

(2) By compactness (Proposition 8.2 (c)), one can extract a subsequence, still denoted by $\tilde{u}_p$, such that

$$\tilde{u}_p \to \tilde{u} \quad \text{weakly in } V_{\rho,\varepsilon}(1), \quad \text{strongly in } L^2(B_R)^3\,. \tag{8.60}$$

The eigenvalue problem (8.53) can be rewritten, using $\tilde{u}_p$ instead of u_p:

$$\begin{aligned} & \int_{\mathbb{R}^2} \varepsilon c^2 \operatorname{rot} \boldsymbol{u}_p \cdot \operatorname{rot} \boldsymbol{v}\ dx + \frac{1}{\beta_p^2} \int_{\mathbb{R}^2} \varepsilon c^2 \nabla \tilde{u}_{3p} \cdot \nabla \tilde{v}_3\ dx + c_\infty^2 \int_{\mathbb{R}^2} \varepsilon \tilde{u}_{3p} \tilde{v}_3\ dx \\ & + \int_{B_R} \varepsilon (c^2 - c_\infty^2) \left(\beta_p^2 \boldsymbol{u}_p \cdot \boldsymbol{v} - \nabla \tilde{u}_{3p} \cdot \boldsymbol{v} - \nabla \tilde{v}_3 \cdot \boldsymbol{u}_p \right) dx \\ & = (\omega_p^2 - \beta_p^2 c_\infty^2) \int_{\mathbb{R}^2} \left(\boldsymbol{u}_p \cdot \boldsymbol{v} + \frac{1}{\beta_p^2} \tilde{u}_{3p} \tilde{v}_3 \right) dx, \quad \text{for all } v \in V_\varepsilon(\beta_p). \end{aligned}$$

This holds in particular if $\tilde{v} \in V_{\rho,\varepsilon}^{comp}(1)$. In such a case, the weight function ρ does not play any role, and we can pass to the limit when $p \to +\infty$ to deduce that, for any $v \in V_{\rho,\varepsilon}^{comp}(1)$,

$$\begin{aligned} & \int_{\mathbb{R}^2} \varepsilon c^2 \operatorname{rot} \boldsymbol{u}\, \operatorname{rot} \boldsymbol{v}\ dx + \frac{1}{\beta^2} \int_{\mathbb{R}^2} \varepsilon c^2 \nabla \tilde{u}_3 \cdot \nabla \tilde{v}_3\ dx + c_\infty^2 \int_{\mathbb{R}^2} \varepsilon \tilde{u}_3 \tilde{v}_3\ dx \\ & + \int_{B_R} \varepsilon (c^2 - c_\infty^2) \left(\beta^2 \boldsymbol{u} \cdot \boldsymbol{v} - \nabla \tilde{u}_3 \cdot \boldsymbol{v} - \nabla \tilde{v}_3 \cdot \boldsymbol{u} \right) dx = 0. \end{aligned} \tag{8.61}$$

Coming back to u and v, we remark that (8.61) is nothing but

$$p(\beta\,;u,v) + c(\beta\,;u,v) = 0 \quad \text{for all } v \in V_{\rho,\varepsilon}^{comp}(\beta)(\beta). \tag{8.62}$$

As $V_{\rho,\varepsilon}^{comp}(\beta)$ is dense in $V_{\rho,\varepsilon}(\beta)$ (cf. Proposition 8.2) and as $p(\beta\,;.,.)$ and $c(\beta\,;.,.)$ are continuous in $V_{\rho,\varepsilon}(\beta)$, (8.62) also holds for any v in $V_{\rho,\varepsilon}(\beta)$. To conclude, it remains to prove that u is nonzero. This comes immediately by passing to the limit in (8.55), using the strong convergence (8.60). □

The Third Threshold Is Strictly Positive

We next show that at most two guided modes can exist for small β. For this, we use the following property of a particular subspace of $H_\rho(\text{rot}, \mathbb{R}^2)$.

LEMMA 8.8. *The space* $P_\varepsilon = \{ \boldsymbol{u} \in H_\rho(rot, \mathbb{R}^2) / rot\boldsymbol{u} = \text{div}(\varepsilon \boldsymbol{u}) = 0 \}$ *has dimension 2.*

Proof. Let $\boldsymbol{u}$ be an element of P_ε. As $\boldsymbol{u} \in L^2_{loc}(\mathbb{R}^2)$ and $\text{rot}\boldsymbol{u} = 0$, by Poincaré's lemma (see [14]), there exists φ in $H^1_{loc}(\mathbb{R}^2)$ such that $\boldsymbol{u} = \nabla \varphi$. It is defined up to an additive constant, and we can impose the condition

$$\int_{|x|=R} \varphi \, d\sigma = 0. \tag{8.63}$$

From the condition $\text{div}(\varepsilon \boldsymbol{u}) = 0$, we deduce that $\text{div}(\varepsilon \nabla \varphi) = 0$. In particular, φ is harmonic in $\{|x| \geq R\}$. Then, we use a Fourier series expansion of φ in polar coordinates ($x_1 = r\cos\theta$, $x_2 = r\sin\theta$):

$$\varphi(r,\theta) = \sum_{n=0}^{+\infty} \varphi_n(r)\cos n\theta + \sum_{n=1}^{+\infty} \tilde{\varphi}_n(r)\sin n\theta. \tag{8.64}$$

Combining the fact that φ is harmonic in $\{|x| \geq R\}$ with

$$\boldsymbol{u} \in P_\varepsilon \implies \int_{\mathbb{R}^2} \rho \,|\nabla\varphi|^2 \, dx < +\infty,$$

we obtain after some calculations

- for $n \geq 2$, $\varphi_n(r) = \varphi_n(R)(R/r)^n$, $\tilde{\varphi}_n(r) = \tilde{\varphi}_n(R)(R/r)^n$, $r \geq R$,
- for $n = 1$, $\varphi_1(r) = a_1 r + b_1 r^{-1}$, $\tilde{\varphi}_1(r) = \tilde{a}_1 r + \tilde{b}_1 r^{-1}$, $r \geq R$,
- for $n = 0$, $\varphi_0(r) = a_0 \text{Log } r + b_0$, $r \geq R$.

Integrating $\text{div}(\varepsilon\nabla\varphi) = 0$ over B_R gives $\int_{|x|=R}(\frac{\partial\varphi}{\partial r})\,d\sigma = 0$ which, combined with (8.63), yields $a_0 = b_0 = 0$. Moreover, for $n \geq 2$,

$$\begin{cases} \varphi'_n(R) + \dfrac{n}{R}\varphi_n(R) = 0, & n \geq 2, \\ \tilde{\varphi}'_n(R) + \dfrac{n}{R}\tilde{\varphi}_n(R) = 0, & n \geq 2. \end{cases} \tag{8.65}$$

Chapter 8. Optical Waveguides

For $n = 1$, we can only write

$$\begin{cases} \varphi_1'(R) + \dfrac{1}{R}\varphi_1(R) = 2a_1, \\ \tilde{\varphi}_1'(R) + \dfrac{1}{R}\varphi_1(R) = 2\tilde{a}_1. \end{cases} \qquad (8.66)$$

One deduces from these remarks that the restriction of φ to B_R is solution of the variational problem

$$\begin{cases} \text{Find } \varphi \in V(B_R) = \left\{ \varphi \in H^1(B_R) \, / \, \displaystyle\int_{\Gamma_R} \varphi \, d\sigma = 0 \right\} \text{ such that} \\ \displaystyle\int_{B_R} \varepsilon \nabla \varphi \nabla \psi \, dx + \varepsilon_\infty b(\varphi, \psi) = \pi \varepsilon_\infty (a_1 \psi_1(R) + \tilde{a}_1 \tilde{\psi}_1(R)) \quad \text{for all } \psi \in V(B_R), \end{cases}$$

where, with obvious notation, the bilinear form $b(\varphi, \psi)$ is defined by

$$b(\varphi, \psi) = \pi \sum_{n=1}^{+\infty} n \{ \varphi_n(R) \psi_n(R) + \tilde{\varphi}_n(R) \tilde{\psi}_n(R) \}. \qquad (8.67)$$

The Lax–Milgram lemma shows that φ is entirely determined in B_R (thus in $\mathbb{R}^2$ by unique continuation) by a_1 and $\tilde{a}_1$. This completes the proof. $\square$

Remark 8.5. When ε is constant, one can check that P_ε is nothing but the space of constant two-dimensional vector fields.

Let $\{\boldsymbol{u}^1, \boldsymbol{u}^2\}$ be a basis of P_ε; we define $(\boldsymbol{w}_1, \boldsymbol{w}_2)$ in $H_\varepsilon(\beta)$ by

$$\int_{\mathbb{R}^2} \varepsilon \boldsymbol{u} \boldsymbol{w}_j \, dx = \int_{\mathbb{R}^2} \boldsymbol{u} \boldsymbol{u}^j \, dx \quad \text{for all } u \in H_\varepsilon(\beta) \qquad (8.68)$$

and let F be the two-dimensional subspace of $H_\varepsilon(\beta)$ spanned by $(\boldsymbol{w}_1, \boldsymbol{w}_2)$. In what follows, we shall use a generalized Poincaré's inequality.

LEMMA 8.9. *There exists a constant $C(R) > 0$ such that for any $u = (\boldsymbol{u}, u_3) \in V_\varepsilon(\beta) \cap F^\perp$:*

$$C(R) \int_{B_R} \varepsilon \, |\, \boldsymbol{u} \,|^2 \, dx \leq \int_{\mathbb{R}^2} \left(\varepsilon c^2 \, |\, \mathrm{rot}\, \boldsymbol{u} \,|^2 + \frac{c_\infty^2}{\varepsilon} \, |\, \mathrm{div}\, \varepsilon \boldsymbol{u} \,|^2 \right) dx. \qquad (8.69)$$

Proof. We omit the details (see [20]). The proof can be done by contradiction using a standard compactness argument based on Proposition 8.2. Let us only emphasize that the kernel of the quadratic form on $V_{\rho,\varepsilon}(\beta)$ defined by the right-hand side of (8.69) is nothing but P_ε and thus generated by $\{\boldsymbol{u}^1, \boldsymbol{u}^2\}$. $\square$

THEOREM 8.9. *The third lower threshold β_3^0 is strictly positive.*

Proof. Taking $\eta = \frac{\varepsilon c^2}{2\beta^2}$ in inequality (8.56), we obtain for any u in $V_\varepsilon(\beta)$

$$p(\beta\,;u,u) + c(\beta\,;u,u) \geq \beta^2 c_\infty^2 \int_{\mathbb{R}^2} \varepsilon \frac{(c^2 - c_\infty^2)}{c^2} \, |\, \boldsymbol{u} \,|^2 \, dx$$
$$+ \beta^2 c_\infty^2 \int_{\mathbb{R}^2} \varepsilon \, |\, u_3 \,|^2 \, dx + \int_{\mathbb{R}^2} \varepsilon c^2 \, |\, \mathrm{rot}\, \boldsymbol{u} \,|^2 \, dx \, .$$

From Lemma 8.9, we can write for any $u \in F^\perp \cap V_\varepsilon(\beta)$, since $\beta u_3 = \varepsilon^{-1} \operatorname{div}(\varepsilon \boldsymbol{u})$,

$$\beta^2 c_\infty^2 \int_{\mathbb{R}^2} \varepsilon \mid u_3 \mid^2 dx + \int_{\mathbb{R}^2} \varepsilon c^2 \mid \operatorname{rot} \boldsymbol{u} \mid^2 dx \geq C(R) \int_{B_R} \varepsilon \mid \boldsymbol{u} \mid^2 dx.$$

Therefore, using (8.33) and setting $K = \parallel (1 - c^2/c_\infty^2) \parallel_{L^\infty}$, we get

$$\forall u \in F^\perp, \quad a(\beta\,;u,u) - c_\infty^2 \beta^2 \mid u \mid_\varepsilon^2 \geq (C(R) - K\beta^2 c_\infty^2) \int_{B_R} \varepsilon \mid \boldsymbol{u} \mid^2 dx.$$

In particular, we have

$$\beta \leq c_\infty^{-1} C(R)^{\frac{1}{2}} K^{-\frac{1}{2}} \implies a(\beta\,;u,u) \geq \beta^2 c_\infty^2 \mid u \mid_\varepsilon^2 \quad \text{for all } u \in F^\perp.$$

By the Min-Max Principle, this means that $s_3(\beta) = c_\infty^2 \beta^2$ and thus that the third lower threshold β_3^0 satisfies

$$\beta_3^0 \geq c_\infty^{-1} C(R)^{\frac{1}{2}} K^{-\frac{1}{2}} > 0. \quad \Box \qquad (8.70)$$

Remark 8.6. It is possible to derive, respectively, sufficient conditions or necessary conditions for the existence of *one* ($\beta_1^0 = 0$) or *two* ($\beta_2^0 = 0$) guided modes at arbitrary low frequencies. Such conditions express the fact that the propagation velocity is less (or more) than its value at infinity in some appropriate *mean value*. We refer to [20] for more details.

A Case Where Lower and Upper Thresholds Coincide

This corresponds to the assumption

$$c(x) \leq c_\infty \quad \text{a.e. } x \in \mathbb{R}^2. \qquad (8.71)$$

THEOREM 8.10. *If (8.71) holds, the function $\beta \to s_m(\beta) - c_\infty^2 \beta^2$ is a decreasing function of β for any $m \geq 1$. Thus, for any $m \geq 1$, $\beta_m^0 = \beta_m^*$.*

Proof. Note that, using (8.33),

$$s_m(\beta) - c_\infty^2 \beta^2 = \inf_{E \in \mathcal{V}_m^\varepsilon(\beta)} \sup_{u \in E \setminus \{0\}} \frac{p(\beta\,;u,u) + c(\beta\,,u,u)}{\mid u \mid_\varepsilon^2}.$$

Moreover, when $u = (\boldsymbol{u}, u_3)$ describes the space $V_\varepsilon(\beta)$, $\tilde{u} = (\boldsymbol{u}, \beta u_3)$ describes the space $V_\varepsilon(1)$ and one has

$$\begin{cases} c(\beta\,;u,u) + p(\beta\,;u,u) = \dfrac{A_0(\tilde{u})}{\beta^2} + A_1(\tilde{u}) + \beta^2 A_2(\tilde{u}), \\ \mid u \mid_\varepsilon^2 = \dfrac{B_0(\tilde{u})}{\beta^2} + B_1(\tilde{u}), \end{cases}$$

where we have set

$$\begin{cases} A_0(\tilde{u}) = \displaystyle\int_{\mathbb{R}^2} \varepsilon c^2 \mid \nabla \tilde{u}_3 \mid^2 dx, \quad A_2(\tilde{u}) = \int_{\mathbb{R}^2} \varepsilon(c^2 - c_\infty^2) \mid \tilde{\boldsymbol{u}} \mid^2 dx, \\ A_1(\tilde{u}) = \displaystyle\int_{\mathbb{R}^2} \varepsilon(c^2 \mid \operatorname{rot} \tilde{\boldsymbol{u}} \mid^2 + c_\infty^2 \mid \tilde{u}_3 \mid^2 - 2(c^2 - c_\infty^2) \nabla \tilde{u}_3 . \tilde{\boldsymbol{u}}) \, dx, \\ B_0(\tilde{u}) = \displaystyle\int_{\mathbb{R}^2} \varepsilon \mid \tilde{u}_3 \mid^2 dx, \quad B_1(\tilde{u}) = \int_{\mathbb{R}^2} \varepsilon \mid \tilde{\boldsymbol{u}} \mid^2 dx. \end{cases}$$

Chapter 8. Optical Waveguides

Therefore, we can write

$$\begin{cases} s_m(\beta) - c_\infty^2 \beta^2 = \inf_{E \in \mathcal{V}_m^\varepsilon(1)} \sup_{\tilde{u} \in E \setminus \{0\}} F(\beta, \tilde{u}), \\ F(\beta, \tilde{u}) = \dfrac{A_0(\tilde{u}) + A_1(\tilde{u})\beta^2 + A_2(\tilde{u})\beta^4}{B_0(\tilde{u}) + B_1(\tilde{u})\beta^2}. \end{cases} \quad (8.72)$$

We now use the fact that $s_m(\beta) - c_\infty^2 \beta^2 \leq 0$ to remark that formula (8.72) does not change if we replace $F(\beta, \tilde{u})$ by $G(\beta, \tilde{u}) = \inf(0, F(\beta, \tilde{u}))$:

$$s_m(\beta) - c_\infty^2 \beta^2 = \inf_{E \in \mathcal{V}_m^\varepsilon(1)} \sup_{\tilde{u} \in E \setminus \{0\}} G(\beta, \tilde{u}). \quad (8.73)$$

As the set $\mathcal{V}_m^\varepsilon(1)$ does not depend on β, it suffices to prove that for any $\tilde{u} \in V_\varepsilon(1)$, $\tilde{u} \neq 0$, the function $\beta \to G(\beta, \tilde{u})$ is nonincreasing. For this we need to study the variations of $\beta \to F(\beta, \tilde{u})$. Note that, because of assumption (8.71),

$$\text{for all } \tilde{u} \neq 0, \quad B_1(\tilde{u}) > 0, \quad B_0(\tilde{u}) \geq 0, \quad A_0(\tilde{u}) \geq 0, \quad A_2(\tilde{u}) \leq 0. \quad (8.74)$$

We now distinguish three cases:

(a) $\tilde{u}_3 = 0$. In such a case $A_0(\tilde{u}) = B_0(\tilde{u}) = 0$. It is then immediate that $\beta \to F(\beta, \tilde{u})$ is nonincreasing, and so is $G(\beta, \tilde{u})$.

(b) $\tilde{u}_3 \neq 0$ and $A_2(\tilde{u}) = 0$. Then, $A_0(\tilde{u}) > 0$ and $B_0(\tilde{u}) > 0$ and either $F(\beta, \tilde{u})$ (and then $G(\beta, \tilde{u})$) is decreasing in β or $F(\beta, \tilde{u})$ is increasing in β with $F(0, \tilde{u}) > 0$, in which case $G(\beta, \tilde{u})$ is equal to zero.

(c) $\tilde{u}_3 \neq 0$ and $A_2(\tilde{u}) < 0$. In this case, $F(\beta, \tilde{u}) \to -\infty$ as $\beta \to +\infty$ and has at most one extremum in β^2. As $F(0, \tilde{u}) > 0$, it is not difficult to see that $G(\beta, v)$ is nonincreasing in β. $\square$

Remark 8.7. We do not know any example for which $\beta_m^0 < \beta_m^*$. We conjecture that the equality $\beta_m^0 = \beta_m^*$ is always true, but this remains an open question.

Finiteness of $N(\beta)$

We first prove this under assumption (8.71).

LEMMA 8.10. *Let $\beta_p > 0$ be a converging sequence of solutions of the threshold equation satisfying $\beta_p \neq \beta_q$ for $p \neq q$. If $c_- < c_\infty$ and if assumption (8.71) holds, its limit is necessarily zero.*

Proof. By the definition of the threshold equation (8.52) one can construct a sequence u_p, $u_p \in V_{\rho,\varepsilon}(\beta_p)$, such that

$$\begin{cases} p_\varepsilon(\beta_p; u_p, v) + c_\varepsilon(\beta_p; u_p, v) = 0 \quad \text{for all } v \in V_{\rho,\varepsilon}(\beta_p), \\ \displaystyle\int_{B_R} \varepsilon \, |u_p|^2 \, dx = 1. \end{cases} \quad (8.75)$$

(1) Using the notation of the proof of Theorem 8.8, the function $\tilde{u}_p = (u_p, \tilde{u}_{3,p} = \beta_p u_{3,p})$ is the solution of

$$\tilde{p}(\beta_p; \tilde{u}_p, \tilde{v}) + \tilde{c}(\beta_p; \tilde{u}_p, \tilde{v}) = 0 \quad \text{for all } \tilde{v} \in V_{\rho,\varepsilon}(1). \quad (8.76)$$

Assume that $\beta_p \to \beta^* \neq 0$. As in Theorem 8.8, we deduce the existence of a subsequence of $\tilde{u}_p$, still denoted $\tilde{u}_p$, such that

$$\tilde{u}_p \rightharpoonup \tilde{u} \quad \text{weakly in } V_{\rho,\varepsilon}(1), \quad \text{strongly in } L^2(B_R)^3,$$

and we have at the limit

$$\tilde{p}(\beta^*; \tilde{u}, \tilde{v}) + \tilde{c}(\beta^*; \tilde{u}, \tilde{v}) = 0 \quad \text{for all } \tilde{v} \in V_{\rho,\varepsilon}(1),$$
$$\int_{B_R} \varepsilon \mid \boldsymbol{u} \mid^2 \, dx = 1. \tag{8.77}$$

(2) Now let us take $\tilde{v} = \tilde{u}_q$ in (8.76). We obtain

$$\tilde{p}(\beta_p; \tilde{u}_p, \tilde{u}_q) + \tilde{c}(\beta_p; \tilde{u}_p, \tilde{u}_q) = 0, \tag{8.78}$$

and, inverting the role of p and q, we have

$$\tilde{p}(\beta_q; \tilde{u}_p, \tilde{u}_q) + \tilde{c}(\beta_q; \tilde{u}_p, \tilde{u}_q) = 0. \tag{8.79}$$

Taking the difference of (8.78) and (8.79), we get after division by $\beta_p^2 - \beta_q^2 \neq 0$

$$\frac{1}{\beta_p^2 \beta_q^2} \int_{\mathbb{R}^2} \varepsilon c^2 \nabla \tilde{u}_{3,q} \cdot \nabla \tilde{u}_{3,p} \, dx - \int_{\mathbb{R}^2} \varepsilon(c^2 - c_\infty^2) \boldsymbol{u}_p \cdot \boldsymbol{u}_q \, dx = 0. \tag{8.80}$$

This equality being satisfied for any p and q we can pass to the limit ($p \to +\infty$ first, then $q \to +\infty$) to obtain

$$\frac{1}{(\beta^*)^4} \int_{\mathbb{R}^2} \varepsilon c^2 \mid \nabla \tilde{u}_3 \mid^2 \, dx - \int_{\mathbb{R}^2} \varepsilon(c^2 - c_\infty^2) \mid \boldsymbol{u} \mid^2 \, dx = 0. \tag{8.81}$$

(3) Because of assumption (8.71), (8.81) implies

$$\tilde{u}_3 = 0 \text{ and } \int_{\mathbb{R}^2} \varepsilon(c^2 - c_\infty^2) \mid \boldsymbol{u} \mid^2 \, dx = 0. \tag{8.82}$$

Coming back to (8.77) with $\tilde{v} = \tilde{u}$, we obtain $\text{rot}\,\boldsymbol{u} = 0$, while $\tilde{u}_3 = 0$ implies $\text{div}(\varepsilon \boldsymbol{u}) = 0$. By Poincaré's lemma, there exists $\varphi \in H^1_{loc}(\mathbb{R}^2)$ such that $\boldsymbol{u} = \nabla \varphi$ and $\text{div}(\varepsilon \nabla \varphi) = 0$. But (8.82) implies that $\boldsymbol{u}$, and thus φ (which is only defined up to an additive constant), is zero in the disk D_γ defined by (8.39) thanks to $c_- < c_\infty$. Therefore, by unique continuation, $\boldsymbol{u} = 0$, which contradicts the second equation of (8.77). □

LEMMA 8.11. *Under the assumptions of Lemma 8.8, the sequence β_m^0 tends to $+\infty$ when m tends to $+\infty$.*

Proof. We know (Theorem 8.10) that $\beta_m^0 = \beta_m^* \equiv \beta_m$. The sequence β_m is increasing. Assume that $\beta_m \to \beta^* < +\infty$. As $\beta_3 > 0$ by Theorem 8.9, $\beta^* > 0$. Now, we remark that any $\beta > \beta^*$ is a solution of problem (8.52). Indeed, if $\beta > \beta^*$, $\beta > \beta_m$ for any m. Thus, by the definition of $\beta_m = \beta_m^*$, the operator $A_\varepsilon(\beta)$ has an infinity of eigenvalues $\omega_m(\beta)^2 < c_\infty^2 \beta^2$ which satisfies

$$\lim_{m \to +\infty} \omega_m(\beta)^2 = c_\infty^2 \beta^2.$$

We can then repeat the proof of Theorem 8.8 (with a fixed β instead of a converging sequence β_p) to conclude that β is solution of the threshold equation. Then, we can

Chapter 8. Optical Waveguides

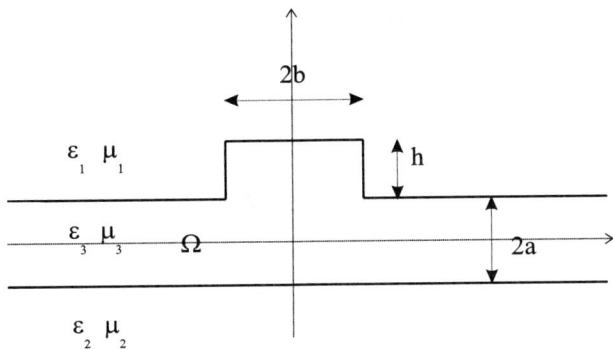

Figure 8.4: The rib waveguide.

construct a strictly decreasing sequence β_p satisfying the assumptions of Lemma 8.9, which contradicts $\beta^* > 0$ and completes the proof. □

We now extend the result to the general case.

THEOREM 8.11. *For any medium, the sequence β_m^0 goes to $+\infty$ when $m \to +\infty$. Consequently the number $N(\beta)$ of guided modes is finite for any $\beta > 0$.*

Proof. Let us introduce $\tilde{\varepsilon}(x)$ and $\tilde{\mu}(x)$ defined by

$$\tilde{\varepsilon}(x) = \varepsilon^+ \quad \text{and} \quad \tilde{\mu}(x) = \mu^+ \quad \text{if } |x| \le R,$$
$$\tilde{\varepsilon}(x) = \varepsilon_\infty \quad \text{and} \quad \tilde{\mu}(x) = \mu_\infty \quad \text{if } |x| > R.$$

By construction $\tilde{\varepsilon}(x) \ge \varepsilon(x), \tilde{\mu}(x) \ge \mu(x)$, and $\tilde{c}(x) \le c_\infty$ almost everywhere. Then, by Lemma 8.10, we have, with obvious notation: $\tilde{\beta}_m^0 \to +\infty$, while, by comparison (Theorem 8.7), $\beta_m^0 \ge \tilde{\beta}_m^0$. □

8.3 Mathematical Analysis of Integrated Optical Guides

Now we suppose that the coefficients ε and μ satisfy hypotheses (8.7) and (8.8). As we will see, the stratified nature of the reference medium will be responsible for original phenomena, compared to the previous case of the optical fiber. The most common example of integrated optical waveguide is the so-called rib or ridge waveguide (cf. Figure 8.4) defined by

$$\varepsilon(x) = \varepsilon_3 \quad \text{and} \quad \mu(x) = \mu_3 \quad \text{if } x \in \Omega,$$
$$\varepsilon(x) = \varepsilon_2 \quad \text{and} \quad \mu(x) = \mu_2 \quad \text{if } x_2 < -a,$$
$$\varepsilon(x) = \varepsilon_1 \quad \text{and} \quad \mu(x) = \mu_1 \quad \text{elsewhere,}$$

where $\Omega = \{x; |x_2| \le a \text{ and } |x_1| > b\} \cup \{x; -a \le x_2 \le a+h \text{ and } |x_1| < b\}$ and

$$c_3 = (\varepsilon_3 \mu_3)^{-1/2} < c_1 = (\varepsilon_1 \mu_1)^{-1/2} \le c_2 = (\varepsilon_2 \mu_2)^{-1/2}.$$

Notice that, contrary to the case of the optical fiber, the region Ω where the velocity is minimum is unbounded. The understanding of the mechanism that allows the confinement of the energy is therefore more complicated here.

Remark 8.8. This problem can be related to that of water-guided waves, which has been extensively studied (see [7], [40], and the references herein); the similarity comes from the dispersive nature of the reference medium. However, the problem we consider here is more involved. Indeed, in hydrodynamics, the problem is scalar and set in a domain that is bounded in the vertical direction.

Let us consider again the operator $A(\beta)$ associated with the formulation for the electric field, defined by (8.27). It has already been proved that it is a self-adjoint operator. To study guided modes, the first step consists in finding the essential spectrum of $A(\beta)$. Existence results are then derived using the Min-Max Principle.

8.3.1 The Essential Spectrum

As for the optical fiber, it will be proved that the essential spectrum of $A(\beta)$ is identical to that of the operator $A_\infty(\beta)$ associated to the reference medium, which is here the perfectly stratified medium defined by

$$\varepsilon(x) = \varepsilon_\infty(x_2) \quad \text{and} \quad \mu(x) = \mu_\infty(x_2).$$

The spectral theory of this operator can be carried out using the Fourier transform in the x_1 direction together with the decoupling between transverse electric (TE) and transverse magnetic (TM) modes.

This way, we will establish the following theorem.

THEOREM 8.12.

$$\sigma_{ess}(A(\beta)) = \sigma_{ess}(A_\infty(\beta)) = [\gamma(\beta), +\infty[,$$

where

$$\gamma(\beta) = \min(\gamma_{TE}(\beta), \gamma_{TM}(\beta))$$

with

$$\gamma_{TE}(\beta) = \inf_{\phi \in H^1(\mathbb{R}) \setminus \{0\}} \frac{\int_\mathbb{R} \frac{1}{\mu_\infty(x_2)} \left(\left| \frac{d\phi}{dx_2} \right|^2 + \beta^2 |\phi|^2 \right) dx_2}{\int_\mathbb{R} \varepsilon_\infty(x_2) |\phi|^2 dx_2} \qquad (8.83)$$

and

$$\gamma_{TM}(\beta) = \inf_{\phi \in H^1(\mathbb{R}) \setminus \{0\}} \frac{\int_\mathbb{R} \frac{1}{\varepsilon_\infty(x_2)} \left(\left| \frac{d\phi}{dx_2} \right|^2 + \beta^2 |\phi|^2 \right) dx_2}{\int_\mathbb{R} \mu_\infty(x_2) |\phi|^2 dx_2}. \qquad (8.84)$$

Remark 8.9. The value $\gamma_{TE}(\beta)$ is the minimum value in the spectrum of the self-adjoint operator $A_{TE}(\beta)$ defined on $L^2(\mathbb{R})$ (equipped with the scalar product $(\phi, \psi) = \int_\mathbb{R} \varepsilon_\infty(x_2) \phi \psi dx_2$) by

$$\begin{cases} D(A_{TE}(\beta)) = \left\{ \phi \in H^1(\mathbb{R}); \; \dfrac{d}{dx_2}\left(\dfrac{1}{\mu_\infty}\dfrac{d\phi}{dx_2}\right) \in L^2(\mathbb{R}) \right\}, \\ A_{TE}(\beta)\phi = \dfrac{1}{\varepsilon_\infty}\left\{-\dfrac{d}{dx_2}\left(\dfrac{1}{\mu_\infty}\dfrac{d\phi}{dx_2}\right) + \dfrac{\beta^2}{\mu_\infty}\phi\right\}. \end{cases}$$

Chapter 8. Optical Waveguides

It is well known that $\beta \to \gamma_{TE}(\beta)$ is nondecreasing on $\mathbb{R}^+$ and that (cf. Appendix A of [12])

$$\begin{cases} \beta^2 \left(c_\infty^-\right)^2 \leq \gamma_{TE}(\beta) \leq \beta^2 c_1^2 & \text{for all } \beta, \\ \gamma_{TE}(\beta) \sim \beta^2 c_1^2 & \text{for } \beta \sim 0, \\ \gamma_{TE}(\beta) \sim \beta^2 \left(c_\infty^-\right)^2 & \text{for } \beta \sim +\infty, \end{cases} \quad (8.85)$$

where

$$c_\infty(x_2) = (\varepsilon_\infty(x_2)\mu_\infty(x_2))^{-1/2}, \quad c_\infty^- = \inf_{x_2 \in \mathbb{R}} c_\infty(x_2), \quad c_1 = (\varepsilon_1 \mu_1)^{-1/2}.$$

Moreover, by the Min-Max Principle applied to $A_{TE}(\beta)$, if

$$\gamma_{TE}(\beta) < \beta^2 c_1^2,$$

then $\gamma_{TE}(\beta)$ is an eigenvalue of $A_{TE}(\beta)$ and the equation $\omega^2 = \gamma_{TE}(\beta)$ is the dispersion relation of the fundamental TE-guided mode of the planar waveguide (cf. [28]).

In the same way, $\gamma_{TM}(\beta)$ is the minimum value in the spectrum of a self-adjoint operator $A_{TM}(\beta)$ (defined like $A_{TE}(\beta)$ by inverting the roles of ε_∞ and μ_∞); it also satisfies properties (8.85), and if

$$\gamma_{TM}(\beta) < \beta^2 c_1^2,$$

it is an eigenvalue of $A_{TM}(\beta)$ and the equation $\omega^2 = \gamma_{TM}(\beta)$ is the dispersion relation of the fundamental TM-guided mode of the planar waveguide.

To establish Theorem 8.12, we are going to demonstrate successively the following two identities:

$$\sigma(A_\infty(\beta)) = [\gamma(\beta), +\infty) \quad \text{and} \quad \sigma_{ess}(A(\beta)) = \sigma_{ess}(A_\infty(\beta)).$$

Let us start by proving the first equality.

PROPOSITION 8.3.
$$\sigma(A_\infty(\beta)) = [\gamma(\beta), +\infty).$$

Proof. Let us denote by $\hat{u}(k_1, x_2)$ the partial Fourier transform of u with respect to the variable x_1 defined by

$$\begin{cases} \hat{u}_1(k_1, x_2) = \dfrac{-i}{\sqrt{2\pi}} \int_\mathbb{R} u_1(x_1, x_2) e^{-ik_1 x_1} dx_1, \\ \hat{u}_j(k_1, x_2) = \dfrac{1}{\sqrt{2\pi}} \int_\mathbb{R} u_j(x_1, x_2) e^{-ik_1 x_1} dx_1, \quad j = 2, 3. \end{cases} \quad (8.86)$$

Classical properties of the Fourier transform give, for all $u \in D(A_\infty(\beta))$,

$$(A_\infty(\beta)u, u) = \int_{\mathbb{R}^2} \mu_\infty^{-1} \left\{ \left| \frac{\partial \hat{u}_3}{\partial x_2} - \beta \hat{u}_2 \right|^2 + |\beta \hat{u}_1 - k_1 \hat{u}_3|^2 + \left| k_1 \hat{u}_2 - \frac{\partial \hat{u}_1}{\partial x_2} \right|^2 \right\} dk_1 \, dx_2$$

and

$$\|u\|_{\varepsilon_\infty}^2 = \int_{\mathbb{R}^2} \varepsilon_\infty(x_2) |\hat{u}|^2 dk_1 \, dx_2.$$

Notice now that using a Fourier transform in the x_1-direction amounts to looking for electromagnetic waves propagating in the direction $k = (k_1, 0, \beta)^t$. Thus, it is natural to perform a rotation in the plane (O, x_1, x_3) so that the direction defined by k is one of the two directions of the new system of coordinates. So, let us set

$$\hat{v}_1 = \frac{1}{\xi}(\beta \hat{u}_1 - k_1 \hat{u}_3), \quad \hat{v}_2 = \hat{u}_2, \quad \hat{v}_3 = \frac{1}{\xi}(k_1 \hat{u}_1 + \beta \hat{u}_3), \tag{8.87}$$

where $\xi = |k| = \sqrt{\beta^2 + k_1^2}$. Then we get

$$(A_\infty(\beta)u, u) = \int_{\mathbb{R}^2} \mu_\infty^{-1} \left(\left|\frac{\partial \hat{v}_1}{\partial x_2}\right|^2 + \xi^2 |\hat{v}_1|^2 + \left|\frac{\partial \hat{v}_3}{\partial x_2} - \xi \hat{v}_2\right|^2 \right) dk_1 \, dx_2. \tag{8.88}$$

Furthermore, we have $|\hat{u}_1|^2 + |\hat{u}_2|^2 + |\hat{u}_3|^2 = |\hat{v}_1|^2 + |\hat{v}_2|^2 + |\hat{v}_3|^2$. On the other hand, the identity $\text{div}_\beta u = 0$ is equivalent to

$$\frac{\partial}{\partial x_2}(\varepsilon_\infty \hat{v}_2) - \xi \varepsilon_\infty \hat{v}_3 = 0.$$

A decoupling between the first component $\hat{v}_1$ and the others, $\hat{v}_2$ and $\hat{v}_3$, appears. It corresponds to the well-known decoupling between TE and TM modes. As a consequence, the spectrum of the operator $A_\infty(\beta)$ is the union of the spectra of the two following operators (remember that $\xi = \xi(\beta) = \sqrt{\beta^2 + k^2}$):

- the self-adjoint operator $\hat{A}_{TE}(\beta)$ of $\hat{H}_\infty$ associated with the bilinear form

$$\hat{a}_{TE}(\beta; \hat{v}_1, \hat{w}_1) = \int_{\mathbb{R}^2} \mu_\infty^{-1} \left(\frac{\partial \hat{v}_1}{\partial x_2} \frac{\partial \hat{w}_1}{\partial x_2} + \xi^2 \hat{v}_1 \hat{w}_1 \right) dk_1 \, dx_2$$

defined on

$$\left\{ \hat{v}_1 \in H^1(\mathbb{R}^2); k_1 \hat{v}_1 \in L^2(\mathbb{R}^2) \right\}.$$

- the self-adjoint operator $\hat{A}_{TM}(\beta)$ of $\hat{H}_\infty(\beta)$ associated with the bilinear form

$$\hat{a}_{TM}(\beta; (\hat{v}_2, \hat{v}_3), (\hat{w}_2, \hat{w}_3)) = \int_{\mathbb{R}^2} \mu_\infty^{-1} \left(\frac{\partial \hat{v}_3}{\partial x_2} - \xi \hat{v}_2 \right) \left(\frac{\partial \hat{w}_3}{\partial x_2} - \xi \hat{w}_2 \right) dk_1 \, dx_2$$

defined on

$$\left\{ (\hat{v}_2, \hat{v}_3) \in \hat{H}_\infty(\beta); \frac{\partial \hat{v}_3}{\partial x_2} - \xi \hat{v}_2 \in \hat{H}_\infty \right\},$$

where $\hat{H}_\infty$ denotes the space $L^2(\mathbb{R}^2)$ equipped with the scalar product

$$(\hat{v}_1, \hat{w}_1) = \int_{\mathbb{R}^2} \varepsilon_\infty \hat{v}_1 \hat{w}_1 \, dk_1 \, dx_2$$

and

$$\hat{H}_\infty(\beta) = \left\{ (\hat{v}_2, \hat{v}_3) \in \hat{H}_\infty^2; \frac{\partial}{\partial x_2}(\varepsilon_\infty \hat{v}_2) - \xi \varepsilon_\infty \hat{v}_3 = 0 \right\}.$$

To conclude, we prove, by standard techniques, that

$$\sigma\left(\hat{A}_{TE}(\beta)\right) = [\gamma_{TE}(\beta), +\infty)$$

Chapter 8. Optical Waveguides

and that (the proof is slightly more complicated; see [33])

$$\sigma\left(\hat{A}_{TM}(\beta)\right) = [\gamma_{TM}(\beta), +\infty).\quad\square$$

Now, we prove the second equality.
PROPOSITION 8.4.

$$\sigma_{ess}(A(\beta)) = \sigma_{ess}(A_\infty(\beta)).$$

Proof. We will prove, for instance, that

$$\sigma_{ess}(A(\beta)) \subset \sigma_{ess}(A_\infty(\beta));$$

the converse inclusion could be established in exactly the same manner.

Let $\sigma \in \sigma_{ess}(A(\beta))$ and $u^n \in D(A(\beta))$ be a singular sequence of $A(\beta)$ associated with σ. Proceeding as in the proof of Lemma 8.6, we can prove that u^n converges weakly to 0 in $V_\varepsilon(\beta)$ and strongly in $L^2(B_R)^3$. In order to build up a singular sequence v^n of $A_\infty(\beta)$ associated to σ, we set

$$v^n(x_1, x_2) = \psi(x_1, x_2) u^n(x_1, x_2), \quad (8.89)$$

where $\psi \in \mathcal{C}^\infty(\mathbb{R}^2)$, $\psi(x) = 0$ if $|x| \leq R$, and $\psi(x) = 1$ if $|x| \geq 2R$. Then

$$\|v^n\|_{H_\varepsilon} \longrightarrow 1 \text{ and } v^n \rightharpoonup 0 \text{ weakly in } H_\varepsilon.$$

We give the proof when ε_∞ and μ_∞ are regular functions of x_2. In this case straightforward calculations (using the fact that $\nabla \psi$ is compactly supported) prove that

$$A_\infty(\beta)v^n - \sigma v^n = \frac{1}{\varepsilon_\infty}\left\{\text{rot}^*_\beta\left(\frac{1}{\mu_\infty}\text{rot}_\beta v^n\right)\right\} - \sigma v^n \longrightarrow 0 \text{ strongly in } H_\varepsilon.$$

However, v^n is not a singular sequence of $A_\infty(\beta)$ since

$$\text{div}_\beta(\varepsilon_\infty v^n) = \varepsilon_\infty \nabla\psi.\boldsymbol{u}^n$$

is not equal to 0 (but notice that it tends to 0 in $L^2(\mathbb{R}^2)$). So we have to add to v^n a corrective term w^n in order to satisfy the divergence-free constraint. This term can be chosen as follows:

$$w^n = -\nabla_\beta \phi_n,$$

where ϕ_n is the unique solution in $H^1(\mathbb{R}^2)$ of

$$\text{div}_\beta(\varepsilon_\infty \nabla_\beta \phi_n) = \text{div}_\beta(\varepsilon_\infty v^n).$$

Then, as $\phi_n \to 0$ in $D(A_\infty(\beta))$, $v^n + w^n$ is a singular sequence of $A_\infty(\beta)$.

The proof is more complicated when ε_∞ and μ_∞ are discontinuous functions. Indeed, in that case, $v^n + w^n$ does not satisfy the required jump conditions in order to belong to the domain of $A_\infty(\beta)$. Again, a corrective term has to be added as explained in [33]. $\square$

8.3.2 The Main Existence Result

In the following, we apply the Min-Max Principle (the min-max associated to $a(\beta; u, u)$ are still denoted $s_m(\beta)$, and the number of eigenvalues in the discrete spectrum by $N(\beta)$) to derive conditions for the existence of eigenvalues below the essential spectrum (these eigenvalues form the discrete spectrum). Notice that these modes propagate more slowly than the fundamental mode of the planar waveguide. The following comparison principle, which extends Theorem 8.7 to stratified waveguides, will be used at many times in the rest of this chapter.

THEOREM 8.13. *Assume that*

$$\varepsilon_1(x) \leq \varepsilon_2(x) \ and \ \mu_1(x) \leq \mu_2(x) \quad a.e. \ x \in \mathbb{R}^2; \tag{8.90}$$

then, with obvious notation,

$$for \ all \ m \geq 1, \quad for \ all \ \beta > 0, \quad s_m^1(\beta) \geq s_m^2(\beta), \tag{8.91}$$

Consequently, if moreover $\gamma^1(\beta) = \gamma^2(\beta)$, then

$$N_1(\beta) \leq N_2(\beta) \quad for \ all \ \beta. \tag{8.92}$$

Let us first point out the following nonexistence result, which is a direct consequence of the previous comparison results.

LEMMA 8.12. *If $\varepsilon(x) \leq \varepsilon_\infty(x_2)$ and $\mu(x) \leq \mu_\infty(x_2)$ almost everywhere, the discrete spectrum of $A(\beta)$ is empty.*

Theorem 8.6 extends easily to stratified waveguides, as shown in the following theorem.

THEOREM 8.14. *Assume that $c_- < c_\infty^-$. Then for any $m \geq 1$, there exists $\beta_m \geq 0$ such that for any $\beta > \beta_m$, $A(\beta)$ admits at least m eigenvalues strictly smaller than $\gamma(\beta)$.*

Proof. Just follow the proof of Theorem 8.6 to prove (8.43) and notice that $c_-^2 < (c_\infty^-)^2 < \gamma(\beta)/\beta^2$ by (8.85). □

The strict inequality $c_- < c_\infty^-$ is essential since $\gamma(\beta)/\beta^2$ tends to $(c_\infty^-)^2$ when β goes to $+\infty$. For instance, this result does not apply to the case of the rib waveguide, described at the beginning of the section, which is such that

$$c_- = c_\infty^- = c_3.$$

The object of the fundamental following result is to establish an existence result of guided modes, which applies in particular to the rib waveguide.

THEOREM 8.15. *Suppose that*

$$\varepsilon(x) \geq \varepsilon_\infty(x_2) \ and \ \mu(x) \geq \mu_\infty(x_2) \quad a.e. \ x \in \mathbb{R}^2 \tag{8.93}$$

and that (m_L denoting the Lebesgue's measure)

$$m_L\{x \in \mathbb{R}^2; \varepsilon(x) - \varepsilon_\infty(x_2) > 0 \ or \ \mu(x) - \mu_\infty(x_2) > 0\} > 0. \tag{8.94}$$

Then $\gamma(\beta) < \beta^2 c_1^2$ implies $s_1(\beta) < \gamma(\beta)$.

In other words, if for a given β, $\gamma(\beta) < \beta^2 c_1^2$, which means that the planar waveguide $(\varepsilon_\infty, \mu_\infty)$ has a guided mode (invariant in the x_1-direction), then $s_1(\beta) < \gamma(\beta)$, which means that the waveguide (ε, μ) also has a guided mode (confined in both transverse directions) for the same value of β.

Chapter 8. Optical Waveguides

Proof.

(a) Let us suppose first that $\varepsilon(x) = \varepsilon_\infty(x_2)$ a.e. in $\mathbb{R}^2$. Both cases $\gamma(\beta) = \gamma_{TE}(\beta)$ and $\gamma(\beta) = \gamma_{TM}(\beta)$ will be considered successively. Suppose first that

$$\gamma(\beta) = \gamma_{TE}(\beta) < \beta^2 c_1^2.$$

That means (cf. Remark 8.9) that there exists a nonvanishing function $\phi \in H^1(\mathbb{R})$ such that

$$-\frac{d}{dx_2}\left(\frac{1}{\mu_\infty}\frac{d\phi}{dx_2}\right) + \frac{\beta^2}{\mu_\infty}\phi = \gamma_{TE}(\beta)\varepsilon_\infty \phi. \tag{8.95}$$

Consider a cut-off function $\psi \in C_0^\infty(\mathbb{R})$ such that $\psi(0) = 1$ and set

$$u^n = \left(\frac{1}{\sqrt{n}}\psi\left(\frac{x_1}{n}\right)\phi(x_2)\,,\;0\,,\;\frac{-1}{\beta n\sqrt{n}}\frac{d\psi}{dx_1}\left(\frac{x_1}{n}\right)\phi(x_2)\right)^t.$$

Clearly $\mathrm{div}_\beta(\varepsilon_\infty u^n) = 0$, so that $u^n \in V_\varepsilon(\beta)$. Then

$$\int_{\mathbb{R}^2} \varepsilon |u^n|^2 dx = \int_{\mathbb{R}^2} \varepsilon_\infty \phi(x_2)^2 \left(\psi\left(\frac{x_1}{n}\right)^2 + \frac{1}{\beta^2 n^2}\frac{d\psi}{dx_1}\left(\frac{x_1}{n}\right)^2\right)\frac{dx_1}{n}dx_2,$$

that is,

$$\int_{\mathbb{R}^2} \varepsilon |u^n|^2 dx = \int_{\mathbb{R}} \varepsilon_\infty \phi(x_2)^2 dx_2 \int_{\mathbb{R}} \psi(x_1)^2 dx_1 + \frac{1}{n^2}K, \tag{8.96}$$

where K is a constant independent of n. It can be proved in the same way that

$$a(\beta; u_n, u_n) = \int_{\mathbb{R}^2} \mu^{-1}|\mathrm{rot}_\beta u^n|^2 dx = I_0 + \frac{1}{n}I_1(n) + \frac{1}{n^2}I_2(n), \tag{8.97}$$

where I_0 and $I_1(n)$ are given by

$$I_0 = \int_{\mathbb{R}} \mu_\infty^{-1}\left(\frac{d\phi}{dx_2}(x_2)^2 + \beta^2 \phi(x_2)^2\right) dx_2 \int_{\mathbb{R}} \psi(x_1)^2 dx_1$$

and

$$I_1(n) = \int_{\mathbb{R}^2} (\mu^{-1} - \mu_\infty^{-1})\left(\frac{d\phi}{dx_2}(x_2)^2 + \beta^2 \phi(x_2)^2\right) \psi\left(\frac{x_1}{n}\right)^2 dx,$$

and where one can check that $I_2(n) \leq C$ for a constant C independent of n. Then, thanks to (8.95), we have

$$I_0 = \gamma(\beta) \int_{\mathbb{R}^2} \varepsilon |u^n|^2 dx, \tag{8.98}$$

while Lebesgue's theorem yields that

$$I_1(n) \to I_1 = \int_{\mathbb{R}^2} (\mu^{-1} - \mu_\infty^{-1})\left(\left(\frac{d\phi}{dx_2}(x_2)\right)^2 + \beta^2 \phi(x_2)^2\right)^2 dx < 0, \tag{8.99}$$

when $n \to +\infty$. I_1 is strictly negative in virtue of (8.93) and (8.94). Combining identities (8.96), (8.97), and (8.98), we obtain

$$\int_{\mathbb{R}^2} \mu^{-1}|\text{rot}_\beta u^n|^2 dx - \gamma(\beta) \int_{\mathbb{R}^2} \varepsilon |u^n|^2 dx = I_1(n) + \frac{1}{n^2}\left(I_2(n) - \gamma(\beta)K\right),$$

which takes a strictly negative value for n great enough. By definition of $s_1(\beta)$, this implies finally that $s_1(\beta) < \gamma(\beta)$.

Suppose now that
$$\gamma(\beta) = \gamma_{TM}(\beta) < \beta^2 c_1^2.$$

Then (cf. Remark 8.9) there exists a nonvanishing function $\phi \in H^1(\mathbb{R})$ such that

$$-\frac{d}{dx_2}\left(\frac{1}{\varepsilon_\infty}\frac{d\phi}{dx_2}\right) + \frac{\beta^2}{\varepsilon_\infty}\phi = \gamma_{TM}(\beta)\mu_\infty \phi. \tag{8.100}$$

Using a cut-off function ψ as above, we set in that case

$$u^n = \left(0,\ \psi\left(\frac{x_1}{n}\right)\frac{\beta}{\varepsilon_\infty(x_2)}\phi(x_2),\ \psi\left(\frac{x_1}{n}\right)\frac{1}{\varepsilon_\infty(x_2)}\frac{d\phi}{dx_2}(x_2)\right)^t.$$

Again $u^n \in V_\varepsilon(\beta)$ since $\text{div}_\beta(\varepsilon_\infty u^n) = 0$. Moreover, thanks to (8.100),

$$\int_{\mathbb{R}^2} \varepsilon_\infty |u^n|^2 dx = \gamma_{TM}(\beta)\int_{\mathbb{R}^2} \mu_\infty(x_2)\phi(x_2)^2\psi\left(\frac{x_1}{n}\right)^2 dx$$

and

$$\int_{\mathbb{R}^2} \mu^{-1}|\text{rot}_\beta u^n|^2 dx = \gamma_{TM}(\beta)^2 \int_{\mathbb{R}^2}\frac{\mu_\infty(x_2)^2}{\mu(x)}\phi(x_2)^2 \psi\left(\frac{x_1}{n}\right)^2 dx + \frac{1}{n}K,$$

where K is a constant independent of n. Consequently,

$$\int_{\mathbb{R}^2} \mu^{-1}|\text{rot}_\beta u^n|^2 dx - \gamma_{TM}(\beta)\int_{\mathbb{R}^2}\varepsilon|u^n|^2 dx$$
$$= \gamma_{TM}(\beta)^2 \int_{\mathbb{R}^2}\frac{\mu_\infty}{\mu}(\mu_\infty - \mu)\phi(x_2)^2\psi\left(\frac{x_1}{n}\right)^2 dx + \frac{1}{n}K,$$

which takes a strictly negative value for n great enough.

(b) If $\mu(x) = \mu_\infty(x_2)$ almost everywhere in $\mathbb{R}^2$, the theorem can be proved in the same way by using the H-formulation.

(c) In the general case, suppose for example that
$$\{x \in \mathbb{R}^2;\ \varepsilon(x) - \varepsilon_\infty(x_2) > 0\}$$
has a strictly positive measure and consider the medium defined by
$$\tilde{\varepsilon} = \varepsilon \quad \text{and} \quad \tilde{\mu} = \mu_\infty.$$

By (b), $\tilde{s}_1(\beta) < \gamma(\beta)$. Then, the theorem derives from the comparison result (see Theorem 8.13). $\square$

Chapter 8. Optical Waveguides

This permits us to exhibit examples of waveguides such that $c_- = c_\infty^-$ and such that the fundamental guided mode (associated to the first eigenvalue) exists at every positive frequency.

COROLLARY 8.1. *Suppose*
$$c_1 = c_2$$
and suppose that one of these two conditions hold:
$$\int_\mathbb{R} \varepsilon_\infty \left(c_1^2 - c_\infty^2\right) dx_2 > 0 \quad or \quad \int_\mathbb{R} \mu_\infty \left(c_1^2 - c_\infty^2\right) dx_2 > 0.$$
Then, if (8.93) and (8.94) hold,
$$s_1(\beta) < \gamma(\beta) \quad \text{for all } \beta > 0.$$

Proof. Indeed, if the above conditions hold, the planar waveguide admits a guided mode for every $\beta > 0$ (cf. Appendix A of [12] or Annex A of [33]). □

Remark 8.10. This last result applies in particular to the case of the rib waveguide, defined at the beginning of this section, if $c_1 = c_2$.

8.3.3 A Low-Frequency Nonexistence Result

Suppose now that $c_1 > c_2$. Then (cf. [12, Appendix A, Proposition A.6]) the planar waveguide $(\varepsilon_\infty, \mu_\infty)$ has no guided modes at low frequency.

LEMMA 8.13. *If $c_1 > c_2$ then $\exists \beta^* > 0$ such that $\gamma(\beta) = \beta^2 c_1^2$ if $\beta < \beta^*$.*

We will see that the same phenomenon occurs for the waveguide (ε, μ).

THEOREM 8.16. *If $c_1 > c_2$, then there exists a $\beta^* > 0$ such that*
$$s_1(\beta) = \gamma(\beta) = \beta^2 c_1^2 \quad \text{if } \beta < \beta^*.$$

That means that the operator $A(\beta)$ has an empty discrete spectrum at low frequency.

Proof. Let us introduce $\tilde{\varepsilon}$ and $\tilde{\mu}$ defined by
$$\begin{aligned}
\tilde{\varepsilon}(x_2) &= \varepsilon_1 \quad \text{and} \quad \tilde{\mu}(x_2) = \mu_1 \quad \text{if} \quad x_2 > a, \\
\tilde{\varepsilon}(x_2) &= \varepsilon_+ \quad \text{and} \quad \tilde{\mu}(x_2) = \mu_+ \quad \text{if} \quad |x_2| < a, \\
\tilde{\varepsilon}(x_2) &= \varepsilon_2 \quad \text{and} \quad \tilde{\mu}(x_2) = \mu_2 \quad \text{if} \quad x_2 < -a.
\end{aligned}$$

Then, Lemma 8.13 applies to the planar waveguide $(\tilde{\varepsilon}, \tilde{\mu})$, and therefore
$$\exists \beta^* > 0 \text{ such that } \tilde{\gamma}(\beta) = \beta^2 c_1^2 \text{ if } \beta < \beta^*.$$

On the other hand,
$$\tilde{\varepsilon}(x_2) > \varepsilon(x) \text{ and } \tilde{\mu}(x_2) > \mu(x) \text{ for a.e. } x \in \mathbb{R}^2,$$

and the comparison result (cf. Theorem 8.13) gives
$$\tilde{s}_1(\beta) < s_1(\beta).$$

To conclude we notice that, by Proposition 8.3,
$$\tilde{s}_1(\beta) = \tilde{\gamma}(\beta)$$
for all β. □

Remark 8.11. In particular, if $c_1 > c_2$, the rib waveguide has no guided modes at low frequency. However, if $c_- < c_1$, it has at least one guided mode at high frequency thanks to Theorem 8.16.

8.3.4 Example of Embedded Eigenvalues

Let us now prove that existence of eigenvalues σ of $A(\beta)$ *embedded* in the essential spectrum, i.e., such that $\sigma > \gamma(\beta)$, can occur. Such an eigenvalue corresponds to a guided mode whose phase velocity ω/β is greater than the velocity of the fundamental guided mode of the planar waveguide $(\varepsilon_\infty, \mu_\infty)$. Notice that results of the same type have been already established by a similar method for different problems (acoustical closed waveguides [44], weakly guiding optical stratified waveguides [13], electromagnetic gratings [8], or elastic waveguides [23]).

In this subsection, we consider a waveguide that is symmetrical with respect to $x_2 = 0$:

$$\varepsilon(x_1, x_2) = \varepsilon(x_1, -x_2) \quad \text{and} \quad \mu(x_1, x_2) = \mu(x_1, -x_2).$$

Then $H_\varepsilon(\beta)$ admits the following orthogonal decomposition:

$$H_\varepsilon(\beta) = H_\varepsilon^1(\beta) + H_\varepsilon^2(\beta),$$

where $H_\varepsilon^1(\beta)$ is the subspace of $H_\varepsilon(\beta)$ composed of symmetrical fields u:

$$H_\varepsilon^1(\beta) = \{u \in H_\varepsilon(\beta); u_j(x_1, x_2) = (-1)^j u_{j+1}(x_1, -x_2), j = 1, 2, 3\}$$

and $H_\varepsilon^2(\beta)$ is the subspace composed of skew-symmetrical fields u:

$$H_\varepsilon^2(\beta) = \{u \in H_\varepsilon(\beta); u_j(x_1, x_2) = (-1)^j u_j(x_1, -x_2), j = 1, 2, 3\}.$$

Notice that $\text{rot}_\beta u$ is skew-symmetrical (respectively, symmetrical) if u is symmetrical (respectively, skew-symmetrical), so that one easily verifies the two following inclusions:

$$A(\beta)\left(H_\varepsilon^1(\beta)\right) \subset H_\varepsilon^1(\beta) \quad \text{and} \quad A(\beta)\left(H_\varepsilon^2(\beta)\right) \subset H_\varepsilon^2(\beta).$$

This allows us to define the restrictions of the operator $A(\beta)$ to the spaces $H_\varepsilon^1(\beta)$ and $H_\varepsilon^2(\beta)$: they are denoted by $A_1(\beta)$ and $A_2(\beta)$. Clearly, $A_1(\beta)$ and $A_2(\beta)$ are self-adjoint, $\sigma(A(\beta)) = \sigma(A_1(\beta)) \cup \sigma(A_2(\beta))$ and $\sigma_{ess}(A(\beta)) = \sigma_{ess}(A_1(\beta)) \cup \sigma_{ess}(A_2(\beta))$.

The spectral theory of operators $A_1(\beta)$ and $A_2(\beta)$ can be carried out exactly in the same way as that of $A(\beta)$. The main point here is that these two operators do not in general have the same essential spectrum.

Suppose, for instance, that $(\varepsilon_\infty, \mu_\infty)$ is a three-layer planar waveguide such that $\mu_\infty(x_2) = \mu_0$ for all $x_2 \in \mathbb{R}$,

$$\varepsilon_\infty(x) = \varepsilon^+ \quad \text{if } |x_2| < a \quad \text{and} \quad \varepsilon_\infty(x) = \varepsilon_- \quad \text{if } |x_2| > a,$$

with $c_-^2 = (\varepsilon^+ \mu_0)^{-1} < c_+^2 = (\varepsilon_- \mu_0)^{-1}$. In that case, the dispersion relation of the guided modes of the planar waveguide has a simple expression, and one can easily show that

$$\gamma_{TE}(\beta) < \gamma_{TM}(\beta) < \beta^2 c_+^2 \quad \text{for all } \beta > 0.$$

Then calculations similar to those of section 8.3.1 lead to the following identities:

$$\sigma_{ess}(A_1(\beta)) = [\gamma_{TE}(\beta), +\infty), \quad \sigma_{ess}(A_2(\beta)) = [\gamma_{TM}(\beta), +\infty).$$

Chapter 8. Optical Waveguides

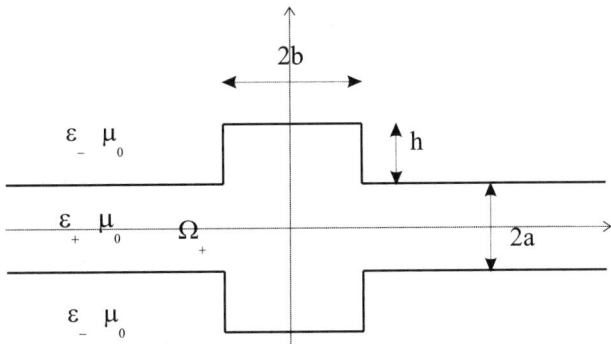

Figure 8.5: A symmetric waveguide.

Indeed, the electric field of the fundamental TE mode of the planar waveguide is symmetrical, while that of the fundamental TM-guided mode is skew-symmetrical. Suppose now that $A_2(\beta)$ has an eigenvalue σ below its essential spectrum which is located between $\gamma_{TE}(\beta)$ and $\gamma_{TM}(\beta)$:

$$\gamma_{TE}(\beta) < \sigma < \gamma_{TM}(\beta). \tag{8.101}$$

Then σ is also an eigenvalue of $A(\beta)$, and by (8.101), it belongs to its essential spectrum. It is what is usually called *an embedded eigenvalue*.

Let us prove finally that $A_2(\beta)$ may have an eigenvalue σ satisfying to (8.101). Suppose, for example, that ε and μ are defined as follows (cf. Figure 8.5):

$$\varepsilon(x) = \varepsilon^+ \quad \text{and} \quad \mu(x) = \mu_0 \quad \text{if } x \in \Omega_+,$$
$$\varepsilon(x) = \varepsilon_- \quad \text{and} \quad \mu(x) = \mu_0 \quad \text{if } |x_2| < -a,$$

where $\Omega_+ = \{x; |x_2| \leq a \text{ and } |x_1| > b\} \cup \{x; |x_2| \leq a+h \text{ and } |x_1| < b\}$.

Then, by Corollary 8.1 extended to the operator $A_2(\beta)$, this operator admits at least one eigenvalue σ such that $\sigma < \gamma_{TM}(\beta)$. Then one can show easily that $\sigma = \sigma(h) \to \gamma_{TM}(\beta)$ when $h \to 0$. In particular, (8.101) holds for h small enough.

8.3.5 Some Remarks on the High-Frequency Behavior

If $c_- < c_\infty^-$, it results from Theorem 8.14 that the number of guided modes $N(\beta)$ tends to infinity with β, as in the case of the optical fiber. What happens if on the contrary, $c_- = c_\infty^-$ (as for the rib waveguide)?

Let us mention first the high-frequency results obtained in [12] for an equivalent scalar model which corresponds to the so-called weak guiding assumption. For this model and if $c_- = c_\infty^-$, it has been proved that the number of guided modes $N(\beta)$ is a bounded function of β. Explicit bounds are derived by comparison with a Neumann problem set in a vertical band containing the perturbation of the reference medium. The function $N(\beta)$ may even vanish at high frequency. That happens, for example, if the reference medium is a three-layer medium, $\mu = \mu_\infty$ is a constant function, and ε is defined by

$$\varepsilon(x) = \varepsilon^* \text{ if } x \in K \text{ and } \varepsilon(x) = \varepsilon(x_2) \text{ elsewhere,}$$

where K is a bounded domain contained in the upper layer and

$$\varepsilon_1 < \varepsilon^* < \varepsilon^+.$$

Unfortunately, up to now, these results have not been extended to the vector model (we conjecture that they remain true). Indeed, the main tool that is the Neumann comparison principle does not work for Maxwell's equations.

8.4 Numerical Methods

8.4.1 Generalities

For the numerical approximation, we have to distinguish two successive steps:

- Reduce the problem initially posed in $\mathbb{R}^2$ to an equivalent one posed in a bounded interior domain Ω_i enclosing the perturbed region.

- Find a finite-dimensional approximation of the new problem, typically using a finite element approach.

Each step presents specific difficulties:

- The first step consists in writing a transparent boundary condition on the artificial boundary $\Gamma = \partial\Omega_i$, in practice an exact relationship between the Neumann- and Dirichlet-type data of the solution on $\partial\Omega_i$. Of course, such a condition depends on the nature of the medium exterior to Ω_i (i.e., on the reference medium) and should be easy to handle numerically (in particular, compatible with a variational formulation). Therefore, it is clear that the case of the optical fiber is much easier than the integrated optical waveguide.

- For the second step, one has to deal with the famous problem of spurious modes that may appear with naive approaches. Here, the difficulty is not linked to the nature of the reference medium but to the divergence-free condition.

To our knowledge, the first difficulty has essentially been treated in a rigorous way only in the case of the optical fiber. The attempts concerning the integrated optics waveguides concern a simplified scalar model whose use is justified under the so-called weak guiding assumptions (see [27], [31]). We shall restrict our attention here to the case of the optical fiber.

There are two natural ways to write a transparent boundary condition based on two different representations of the solution in the exterior domain Ω_e. The first one makes use of an integral representation of the solution in Ω_e in terms of boundary data [19], [18]. It only requires the Green's function of the exterior medium, in this case a homogeneous medium, which is known analytically. (Note that this approach could be applied in principle to the integrated optics case, but the computation of the Green's function is much more complicated [15], [10], [30].) However, this method does not naturally preserve the symmetry of the initial problem. That is why we have chosen to privilege a second approach based on a Fourier series expansion in polar coordinates of the solution in the exterior domain (section 3.2).

This permits us to give an explicit expression of the Dirichlet-to-Neumann–type operator, provided that the artificial boundary is chosen to be circular. Note that this technique is not applicable to the integrated optics case since there is no separation of variables in polar coordinates in a stratified medium.

The reduction to a bounded domain is obviously necessary for numerical reasons. It also has interesting mathematical properties. In particular, it permits us to get rid of the continuous spectrum present in the initial problem. Indeed, the new problem consists in finding the eigenvalues of an operator having a compact resolvent and thus a pure point spectrum, as for the case of the closed waveguide. On the other hand, one has to pay the price of an additional difficulty: the new operator depends on the eigenvalue one is looking for. This corresponds to some artificial nonlinearity, which leads to solving fixed-point equations.

Remark 8.12. In fact, the use of tranparent boundary conditions is not the only possible way to reduce effective calculations to a bounded domain. In [39], Urbach proposed using a "volume" integral equation which is the equivalent for the waveguide problem of the Lipman–Schwinger equation for the scattering problem. Here also, he finally had to solve a "fixed-point eigenvalue" problem. The drawback of this type of method is that the matrix one has to deal with after space discretization is dense.

Concerning the finite element approximation, it is now well known that an efficient approach for avoiding spurious modes consists in using edge-type elements which naturally lead to approximation spaces for $H(\text{rot}; \Omega_i)$ [9], [3]. The method we shall present in section 8.4.3 belongs to this class of methods. Let us emphasize that, in comparison with more standard problems such as cavity or closed waveguide problems, there is an additional difficulty due to the treatment of the nonstandard boundary condition.

An alternative to the use of edge-type elements consists in working with the so-called regularized formulation of Maxwell's equations [16], [2], [26]. In its principle, this method permits us to work within the H^1 functional framework and thus to use nodal finite elements. It consists in replacing the bilinear form $a(\beta, u, v)$ by the following one:

$$a_s(\beta; u, v) = \int_{\mathbb{R}^2} \left(\mu^{-1} \text{rot}_\beta u . \text{rot}_\beta v + s \, \text{div}_\beta(\varepsilon u) \, \text{div}_\beta(\varepsilon v) \right) dx,$$

where s is a positive real number, with the domain

$$W_\varepsilon = \{ u \in V_\varepsilon; \text{div}(\varepsilon \mathbf{u}) \in L^2(\mathbb{R}^2) \}.$$

With an appropriate choice of s, it can be proved that the new problem is equivalent to the previous one, so that one can speak of an *exact penalty method*. If ε is a regular function, W_ε is equal to the space $H^1(\mathbb{R}^2)^3$. On the contrary, if ε is a discontinuous function, and if the discontinuities arise on irregular polygonal curves, the space W_ε contains very singular fields that do not have the piecewise H^1 regularity. This gives rise to serious numerical difficulties, which are currently being studied [6].

8.4.2 Reduction to a Bounded Domain

The idea here is to reduce the numerical computations to a ball $\Omega_i = B_R$ containing the perturbation and to exploit the invariance by rotation of the exterior domain

$\Omega_e = \mathbb{R}^2 \backslash B_R$ to write an explicit *transparent boundary condition* on $\partial \Omega_i = \Gamma_R$. In what follows, we shall denote by n the unit vector normal to Γ_R, outgoing with respect to Ω_i. The same notation will be used for the vector of $\mathbb{R}^3$, which is normal to the cylinder $\Omega_i \times R$.

A Transmission Problem

We begin by pointing out a transmission problem equivalent to the initial eigenvalue problem:

$$(P) \quad \begin{cases} \text{Find } u \in D(A(\beta)), \ 0 < \omega^2 < \beta^2 c_\infty^2, \\ A(\beta)u = \omega^2 u, \quad u \neq 0. \end{cases} \quad (8.102)$$

LEMMA 8.14. *The eigenpair (ω^2, u) is a solution of (P) if and only if u_i and u_e, the respective restrictions of u to Ω_i and Ω_e, are solutions of*

$$(P^i) \quad \begin{cases} (\omega^2, u_i) \in \mathbb{R}^{+\star} \times H(rot_\beta, \Omega_i), \\ rot_\beta{}^\star(\mu^{-1} rot_\beta u_i) = \omega^2 \varepsilon u_i, \quad \text{div}_\beta(\varepsilon u_i) = 0, \end{cases}$$

$$(P^e) \quad \begin{cases} (\omega^2, u_e) \in \mathbb{R}^{+\star} \times H(rot_\beta, \Omega_e), \\ rot_\beta{}^\star(\mu^{-1} rot_\beta u_e) = \omega^2 \varepsilon u_e, \quad \text{div}_\beta(u_e) = 0, \end{cases}$$

along with the following transmission relations:

$$\begin{aligned} &\text{(a)} \quad n \wedge (u_i \wedge n) = n \wedge (u_e \wedge n) \quad &\text{on } \Gamma_R, \\ &\text{(b)} \quad rot_\beta u_i \wedge n = rot_\beta u_e \wedge n \quad &\text{on } \Gamma_R, \\ &\text{(c)} \quad \varepsilon \boldsymbol{u_i} \cdot n = \varepsilon \boldsymbol{u_e} \cdot n \quad &\text{on } \Gamma_R. \end{aligned} \quad (8.103)$$

In what follows we shall use polar coordinates and the polar representation of transverse vector fields: $\boldsymbol{u} = (u_r, u_\theta)$. On Γ_R, $u_i \cdot n = u_r$ (the *normal trace* of u) while $n \wedge (u_i \wedge n) = (u_\theta, u_3)$ (the *tangential trace* of u). By standard trace theorems [14], we can introduce the continuous linear mappings ($\Omega = \Omega_i$ or Ω_e):

$$u \in H(rot_\beta, \Omega) \to n \wedge (u_i \wedge n)|_{\Gamma_R} = (u_\theta, u_3)|_{\Gamma_R} \in H^{-\frac{1}{2}}(\Gamma_R) \times H^{\frac{1}{2}}(\Gamma_R),$$

$$u \in H(\text{div}_\beta, \Omega) \to (u_i \cdot n)|_{\Gamma_R} = u_r|_\Gamma \in H^{-\frac{1}{2}}(\Gamma_R).$$

The next step is to study and find explicit resolution of the exterior problem in Ω_e.

LEMMA 8.15. *Let ω and β be such that $\beta^2 - \omega^2/c_\infty^2 > 0$. Then, for each $\varphi = (\varphi_\theta, \varphi_3) \in H^{-\frac{1}{2}}(\Gamma) \times H^{\frac{1}{2}}(\Gamma)$, there is a unique field $u_{e,\varphi} \in H(rot_\beta, \Omega_e)$ solution of the problem:*

$$(P^e_{\omega,\beta}) \quad \begin{cases} rot_\beta^\star(rot_\beta u_{e,\varphi}) = \dfrac{\omega^2}{c_\infty^2} u_{e,\varphi}, \quad u_{e,\varphi} \in H(rot_\beta, \Omega_e), \\ n \wedge (u_e \wedge n) = \varphi \text{ on } \Gamma_R. \end{cases}$$

Chapter 8. Optical Waveguides

This solution is given by $(\alpha = (\beta^2 - \omega^2/c_\infty^2)^{\frac{1}{2}})$

$$\begin{cases} u_r = \sum_{n\in\mathbb{Z}} \left\{ \frac{\beta}{\alpha}\varphi_3^n \frac{K_n'(\alpha r)}{K_n(\alpha R)} - \left(i\varphi_\theta^n + \frac{\beta n}{\alpha^2 R}\varphi_3^n\right) \frac{n}{\alpha r}\frac{K_n(\alpha r)}{K_n'(\alpha R)} \right\} e^{in\theta}, \\ u_\theta = \sum_{n\in\mathbb{Z}} \left\{ \left(\varphi_\theta^n - i\frac{\beta n}{\alpha^2 R}\varphi_3^n\right) \frac{K_n'(\alpha r)}{K_n'(\alpha R)} - i\frac{\beta n}{\alpha^2 r}\varphi_3^n \frac{K_n(\alpha r)}{K_n(\alpha R)} \right\} e^{in\theta}, \\ u_3 = \sum_{n\in\mathbb{Z}} \varphi_3^n \frac{K_n(\alpha r)}{K_n(\alpha R)} e^{in\theta}, \end{cases}$$
(8.104)

where the functions K_n are the usual modified Bessel's functions [1].

Proof. The existence and uniqueness result can de deduced from a coercivity result analogous to the one of Lemma 8.3. The solution is calculated by separation of variables, which leads to a system of modified Bessel's equations coupled by the boundary conditions. We omit the details (see [22]). □

The Boundary Operators $T_R(\omega,\beta)$ and $S_R(\omega,\beta)$

Thanks to trace theorems, we can define

$$\begin{aligned} T_R(\omega,\beta) &\in \mathcal{L}\left(H^{-\frac{1}{2}}(\Gamma_R) \times H^{\frac{1}{2}}(\Gamma_R), H^{\frac{1}{2}}(\Gamma_R) \times H^{-\frac{1}{2}}(\Gamma_R)\right), \\ S_R(\omega,\beta) &\in \mathcal{L}\left(H^{-\frac{1}{2}}(\Gamma_R) \times H^{\frac{1}{2}}(\Gamma_R), H^{-\frac{1}{2}}(\Gamma_R)\right), \end{aligned}$$
(8.105)

such that

$$T_R(\omega,\beta)\varphi = \mathrm{rot}_\beta u_{e,\varphi} \wedge n|_{\Gamma_R}, \quad S_R(\omega,\beta)\varphi = u_{e,\varphi} \cdot n|_{\Gamma_R}.$$
(8.106)

It is possible to compute these two operators via series expansions in polar coordinates. If $\langle .,. \rangle$ stands for the duality product on Γ_R, we have

$$\text{for all } f \in \mathcal{D}'(\Gamma_R), \quad f = \sum_{n\in\mathbb{Z}} f^n e^{in\theta}, \quad f^n = \left\langle f, \frac{1}{2\pi R}e^{in\theta} \right\rangle.$$

THEOREM 8.17.

$$((T_R(\omega,\beta)\varphi)_\theta, (T_R(\omega,\beta)\varphi)_3)^t = \sum_{n\in\mathbb{Z}} [T_R^n(\omega,\beta)](\varphi_\theta^n, \varphi_3^n)^t e^{in\theta},$$
(8.107)

$$S_R(\omega,\beta)(\varphi) = \sum_{n\in\mathbb{Z}} S_R^n(\omega,\beta) \cdot (\varphi_\theta^n, \phi_3^n)^t\, e^{in\theta},$$
(8.108)

where, setting $\alpha = (\beta^2 - \omega^2/c_\infty^2)^{\frac{1}{2}}, T_R^n(\omega,\beta)$ denotes the 2×2 hermitian matrix

$$T_R^n(\omega,\beta) = \begin{bmatrix} \alpha\dfrac{K_n(\alpha R)}{K_n'(\alpha R)} & \dfrac{i\beta n}{\alpha R}\dfrac{K_n(\alpha R)}{K_n'(\alpha R)} \\ -\dfrac{i\beta n}{\alpha R}\dfrac{K_n(\alpha R)}{K_n'(\alpha R)} & \dfrac{\beta^2-\alpha^2}{\alpha}\dfrac{K_n'(\alpha R)}{K_n(\alpha R)} - \dfrac{\beta^2 n^2}{\alpha^3 R^2}\dfrac{K_n(\alpha R)}{K_n'(\alpha R)} \end{bmatrix}$$

and where $S_R^n(\omega, \beta)$ is the vector defined by

$$S_R^n(\omega, \beta) = \left(\frac{in}{\alpha R} \frac{K_n(\alpha R)}{K_n'(\alpha R)}, \frac{\beta}{\alpha} \left(\frac{n^2}{\alpha^2 R^2} \frac{K_n(\alpha R)}{K_n'(\alpha R)} - \frac{K_n'(\alpha R)}{K_n(\alpha R)} \right) \right).$$

Proof. This is a straightforward consequence of Lemma 8.15. The difficulties are purely computational [22]. □

The Interior Problem

Let us first introduce the functional spaces:

$$\begin{aligned} H_\varepsilon(\beta; \Omega_i) &= \{\, u_i \in L^2(\Omega_i)^3 \,;\, \mathrm{div}_\beta(\varepsilon u_i) = 0 \,\}, \\ V_\varepsilon(\beta; \Omega_i) &= \{\, u_i \in H(\mathrm{rot}_\beta, \Omega_i) \,;\, \mathrm{div}_\beta(\varepsilon u_i) = 0 \,\}, \end{aligned} \quad (8.109)$$

which are Hilbert spaces with respective norms:

$$\|u\|_{\varepsilon,\Omega_i}^2 = \int_{\Omega_i} \varepsilon |u|^2 \, dx,$$

$$\|u\|_{H(\mathrm{rot}_\beta, \Omega_i)}^2 = \int_{\Omega_i} (|\mathrm{rot}_\beta u|^2 + |u|^2) \, dx.$$

These are nothing but the spaces described by the restrictions to Ω_i of elements in $H_\varepsilon(\beta)$ and $V_\varepsilon(\beta)$, respectively. Taking into account the transmission conditions (8.103), we see that the restriction u_i of u, the solution of (P), is the solution of

$$(\widetilde{P}_{\omega,\beta}^i) \quad \begin{cases} \text{Find } (\omega^2, u_i) \in \mathbb{R}^{\star+} \times V_\varepsilon(\beta, \Omega_i) \setminus \{0\} \text{ such that} \\ \mathrm{rot}_\beta^\star(\mu^{-1} \mathrm{rot}_\beta u_i) = \omega^2 \varepsilon u_i \quad \text{in } \Omega_i, & (8.110) \\ \mathrm{rot}_\beta u_i \wedge n|_{\Gamma_R} = T_R(\omega, \beta)(n \wedge (u_i \wedge n)|_{\Gamma_R}), & (8.111) \\ u_i \cdot n|_{\Gamma_R} = S_R(\omega, \beta)(n \wedge (u_i \wedge n)|_{\Gamma_R}), & (8.112) \end{cases}$$

where (8.111) and (8.112) are the *transparent boundary conditions* we were looking for. More precisely, we have an equivalence result (the proof is trivial):

THEOREM 8.18. *Assume $0 < \omega^2 < \beta^2 c_\infty^2$.*

- *Let (ω^2, u) be a solution of (P), then $(\omega^2, u_i := u_{/\Omega_i})$ is a solution of $(\widetilde{P}_{\omega,\beta}^i)$.*

- *Conversely, let (ω^2, u_i) be a solution of $(\widetilde{P}_{\omega,\beta}^i)$; then u_i has a unique extension u to $\mathbb{R}^2$ such that (ω^2, u) is solution of (P).*

We next reinterpret $(\widetilde{P}_{\omega,\beta}^i)$ as a generalized eigenvalue problem, via a variational formulation that we shall also use for the discretization. Let us emphasize the difference in nature between the two boundary conditions (8.111) and (8.112). Indeed, (8.111) will appear as a natural boundary condition while (8.112) is an essential one.

Variational Formulation of the Interior Problem

Because of (8.112), we see that we have to work in the closed subspace of space $V_\varepsilon(\beta; \Omega_i)$:

$$V_\varepsilon(\omega, \beta; \Omega_i) = \{ u_i \in V_\varepsilon(\beta; \Omega_i) \,/\, \boldsymbol{u_i} \cdot n|_{\Gamma_R} = S_R(\omega, \beta)(n \wedge (u_i \wedge n)|_{\Gamma_R}) \}.$$

This space has the following fundamental property.

Chapter 8. Optical Waveguides   313

THEOREM 8.19. *The embedding $V_\varepsilon(\omega, \beta; \Omega_i) \subset H_\varepsilon(\beta; \Omega_i)$ is dense and compact.*

Proof. This result is nontrivial and the proof rather complicated. We refer the reader to [22] for the details and only mention some features of the proof.

(i) Once again, the compactness result is derived from Weber's result. The fact that condition (8.112) is included in the definition of the space plays a fundamental role: one cannot expect any compactness without any control of traces on the boundary. Such a control is given by (8.112).

(ii) The proof of the density result is not direct at all. We first characterize $H_\varepsilon(\beta, \Omega_i)$ as $H_\varepsilon(\beta, \Omega_i) = \{ \varepsilon^{-1}\text{rot}_\beta^\star v \ ; \ v \in H(\text{rot}_\beta, \Omega)\}$. Then, we construct a self-adjoint operator $B_{\omega,\beta}$ in $L^2(\Omega_i)^3$, whose domain $\mathcal{D}(B_{\omega,\beta})$ satisfies $u \in \mathcal{D}(B_{\omega,\beta}) \Longrightarrow \varepsilon^{-1}\text{rot}_\beta^\star u \in V_{\omega,\beta}(\Omega_i)$. The domain $\mathcal{D}(b_{\omega,\beta})$ of the bilinear form $b_{\omega,\beta}$ associated to $B_{\omega,\beta}$ is $H(\text{rot}_\beta, \Omega_i)$. Therefore, by classical properties of self-adjoint operators, $\mathcal{D}(B_{\omega,\beta})$ is dense in $\mathcal{D}(b_{\omega,\beta})$, i.e., in $H(\text{rot}_\beta, \Omega_i)$. One then concludes by using the fact that the continuous image of a dense subspace is dense in the image of the whole space (we use the mapping $u \to \varepsilon^{-1}\text{rot}_\beta^\star u$).
□

Now, let us recall a Green's formula, valid for any (u, v) in $H(\text{rot}_\beta, \Omega)^2$ such that $\text{rot}_\beta(\mu^{-1}\text{rot}_\beta u) \in L^2(\Omega)^3$ ($\Omega = \Omega_i$ or Ω_e):

$$\int_\Omega \text{rot}_\beta(\mu^{-1}\text{rot}_\beta u) \cdot v \, dx = \int_\Omega \mu^{-1}\text{rot}_\beta u \cdot \text{rot}_\beta v \, dx \\ + \langle \mu^{-1}\text{rot}_\beta u \wedge n, \ n \wedge (v \wedge n) \rangle, \quad (8.113)$$

and define the symmetric bilinear form on $V_\varepsilon(\omega, \beta; \Omega_i)$:

$$a_{\omega,\beta}(u_i, v_i) = \int_{\Omega_i} \mu^{-1}\text{rot}_\beta u_i \cdot \text{rot}_\beta v_i \, dx + \mu_\infty^{-1} b_R(\omega, \beta; u_i, v_i),$$

where $b_R(\omega, \beta; u, v)$ is the bilinear form associated to $T_R(\omega, \beta)$:

$$b_R(\omega, \beta; u, v) = \langle T_R(\omega, \beta)(n \wedge (u \wedge n)|_{\Gamma_R}), \ n \wedge (v \wedge n)|_{\Gamma_R} \rangle, \\ = 2\pi R \sum_{n \in \mathbb{Z}} [T_R^n(\omega, \beta)](\boldsymbol{u}_\theta^n, u_3^n)^t \cdot (\boldsymbol{u}_\theta^n, u_3^n)^t.$$

We introduce the variational problem

$$(P_{\omega,\beta}^i) \quad \begin{cases} \text{Find } (\omega^2, u_i) \in \mathbb{R}^{\star+} \times V_\varepsilon(\omega, \beta; \Omega_i) \text{ such that} \\ a_{\omega,\beta}(u_i, v_i) = \omega^2 (u_i, v_i)_\varepsilon \quad \text{for all } v_i \in V_\varepsilon(\omega, \beta; \Omega_i). \end{cases}$$

THEOREM 8.20. *The two problems $(P_{\omega,\beta}^i)$ and $(\widetilde{P}_{\omega,\beta}^i)$ are equivalent.*

Proof. Going from $(\widetilde{P}_{\omega,\beta}^i)$ to $(P_{\omega,\beta}^i)$ is easy: it suffices to multiply equation (8.110) by a test field v_i, integrate over Ω_i, use Green's formula (8.113) and boundary condition (8.111). The reciprocity is much less obvious but is a consequence (see [22]) of a generalized Helmholtz decomposition for vector fields in $V_\varepsilon(\beta; \Omega_i)$.

LEMMA 8.16. *For any $u_i \in V_\varepsilon(\beta; \Omega_i)$, there exists $(v_i, \phi) \in V_\varepsilon(\omega, \beta; \Omega_i) \times H^1(\mathbb{R}^2)$ such that*

$$u_i = v_i + \nabla_\beta \phi|_{\Omega_i}.$$

The proof of this result relies on the resolution on a nonstandard boundary value problem for ϕ in Ω_i, a problem which is obtained by writing that $\text{div}_\beta(\varepsilon u_i) = 0$. The main difficulties consist in writing appropriate boundary conditions so that v_i satisfies (8.111) and in proving the wellposedness of the resulting boundary value problem (see [22] for the details of this difficult proof). □

A Fixed Point Eigenvalue Problem

The space $V_\varepsilon(\omega, \beta; \Omega_i)$ being dense in $H_\varepsilon(\beta, \Omega_i)$, we can associate to the bilinear form $a_{\omega,\beta}$ the unbounded operator $A_{\omega,\beta}$ in $H_\varepsilon(\beta, \Omega_i)$, defined by

$$\begin{cases} \mathcal{D}(A_{\omega,\beta}) = \{u_i \in V_\varepsilon(\omega, \beta; \Omega_i) \,;\, \text{rot}_\beta^\star(\mu^{-1}\text{rot}_\beta u_i) \in L^2(\Omega_i)^3, \\ \qquad \text{rot}_\beta u_i \wedge n|_{\Gamma_R} = T_R(\omega, \beta)(n \wedge (u_i \wedge n)|_{\Gamma_R}) \}, \\ A_{\omega,\beta} u_i = \varepsilon^{-1}\text{rot}_\beta^\star(\mu^{-1}\text{rot}_\beta u_i) \quad \text{for all } u_i \in \mathcal{D}(A_{\omega,\beta}). \end{cases} \quad (8.114)$$

A fundamental property of $a_{\omega,\beta}$ is given by the following lemma.

LEMMA 8.17. *The bilinear form $a_{\omega,\beta}$ is coercive on $V_\varepsilon(\omega, \beta; \Omega_i)$, that is, there exist $\alpha > 0$ and $C > 0$ such that*

$$\text{for all } u \in V_\varepsilon(\omega, \beta; \Omega_i), \quad a_{\omega,\beta}(u, u) + \alpha \|u\|_{\varepsilon,\Omega_i}^2 \geq C \|u\|_{H(\text{rot}_\beta, \Omega_i)}^2. \quad (8.115)$$

Proof. The proof clearly illustrates the importance of the condition (8.112). Let us consider $u_i \in V_\varepsilon(\omega, \beta; \Omega_i)$, set $\varphi = n \wedge (u_i \wedge n)|_{\Gamma_R}$, and introduce the $u_{e,\varphi}$ solution of $(P^e_{\omega,\beta})$. Thanks to Green's formula (8.113), it is not difficult to obtain the identity ($\|\cdot\|_{0,\Omega}$ stands for the $L^2(\Omega)$-norm)

$$b_R(\omega, \beta; u_i, u_i) = \|\text{rot}\boldsymbol{u}_{e,\varphi}\|_{0,\Omega_e}^2 + \left(\beta^2 - \frac{\omega^2}{c_\infty^2}\right) \|\boldsymbol{u}_{e,\varphi}\|_{0,\Omega_e}^2 + \|\nabla u_{e,\varphi 3}\|_{0,\Omega_e}^2$$
$$+ \left(2\beta^2 - \frac{\omega^2}{c_\infty^2}\right) \|u_{e,\varphi 3}\|_{0,\Omega_e}^2 + 2\beta \langle \boldsymbol{u}_e \cdot n, u_{e,\varphi 3}\rangle.$$

By trace theorems, we first have

$$|\langle \boldsymbol{u}_{e,\varphi} \cdot n, u_{e,\varphi 3}\rangle| \leq C \|\boldsymbol{u}_e \cdot n\|_{H^{-\frac{1}{2}}(\Gamma_R)} \|u_{e,\varphi 3}\|_{H^1(\Omega_e)},$$

and using $\text{div}(\varepsilon u_i) = \beta \varepsilon u_{i3}$, we also have

$$\varepsilon_\infty \|\boldsymbol{u}_i \cdot n\|_{H^{-\frac{1}{2}}(\Gamma_R)} \leq \tilde{C} \, (\|\varepsilon \boldsymbol{u_i}\|_{0,\Omega_i}^2 + \|\beta \varepsilon u_{i3}\|_{0,\Omega_i}^2)^{\frac{1}{2}} \leq C(\beta, \varepsilon) \|u_i\|_{\varepsilon,\Omega_i}.$$

By (8.112) and the definition of $S_R(\omega, \beta)$, $u_i \in V_\varepsilon(\omega, \beta; \Omega_i)$ means that $\boldsymbol{u}_i \cdot n|_{\Gamma_R} = \boldsymbol{u}_e \cdot n|_{\Gamma_R}$. Thus

$$2\beta |\langle \boldsymbol{u}_{e,\varphi} \cdot n, u_{e,\varphi 3}\rangle| \leq 2C(\beta, \varepsilon) \|u_i\|_{\varepsilon,\Omega_i} \|u_{e,\varphi 3}\|_{H^1(\Omega_e)},$$

with another constant $C(\beta, \varepsilon)$. By Young's inequality, we obtain for any $\eta > 0$

$$b_R(\omega, \beta; u_i, u_i) \geq \left(2\beta^2 - \frac{\omega^2}{c_\infty^2} - \eta \, C(\beta, \varepsilon)\right) \|u_{e,\varphi 3}\|_{0,\Omega_e}^2$$
$$- \eta^{-1} C(\beta, \varepsilon) \|u_i\|_{\varepsilon,\Omega_i}^2 + (1 - \eta \, C(\beta, \varepsilon)) \|\nabla u_{e,\varphi 3}\|_{0,\Omega_e}^2.$$

Chapter 8. Optical Waveguides

Choosing $\eta > 0$ such that

$$\eta C(\beta, \varepsilon) < \min\left(1, 2\beta^2 - \frac{\omega^2}{c_\infty^2}\right),$$

we obtain

$$b_R(\omega, \beta; u_i, u_i) \geq -\eta^{-1} C(\beta, \varepsilon) \|u_i\|_{\varepsilon, \Omega_i}^2, \qquad (8.116)$$

from which we deduce that, if $\alpha = 2(\varepsilon_\infty \eta)^{-1} C(\beta, \varepsilon)$,

$$a_{\omega,\beta}(u_i, u_i) + \alpha \|u_i\|_{\varepsilon,\Omega_i}^2 \geq \|\mathrm{rot}_\beta u_i\|_{0,\Omega_i}^2 + (\varepsilon_\infty \eta)^{-1} C(\beta, \varepsilon) \|u_i\|_{\varepsilon,\Omega_i}^2,$$

which proves the coercivity of $a_{\omega,\beta}$. □

As a consequence of Theorem 8.19 and Lemma 8.17, we state the following theorem.

THEOREM 8.21. *The operator $A_{\omega,\beta}$ is self-adjoint, bounded from below, and has a compact resolvent. Its spectrum is then a pure point spectrum:*

$$\sigma(A_{\omega,\beta}) = \{\lambda_n(\omega, \beta) \, ; \, n \in \mathbb{N}\}, \qquad (8.117)$$

where $\lambda_n(\omega, \beta)$ is an increasing sequence tending to $+\infty$ and each $\lambda_n(\omega, \beta)$ depends continuously on ω and β.

Combining Theorems 8.18, 8.20, and 8.21, we can now give the characterization of guided modes that we shall use for the numerical approximation.

THEOREM 8.22. *The triplet (u, ω, β) represents a guided mode if and only if:*

(a) *(ω, β) is the solution of one of the equations $\lambda_n(\omega, \beta) = \omega^2$, $n \in \mathbb{N}$;*

(b) *$u_i = u|_{\Omega_i}$ is an eigenvector of $A_{\omega,\beta}$ associated to the eigenvalue $\lambda_n(\omega, \beta)$.*

8.4.3 Finite Element Discretization Using Edge-Type Elements

General Considerations

We deduce from the previous section a two-step algorithm for the computation of the guided modes:

(i) Step 1: Compute the eigenvalues $\lambda_n(\omega, \beta), n \geq 0$ of the operator $A_{\omega,\beta}$.

(ii) Step 2: Solve the fixed-point equations $\lambda_n(\omega, \beta) = \omega^2, n \geq 0$.

The difficult step of the algorithm is the first one, which requires a numerical approximation. In the method we present here, we have to introduce two approximation parameters:

- h: step size of the mesh of $\Omega_i = B_R$ for the spatial discretization.

- N: truncation order for the series appearing in the definition of $(P_{\omega,\beta}^i)$.

The approximation procedure requires

- an approximation space $V_{\omega,\beta,h}^N$ for $V_\varepsilon(\omega,\beta;\Omega_i)$,
- an approximate bilinear form $a_{\omega,\beta}^N(u,v)$ defined on $V_{\omega,\beta,h}^N$.

The problem to be solved for Step 1 can then be written

$$\begin{cases} \text{Find } (u_h^N, \lambda^{h,N}) \in V_{\omega,\beta,h}^N \setminus \{0\} \times \mathbb{R}^{\star+}, \\ a_{\omega,\beta}^N(u_h^N, v_h) = \lambda^{h,N}(u^N, v_h)_\varepsilon \quad \text{for all } v_h \in V_{\omega,\beta,h}^N. \end{cases} \quad (8.118)$$

Of course, when h and N^{-1} tend to 0, the eigenpairs $(u_h^N, \lambda^{h,N})$ solutions of (8.118) are approximations of the eigenpairs of $A_{\omega,\beta}$.

Construction of the Space $V_{\omega,\beta,h}^N$

In what follows, τ_h represents a "conforming triangulation" of $\Omega_i = B_R$ by disjoint triangles $\{K\}$, with mesh size $h = \max \operatorname{diam}(K)$. For simplicity, we assimilate here the edges with vertices on Γ_R with the corresponding portions of the circle (see [22] or [32] for a more rigorous presentation). The first idea is to look for $V_{\omega,\beta,h}^N$ as a subspace of

$$R_h \times P_h \subset H(\operatorname{rot}_\beta, \Omega_i) = H(\operatorname{rot}, \Omega_i) \times H^1(\Omega_i) \quad (\supset V_\varepsilon(\beta;\Omega_i)),$$

where we use for R_h (respectively, P_h) standard edge (respectively, Lagrange) finite elements:

$$R_h = \{v \in H(\operatorname{rot}, \Omega_i) \;/\; \forall K \in \tau_h, v|_K = \alpha + \gamma(-x_2, x_1)^t,\; \alpha \in \mathbb{R}^2, \gamma \in \mathbb{R}\},$$
$$P_h = \{v_3 \in H^1(\Omega_i) \;/\; \forall K \in \tau_h, v_3|_K \in P_1(K)\}.$$

One of the main interests in the use of such elements is to avoid the apparition of *spurious modes* as with classical Lagrange elements [9], [3]. The second idea comes from the observation that in $V_\varepsilon(\omega,\beta;\Omega_i)$, the third component of u is completely determined by the first two, thanks to the divergence-free condition

$$(DFC) \quad \beta \varepsilon u_3 = \operatorname{div}(\varepsilon \boldsymbol{u}).$$

One would like for the approximation space $V_{\omega,\beta,h}^N$ to also satisfy such a property. The most natural idea would be to define it as a space of vector fields in $R_h \times P_h$ satisfying (DFC). Unfortunately, this is not possible since

$$\{u \in R_h \times P_h \;/\; \beta \varepsilon u_3 = \operatorname{div}(\varepsilon \boldsymbol{u})\} = R_h \times P_h \cap V_\varepsilon(\omega,\beta) = \{0\}.$$

That is why we have to take into account the divergence-free condition (DFC) in a *weak sense*. More precisely, any field u in $V_\varepsilon(\omega,\beta;\Omega_i)$ satisfies (multiply (DFC) by $\varphi \in H^1(\Omega_i)$, integrate by parts over Ω_i, and use boundary condition (8.112))

$$(WDFC) \quad \beta \int_{B_R} \varepsilon u_3 \varphi \, dx + \int_{B_R} \varepsilon \boldsymbol{u} \cdot \nabla \varphi \, dx = s_R(\omega,\beta;u,\varphi) \quad \forall \varphi \in H^1(B_R),$$

where $s_R(\omega,\beta;.,.)$ is the bilinear form on $V_\beta \times H^1(\Omega_i)$ associated to $S_R(\omega,\beta)$ (we set $(ia^n, b^n)^t = S_R^n(\omega,\beta)$, where $S_R^n(\omega,\beta)$ has been defined in Theorem 8.17):

$$s_R(\omega,\beta;u,\varphi) = \langle S_R(\omega,\beta)(u \wedge n) \wedge n, \varphi \rangle_{\Gamma_R}$$
$$= -2\pi R \sum_{n \in \mathbb{Z}} b^n u_3^n \overline{\varphi}^n - 2i\pi R \sum_{n \in \mathbb{Z}} a^n \boldsymbol{u}_\theta^n \overline{\varphi}^n. \quad (8.119)$$

Chapter 8. Optical Waveguides

Then, we define
$$V_{\omega,\beta,h}^N = \{u_h \in R_h \times P_h \text{ such that (DDFC) holds}\}, \qquad (8.120)$$
where (DDFC) is the *discrete divergence-free condition*:

$$(DDFC) \quad \forall \varphi_h \in P_h, \quad \beta \oint_{\Omega_i} \varepsilon u_{3,h} \varphi_h dx + \int_{\Omega_i} \boldsymbol{u}_h \cdot \nabla \varphi_h dx = s_R^N(\omega,\beta;u_h,\varphi_h),$$

where we have set $\mathcal{M}_h$ denoting the set of the vertices of τ_h:

$$\begin{cases} \oint_{\Omega_i} f dx = \frac{1}{3} \sum_{K \in \tau_h} \text{meas}(K) \sum_{M \in \mathcal{M}_h} f(M) & \text{(a quadrature formula)}, \\ s_R^N(\omega,\beta;u_h,\varphi_h) = -2\pi R \sum_{|n| \leq N} b^n u_3^n \overline{\varphi}^n + 2i\pi R \sum_{|n| \leq N} a^n \boldsymbol{u}_\theta^n \overline{\varphi}^n. \end{cases}$$

We have to check that (DDFC) entirely defines u_3 as a function of $\boldsymbol{u}$. This is the object of the next lemma.

LEMMA 8.18. *The field $\boldsymbol{u}_h$ being given in R_h, there exists a unique element $u_{3,h} \in P_h$ which satisfies (DDFC), which we shall denote by $u_{3,h} \equiv (\beta\varepsilon)^{-1} \text{div}_h(\varepsilon\boldsymbol{u}_h)$.*

Proof. If we first write (DDFC) with $\varphi_h = \lambda_i$, where λ_i is a Lagrange P_1 basis function of P_h associated to an interior vertex M_i, we simply obtain

$$\beta \oint_{B_R} \varepsilon u_{3,h} \lambda_i dx + \int_{B_R} \varepsilon \boldsymbol{u}_h \cdot \nabla \lambda_i dx = 0,$$

which completely defines $u_{3,h}$ at the vertex i. It remains to compute $u_{3,h}$ at each vertex of the boundary. It is easy to see that we are led to invert the $N_b \times N_b$ symmetric matrix Q_R^h (N_b being the number of vertices on Γ_R) given by

$$Q_R^{ij} = \beta \oint_{\Omega_i} \varepsilon \lambda_i \lambda_j dx + 2\pi R \sum_{n \in \mathbb{Z}} b^n \lambda_i^n \overline{\lambda_j^n}, \quad 1 \leq i,j \leq N_b. \qquad (8.121)$$

Then, we note that, for any $\phi = (\phi_i) \in \mathbb{R}^{N_b}$,

$$\varphi = \sum_{i \in N_b} \phi_i \lambda_i \in P_h \Rightarrow (Q_R^h \phi, \phi) = \beta \oint_{\Omega_i} \varepsilon |\varphi|^2 dx + 2\pi R \sum_{n \in \mathbb{Z}} b^n |\varphi^n|^2.$$

Therefore, the positivity of the coefficients b^n (see [22]) implies that the matrix Q_R^h is positive definite, which ends the proof. □

Remark 8.13. The space $V_{\omega,\beta,h}^N$ is not included in $V_\varepsilon(\beta,\Omega_i)$: this is a *nonconforming method*. Moreover, the use of the numerical integration formula is very important: it avoids the inversion of a mass matrix for computing u_{h3}. This also ensures that the support of basis functions of $V_{\omega,\beta,h}^N$ remains small: the discrete divergence operator div_h is still local in some sense.

In practice, it is useful to describe a basis of the space $V_{\omega,\beta,h}^N$. Letting $\mathcal{A}_h$ be the set of the $\mathcal{N}_a$ edges of the mesh τ_h and $a \in \mathcal{A}_h$, we denote by $\boldsymbol{w}_a$ the usual basis function of the space R_h associated with the edge a. The dimension of $V_{\omega,\beta,h}^N$ is $\mathcal{N}_a$, and one constructs a basis by considering the vector fields

$$w_a = (\boldsymbol{w}_a, w_{3,a}), \quad \text{where } w_{3,a} = (\beta\varepsilon)^{-1} \text{div}_h(\varepsilon \boldsymbol{w}_a). \qquad (8.122)$$

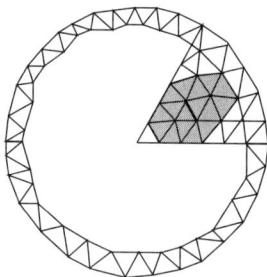 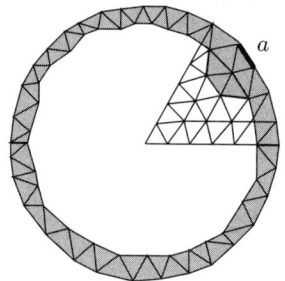

Figure 8.6: Interior edge.   Figure 8.7: Boundary edge.

One then naturally distinguishes two types of basis functions, which essentially differ by the support of the third component $w_{3,a}$ (see Figures 8.6 and 8.7):

- the basis functions associated to an interior edge,
- the basis functions associated to an edge on the boundary.

Finally, the finite-dimensional algorithm is

- Step 1: Solve the eigenvalue problem

$$a_{\omega,\beta}^N(u_h^N, v_h) = \lambda^{h,N} (u_h^N, v_h)_\varepsilon \quad \text{for all } v_h \in V_{\omega,\beta,h}^N, \ u_h^N \in V_{\omega,\beta,h}^N. \tag{8.123}$$

- Step 2: $\lambda_n^{h,N}(\omega, \beta)$ being the eigenvalues of step 1, solve

$$\lambda_n^{h,N}(\omega, \beta) = \omega^2, \quad 0 < \omega < \beta c_\infty, \quad 1 \leq n \leq \mathcal{N}_a. \tag{8.124}$$

Remark 8.14.

- The matrix $\mathbb{A}_{\omega,\beta,h}^N$ associated with $a_{\omega,\beta}^N$ in $V_{\omega,\beta,h}^N$ will contain a dense block corresponding to the degrees of freedom on the boundary.

- To compute the dispersion curves $\beta \to \omega(\beta)$, it is useful to couple the previous algorithm to a continuation method with respect to β (see [32]).

- It is also useful to determine the starting points of the dispersion curves, i.e., the thresholds; this would require us to design a numerical method for solving the threshold equation.

About the Approximation Theory

To our knowledge the theory of the convergence of the approximation presented in the previous paragraph is not yet complete. Let us, however, mention some results in this direction.

- In [21], the convergence of the finite element approximation is achieved in the case of a closed waveguide. In this case, the only approximation parameter is h. The proof is based on the theory of collectively compact operators of Anselone. It combines a discrete compactness property that is an adaptation of a result by Kikuchi with the use of Strang's lemma to deal with the quadrature error and the nonconformity (see [21] for references).

- In [5], the complete theory is developed for the weakly guiding scalar problem. In particular, it is proved that, as far as the series truncation is concerned, the error decreases more rapidly that any negative power of N. By adapting the techniques of [34] (see also [31]), it is possible to show that the decay is in fact exponential with respect to N (the so-called *spectral accuracy*).

8.4.4 Numerical Results

To conclude this section, we present some numerical results of the simulation of a simple test case which corresponds to a step-index optical fiber. As we said previously, an analytical solution is available, which permits us to validate the simulation code. The data of the physical model are the following:

$$\begin{cases} \varepsilon(x) = 2, & \mu(x) = 1 \quad \text{if } |x| \leq 1, \\ \varepsilon(x) = 1, & \mu(x) = 1 \quad \text{if } |x| \geq 1. \end{cases} \tag{8.125}$$

The parameters for the discretization are

$$\begin{cases} \text{The size of the computational domain } R=1.5, \\ \text{The truncation order of the series } N=100. \end{cases} \tag{8.126}$$

The chosen value for N is more than sufficient to get a good accurary. Finally, for the discretization, we have used a mesh made of about 2,000 triangles, with 1,392 nodes: h is approximately 0.08.

In our first computation, we choose the propagation constant $\beta = 2.6$. We compute the fundamental mode (the first one), which gives $\omega = 2.2108$, to be compared with the exact value $\omega = 2.2097$ computed analytically. We represent in Figure 8.8 the distribution of the TE in the cross section of the fiber: on the left we represent the field in the computational domain. We see that it is not zero on the boundary. On the right, we observe the field in a larger domain: this gives an illustration of the concentration of the energy of the mode near the core of the fiber. In our second computation, we take $\beta = 6$, which happens to be larger than the tenth threshold, and we calculate the mode of order 10. We get $\omega = 5.3073$, the exact value being $\omega = 5.2941$. We represent in Figure 8.9 the distribution of the TE field using the same convention than in Figure 8.8. We observe a more complicated structure of the mode, which is characteristic of higher order modes.

Appendix

Let us briefly recall some fundamental notions about the spectral theory of self-adjoint operators (see [35] or [37] for the proofs). In the following, H denotes a complex Hilbert space, (u, v) the scalar product in H, and $|u|$ the associated norm.

First Mode : $\omega = 2.2108$

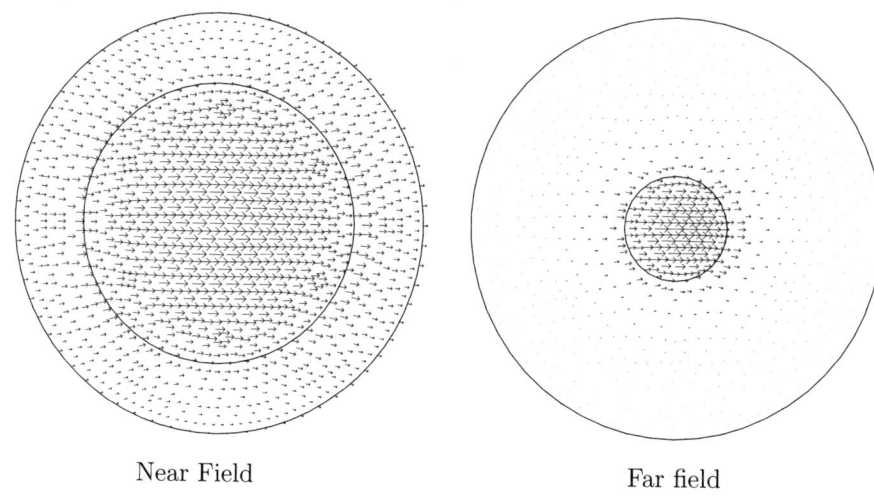

Figure 8.8: Optical fiber, $\beta = 2.6$. TE field.

10 Mode : $\omega = 5.3073$

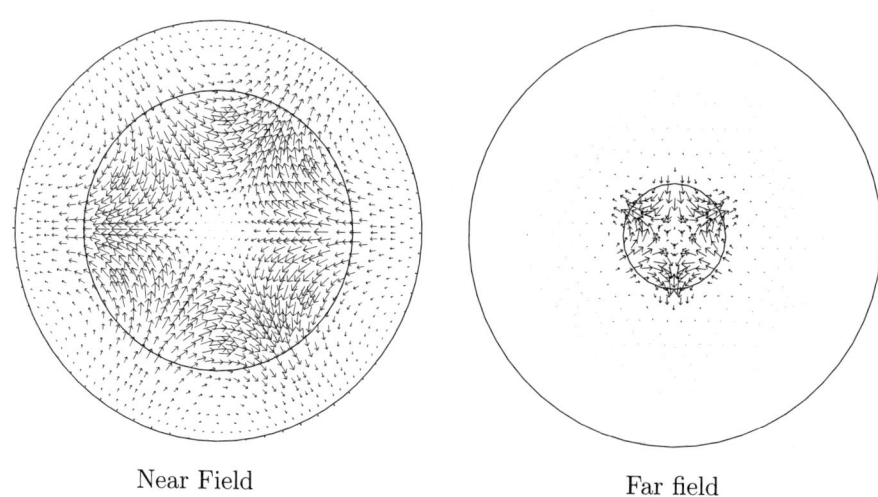

Figure 8.9: Optical fiber, $\beta = 6$. TE field.

Chapter 8. Optical Waveguides

An operator A in H is by definition a linear mapping from a dense subspace $D(A)$ of H into H. It is said to be unbounded if and only if

$$\exists u_n \in D(A) \text{ such that } |u_n| = 1 \text{ and } |Au_n| \to +\infty.$$

For example, a linear partial differential operator defines an unbounded operator in $L^2(\mathbb{R}^N)$.

The adjoint A^* of A is defined as the operator with domain

$$D(A^*) = \{v \in H; \exists C > 0 \text{ such that } |(Au,v)| \leq C|u| \text{ for all } u \in D(A)\}$$

such that, for all $v \in D(A^*)$ and $u \in D(A)$, $(Au,v) = (u, A^*v)$. By definition, A is said to be *self-adjoint* if $A = A^*$, which implies in particular that the domains are the same. A self-adjoint operator is in particular *symmetric*, i.e.,

$$\forall (u,v) \in D(A) \times D(A), \quad (Au,v) = (u, Av),$$

but the converse is false in general. If A is symmetric, $(Au, u) \in \mathbb{R}$ for all $u \in D(A)$. A symmetric operator A is said to be positive if $(Au, u) \geq 0$ for all $u \in D(A)$.

The *resolvent set* $\rho(A)$ of A is the set of all complex numbers λ such that $A - \lambda I$ is a bijective mapping from $D(A)$ into H. By the Banach theorem, this implies in particular that $(A - \lambda I)^{-1}$ is a bounded operator in H. The resolvent set is an open subset of $\mathbb{C}$; its complement, a closed subset of $\mathbb{C}$, is called by definition the *spectrum* of A and denoted $\sigma(A)$.

In the particular case of a self-adjoint operator, the following characterization of the spectrum holds: $\lambda \in \sigma(A)$ if and only if there exists a sequence u_n in $D(A)$ such that $|u_n| = 1$ and $Au_n - \lambda u_n$ tends to 0 strongly in H. This proves in particular that $\sigma(A) \subset \mathbb{R}$ since $(Au_n, u_n) \to \lambda$ and $(Au_n, u_n) \in \mathbb{R}$. If, moreover, A is a positive operator, then $\sigma(A) \subset \mathbb{R}^+$.

The spectrum of a self-adjoint operator A is composed of two distinct parts, the *discrete spectrum* and the *essential spectrum*. The discrete spectrum denoted by $\sigma_{disc}(A)$ is the set of eigenvalues of A that are isolated in the spectrum and have a finite multiplicity. The essential spectrum, which is the complement of the discrete spectrum, is denoted $\sigma_{ess}(A)$; it is a closed subset of $\mathbb{R}$. The essential spectrum has the following characterization: $\lambda \in \sigma_{ess}(A)$ if and only if there exists a sequence u_n in $D(A)$ such that $|u_n| = 1$, u_n converges weakly to 0 in H, and $Au_n - \lambda u_n$ tends to 0 strongly in H. Such a sequence is called *a singular sequence*.

A fundamental example of the applications is that of an operator associated to a sesquilinear form. More precisely, let us consider now a dense subspace V of H, such that V equipped with the norm $|u|_V$ is a Hilbert space and such that the embedding $(V, |\cdot|_V)$ into $(H, |\cdot|)$ is continuous. Let $a(u,v)$ be a sesquilinear hermitian form which is coercive on $V \times V$. Then the operator A associated to the form a (which is defined as the operator with domain $D(A) = \{u \in V; \exists C > 0 \text{ such that } |a(u,v)| \leq C|v| \text{ for all } v \in V\}$ such that, for all $u \in D(A)$ and $v \in V$, $(Au,v) = a(u,v)$) is a self-adjoint positive operator. Moreover, we can set $s_e = \min(\sigma_{ess}(A))$ and for $m \geq 1$,

$$s_m = \inf_{E \in \mathcal{V}_m} \sup_{u \in E \setminus \{0\}} \frac{a(u,u)}{|u|^2} = \sup_{F \in \mathcal{H}_{m-1}} \inf_{u \in F^\perp \cap V \setminus \{0\}} \frac{a(u,u)}{|u|^2},$$

where $\mathcal{V}_m$ is the set of m-dimensional subspaces of V, $\mathcal{H}_m$ is the set of m-dimensional subspaces of H, and, for any subset F of H, $F^\perp = \{u \in H \mathbin{/} (u,v) = 0 \ \forall v \in F\}$.

The sequence s_m is a nondecreasing converging sequence and its limit is equal to s_e; moreover, the Min-Max Principle says that, if $N(A)$ denotes the (possibly infinite) number of eigenvalues of A located below the essential spectrum (each of them counted a number of times equal to its multiplicity), then

$$\begin{cases} m \leq N \iff s_m < s_e, \\ m > N \iff s_m = s_e \ . \end{cases}$$

References

[1] M. Abramowitz and I. A. Stegun, *Handbook of Mathematical Functions*, Dover, New York, 1972.

[2] A. Bamberger and A. S. Bonnet, *Mathematical analysis of the guided modes of an optical fiber*, SIAM J. Math. Anal., 21 (1990), pp. 1487–1510.

[3] A. Bermudez and D. Pedreira, *Mathematical analysis of a finite element method without spurious solutions for the computation of dielectric waveguides*, Numer. Math., 61 (1992), pp. 39–57.

[4] A. S. Bonnet and R. Djellouli, *High frequency asymptotics of guided modes in optical fibers*, IMA J. Appl. Math., 52 (1994), pp. 271–287.

[5] A. S. Bonnet-Ben Dhia and N. Gmati, *Spectral approximation of a boundary condition for an eigenvalue problem*, SIAM J. Numer. Anal., 32 (1995), pp. 1263–1279.

[6] A. S. Bonnet-Ben Dhia, C. Hazard, and S. Lohrengel, *A singular field method for the solution of Maxwell's equations*, SIAM J. Appl. Math., 59 (1999), pp. 2028–2044.

[7] A. S. Bonnet-Ben Dhia and P. Joly, *Mathematical analysis of guided water waves*, SIAM J. Appl. Math., 53 (1993), pp. 1507–1550.

[8] A. S. Bonnet-Ben Dhia and F. Starling, *Guided waves by electromagnetic gratings and non-uniqueness examples for the diffraction problem*, Math. Methods Appl. Sci., 17 (1994), pp. 305–338.

[9] A. Bossavit, *Solving Maxwell equations in a closed cavity and the question of spurious modes*, IEEE Trans. Magnetics, 26 (1990), pp. 702–705.

[10] P. M. Cutzach and C. Hazard, *Existence, uniqueness, and analyticity properties for electromagnetic scattering in a two-layered medium*, Math. Methods Appl. Sci., 21 (1998), pp. 433–461.

[11] R. Dautray and J. L. Lions, *Analyse mathématique et calcul numérique pour les sciences et les techniques*, Tome 2, Masson, Paris, 1985.

[12] A. S. Bonnet-Ben Dhia, G. Caloz, and F. Mahé, *Guided modes of integrated optical guides. A mathematical study*, IMA J. Appl. Math., 60 (1998), pp. 225–261.

[13] A. S. Bonnet-Ben Dhia and F. Mahé, *A guided mode in the range of the radiation modes for a rib waveguide*, J. Opt., 28 (1997), pp. 41–43.

[14] V. Girault and P. A. Raviart, *Finite Element Methods for Navier-Stokes Equations, Theory and Algorithms*, Springer-Verlag, New York, 1986.

[15] A. S. Bonnet-Ben Dhia and N. Gmati, *Computation of the modes of dielectric waveguides by finite elements coupled with an integral representation*, in Numerical Methods in Engineering '92, Ch. Hirsch et al., eds., Elsevier, New York, 1992, pp. 73–77.

[16] C. Hazard and M. Lenoir, *On the solution of time-harmonic scattering problems for Maxwell's equations*, SIAM J. Math. Anal., 27 (1996), pp. 1597–1630.

[17] R. G. Hunsperger, *Integrated Optics: Theory and Technology*, Optical Sciences 33, Springer-Verlag, New York, 1991.

[18] A. Jami and M. Lenoir, *A variational formulation for exterior problems in linear hydrodynamics*, Comput. Methods Appl. Mech. Engrg, 16 (1978), pp. 341–359.

[19] C. Johnson and J. C. Nédélec. *On the coupling of boundary integral and finite element methods*, Math. Comp., 35 (1980), pp. 1036–1079.

[20] P. Joly and C. Poirier, *Mathematical analysis of electromagnetic open waveguides*, Math. Model. Numer. Anal., 29 (1995), pp. 505–575.

[21] P. Joly and C. Poirier, *A new non-conforming finite element method for the computation of electromagnetic guided waves. I. Mathematical analysis*, SIAM J. Numer. Anal., 33 (1996), pp. 1494–1525.

[22] P. Joly and C. Poirier, *A numerical method for the computation of electromagnetic modes in optical fibers*, Math. Methods Appl. Sci., 22 (1999), pp. 389–447.

[23] P. Joly and R. Weder, *New results for guided waves in elastic heterogeneous media*, Math. Methods Appl. Sci., 15 (1992), pp. 395–409.

[24] I. P. Kaminow and T. L. Koch, *Optical Fiber Telecommunications* III, Academic Press, New York, 1997.

[25] T. Kato, *Perturbation Theory for Linear Operators*, Springer-Verlag, New York, 1976.

[26] R. Leis, *Zur theorie elektromagnetischer schwingungen in anisotropen medien*, Math. Z., 106 (1968), pp. 213–224.

[27] F. Mahé, *Etude mathématique et numérique de la propagation d'ondes electromagnétiques dans les microguides de l'optique intégrée*, Rapport de Recherche n°276, E.N.S.T.A., Paris, 1993.

[28] D. Marcuse, *Theory of Dielectric Optical Waveguide*, Academic Press, New York, 1974.

[29] D. Marcuse, *Light Transmission Optics*, Van Nostrand Reinhold, New York, 1982.

[30] J. R. Mosig and F. E. Gardiol, *General integral equation formulation for microstrip antennas and scatterers*, in IEEE Proc. 132H, 1985, pp. 424–432.

[31] D. Gomez and P. Joly, *A Method for computing guided waves in integrated optics. Part I: Mathematical analysis*, SIAM J. Numer. Anal., to appear.

[32] P. Joly and C. Poirier, *A numerical method for the computation of electromagnetic modes in optical fibers*, Math. Methods Appl. Sci., 22 (1999), pp. 389–447.

[33] A. S. Bonnet-Ben Dhia and K. Ramdani, *Mathematical analysis of conductive and superconductive transmission lines*, SIAM J. Appl. Math., 60 (2000), pp. 2087–2113.

[34] J. Razafiarivelo, *Optimisation de la forme de transitions entre guides électromagnétiques par une méthode intégrale d'éléments finis*, Ph.D. thesis, Université de Paris 6, 1996.

[35] M. Reed and B. Simon, *Methods of Modern Physics, Analysis of Operators*, Vol. 4, Academic Press, New York, 1980.

[36] F. Rellich, *Uber das asymptotische verhalten der losungen von $\Delta u + \lambda u = 0$ in unendlichen gebieten*, Uber. Deutsch. Math. Verein, 53 (1943), pp. 57–65.

[37] M. Schechter, *Operator Methods in Quantum Mechanics*, North-Holland, New York, 1981.

[38] A. W. Snyder and J. D. Love, *Optical Waveguide Theory*, Chapman and Hall, London, 1983.

[39] H. P. Urbach, *Analysis of the domain integral operator for anisotropic dielectric waveguides*, SIAM J. Math. Anal., 27 (1996), pp. 204–220.

[40] F. Ursell, *Mathematical aspects of trapping modes in the theory of surface waves*, J. Fluid Mech., 183 (1987), pp. 421–437.

[41] C. Vassalo, *Optical Waveguide Concepts*, Elsevier, New York, 1991.

[42] V. Vogelsang, *On the strong unique continuation principle for inequalities of Maxwell type*, Math. Ann., 289 (1991), pp. 285–295.

[43] C. Weber, *A local compactness theorem for Maxwell's equations*, Math. Methods Appl. Sci., 2 (1967), pp. 12–25.

[44] H. D. Witsch, *Examples of embedded eigenvalues for the Dirichlet Laplacian in perturbed waveguides*, Math. Methods Appl. Sci., 12 (1990), pp. 91–93.

Index

aberrations, 10
abscissae of the discontinuities, 122
absence of embedded impurity eigenvalues, 258
absolutely continuous, 226
absolutely convergent, 116
absorbers, 201
absorbing boundary condition (ABC), 41, 154
acoustical closed waveguides, 306
acoustically soft, 96
admissible class, 59
admissible curves, 59
almost localize, 243
analytic continuation, 72, 80, 84
analytically, 73
Anderson localization, 255, 258
angle of incidence, 55
anisotropic gratings, 112
anisotropic lamellar gratings, 131
antenna arrays, 37
antennas, 180
antireflective (AR) conditions, 24
artificial index modulation, 4
artificial nonlinearity, 309
aspheric aberration correction, 3
asymptotic behavior, 237
asymptotic behavior of the spectrum, 248
asymptotic matching techniques, 63
axial symmetry, 148

Banach theorem, 321
band functions, 217, 219, 222
band gap, 32
band gap structures, 180, 201
Beltrami fields, 185
BEM, 162, 167

BEM for perfectly conducting DOEs, 169
Bessel functions, 98, 311
Bethe–Sommerfeld conjecture, 230
bilinear form, 280, 314
binary optical elements, 4
binary optical structures, 10, 61
binary optics, 1
biperiodic surfaces, 92
Birman–Schwinger method, 251
bisinusoidal gratings, 92
BKK method, 111
Bohren transform, 187, 188
Born approximation, 181
bound states, 245
boundary element, 61
boundary element method (BEM), 31, 143, 183
boundary integral formulation for perfect conductors, 169
boundary integral methods (BIMs), 143, 162
boundary perturbation methods, 72
boundary pseudodifferential operators, 190
boundary variations, 73
bounded cylindrical obstacle, 90
bounded obstacles, 82
branch point singularities, 73
Brillouin zone, 218, 223, 256

Cartesian coordinates, 79, 155
Cauchy's principal value of integration, 166
Cauchy–Schwarz inequality, 199
cavities, 96
cavity mirrors, 29
central difference method, 149

central finite-difference approximation, 144
characteristic length, 89
characterizations of uniqueness, 56
chiral gratings, 180
chiral medium, 179
chiral object, 179
chirality admittance, 179
chirality parameter, 186
chromatic aberration, 2, 3
circular step-index fiber, 274
circularly birefringent, 179
cladding, 275
closed waveguides, 274
coherent multiple scattering, 207
cohomologies, 212
cokernel, 221
compact admissible set, 60
compact imbedding result, 182
compact perturbations, 281
compact resolvent, 274, 281, 315
compactness, 291
compactness lemma, 194
comparison principle, 288, 302
complementary jumps, 122
complete band gap, 207
complex Bloch variety, 224
complex pseudodifferential, 244
concurrent discontinuities, 122
conductivity, 144
conformal transformation, 86
conforming triangulation, 316
conical diffraction, 44
conical mountings, 112
connected structures, 242
constant coefficient partial differential operator, 216
constitutive relations, 209
continuation of multiple eigenvalues, 98
continuous dependence, 48
contraction mapping argument, 64
convergence analysis, 49
convolution operators, 216
corrugated surfaces, 72
coupling FEM/BEM variational formulations, 183

covariant Cauchy–Riemann derivative operator, 228
critical values, 288
crossed gratings, 112
crystalline materials, 37
curl operator, 212
curvilinear tetrahedra, 200
cut-off function, 303
cutoff property, 18
cylindrical coordinates, 146

Dammann gratings, 3
data communications, 13
dense subspace, 321
determinant of infinite order, 116
diamond turning, 4
dielectric coefficient, 39
dielectric permittivity ε, 276
dielectric-air coupling, 245
differential formalisms, 72
diffraction, 37, 111
diffraction efficiencies, 7, 15, 128
diffraction efficiency of the mth order, 7
diffraction gratings, 1, 37, 63, 71, 92
diffraction problem, 51
diffraction theory, 10
diffractive optical element (DOE), 1, 12, 141
diffractive optics, 1–3, 37
diffractive optics technology, 5
diffractive orders, 7, 10
    negative, 7
    positive, 7
    reflective, 7
    transmissive, 7
diffusion equation, 31
direct problem, 30
direct-write electron beam lithography, 4, 27
Dirichlet Laplacian, 248
Dirichlet problem, 238
Dirichlet-to-Neumann (D–N) operator, 190, 238, 309
discrete spectrum, 284, 321
discretization of the continuous problem, 49
dispersion curves, 275, 318

INDEX

dispersion relations, 217, 219, 222, 240, 245, 274, 278, 306
distinct eigenvalues, 98
distributed Bragg reflector (DBR), 29
$\text{div}_\beta u$, 279
divergence-free condition, 316
divergence-type problem, 215
domain of analyticity, 86
domain of validity, 73
Drude–Born–Fedorov constitutive equations, 179, 184
dry etching, 4
dry-plasma etch procedures, 5
dual lattice, 218

E-formulation, 278
edge-emitting lasers, 13
edge-type elements, 309
effective dielectric constant, 25
effective refractive index, 26
effective-medium theory, 31
efficiency, 39
eigenfunctions, 111
eigenvalue branches, 217
eigenvalue problem, 96, 310
eikonal equation, 6
elastic waveguides, 306
electric, 39
electromagnetic gratings, 306
electromagnetic waves, 111
elliptic complex, 212
embedded eigenvalue, 253, 307
embedding, 313
Engquist–Majda wave equation, 154
enhanced convergence, 88
essential spectrum, 249, 281, 282, 298, 306, 307, 321
Euler's number, 166
evanescent, 249
evanescent modes, 39
exact penalty method, 309
existence, 287
existence and size of gaps, 258
existence and uniqueness, 194, 197
existence of spectral gaps, 229, 230
existence of the point spectrum, 217
exponential localization estimates, 226

far-field eigenfunction expansions, 80

Faraday's law, 150
fast Fourier transform (FFT), 51, 161
FDTD approach, 32, 97
Fermi surface, 224, 227, 254
fiber coupler, 15
finite aperture, 31
finite differences, 72
finite element approximations, 37
finite element method, 31, 48, 255
finite elements, 72, 275, 316
finite Laurent's rule, 122
finite periodic structures, 63
finite-difference method (FDM), 143
finite-difference time-domain method (FDTD), 31, 143, 255
finite-dimensional subspaces, 48
finite-element method (FEM), 143, 183
finite-element subspaces, 200
finiteness of the number of gaps, 258
first Hankel function, 77
first spherical Hankel functions, 77
first-order diffraction efficiency, 120
fixed point eigenvalue problem, 314
fixed-point equations, 309
fixed-point iteration, 64
flattened bands, 258
Floquet condition, 218, 252
Floquet modes, 182
Floquet theory, 216
Floquet transform, 217, 220, 252
Floquet–Bloch eigenmodes, 234, 242
Floquet–Bloch solution, 225
Floquet–Bloch theorem, 20, 40
Floquet–Bloch transform, 226
Floquet–Fourier series, 114
focal length, 6
Fourier expansion method, 111
Fourier factorization, 119, 122
Fourier factorization theory, 127
Fourier modal method (FMM), 111, 112
Fourier modes, 147
Fourier series, 111, 114
Fourier series coefficients, 81
Fourier series expansion, 41, 99, 147, 292
Fourier transform, 209, 216, 298
Fraunhofer approximation, 2, 7

Fredholm alternative, 45
Fredholm theory, 45
frequency domain techniques, 143
Fresnel lens, 2
Fresnel phase lenses, 1
Fresnel zone, 2, 4, 6
fundamental cell, 40
fundamental domain, 218
fundamental guided mode, 306

Galerkin-type methods, 257
Gaussian beams, 159, 171
Gelfand transform, 217
generalized gradient, 258
generalized Helmholtz decomposition, 313
generic algorithm, 62
geometric dimensions, 19
geometric perturbations, 71
geometrical optics, 6, 10, 37
geometrical wavefront, 6
Gibbs–Wilbraham phenomenon, 126
global uniqueness result, 56
GMRES, 50
grating efficiencies, 78
grating enhanced second harmonic generation, 63
grating period, 12
grating profile, 58
gratings, 90, 182
gratings with high aspect ratio, 30
grayscale lithography, 4
Green's function, 98, 161
Green's second identity, 163
guided modes, 274, 277, 297
guided waves, 277
guided-mode grating resonance, 23
guiding condition, 274

H-formulation, 278, 304
$H(\mathrm{curl}, \Omega)$, 192
$H(\mathrm{div}, \Omega)$, 192
$H(\mathrm{rot}; \mathbb{R}^2)$, 279
$H_\rho(\mathrm{rot}, \mathbb{R}^2)$, 289
Hankel functions, 166
hard acoustic media, 63
Hardy's inequality, 290
harmonic conversion efficiency, 63

harmonic plane wave, 277
Hausdorff distance, 57, 234
head-mounted display (HMD), 11
head-up displays, 1
helices, 179
Helmholtz equation, 44, 75, 80
Helmholtz type, 215
high dielectric contrast, 232
high modulation rates, 14
high-contrast PBG material, 248
high-contrast periodic medium, 248
high-order boundary perturbation theory, 97
high-resolution pattern generation techniques, 4
higher order aberration correction, 3
Hilbert space, 278, 290
Hill operator, 230, 253
Hodge decomposition, 182, 194, 197
holographic lithography, 30
holomorphic framework, 73
honeycomb structure, 245
hybrid finite element–boundary element methods, 61
hypoelliptic operator, 226

image quality, 10
imbeddings, 193
impurities in PBG materials, 248
impurity eigenvalue, 250
impurity spectra, 258
incidence vector, 40
incident beam of finite width, 63
incident wave, 55
inf-sup condition, 52
infinite eigenvalue problem, 117
infinitely degenerate eigenvalues, 245
integer programming, 62
integral equation approach, 45
integral equations, 72, 90
integrated optical circuits, 273
integrated optical guide, 276
interface least-squares finite element, 52
interface problem, 201
interior problem, 312
inverse diffraction problem, 55
inverse problem, 30, 37, 54, 274

INDEX

inverse rule, 123
irreducible, 225
irreducible Brillouin zone, 240
isobathic procedure, 10
isobathic process, 5
isotropic, 179
isotropic gratings, 112

joint analyticity, 73
jointly analytic, 79
jump conditions, 45

Kato's perturbation theory, 286
kernel, 221
kinoforms, 1
Kirchhoff approximation, 62

L-periodic Sobolev space, 43
L^2-based Sobolev space, 43, 192
Lagrange elements, 316
Laplace operator, 211
Laplace's equation, 98
Laplace–Beltrami operator, 230
large, sparse matrix equation, 50
laser writing, 4
Laurent's multiplication rule, 121
Laurent's rule, 116
Lax–Milgram lemma, 45, 47, 195, 197, 293
leaky modes, 274
least-squares finite element methods, 51
least-squares functional, 53
least-squares minimization problem, 53
Lebesgue's theorem, 303
left-circularly and right-circularly polarized plane waves, 185
Legendre functions, 77
LIGA techniques, 17
linear defect, 254
linear self-adjoint operators, 275
linear systems theory, 10
Liouville–Green transform, 216
Liouville-type theorems, 232
Lipman–Schwinger equation, 309
liquid crystals, 180
local propagation velocity, 276
localized defect, 249, 250
localized perturbation, 252

localized waves, 258
locally Lipschitz, 286
low dielectric filling fractions, 233
low-frequency asymptotics, 180
low-order diffractive gratings, 2, 30
low-order methods, 72
low-pass spectral filter, 18
lower threshold, 288
luminance, 11

magnetic field vectors, 39
magnetic permeability, 39, 276
magnitude of the ratio P/λ, 89
manufacturing constraints, 62
mask alignment error, 8
material tensors, 209
matrix eigenvalue problem, 111
Maxwell's equations, 2, 6, 113, 142, 208, 274, 276
Maxwell's time-dependent curl equations, 144
mean value, 294
Mellin transform, 244
MEMS technology, 16
metallic gratings, 112
method of integral equations, 183
method of moments, 90
method of moments (MoM), 143
method of separation of variables, 42
micro-electrical-mechanical structures (MEMS), 51
microelectromechanical systems (MEMS), 13
microstrip grating, 201
microwave devices, 180
microwave structures, 37
Min-Max Principle, 284, 285, 298, 299, 302, 322
minimum feature size, 2, 5
minimum Fresnel zone, 7
modal method, 111
modes, 273
modulation, 11
modulation transfer function (MTF), 10
Moharam–Gaylord method, 111
multigrid methods, 52
multimode, 14

$\nabla_\beta \varphi$, 279
narrow strip gratings, 201
Nédélec's finite elements, 183, 200
Neumann series, 73
non-Hermitian matrix, 50
nonconforming method, 317
nonconformity, 319
nonellipticity, 212
nonexistence, 284, 302
nonexistence of bound states, 216
nonlinear devices, 63
nonlinear effects, 258
nonlinear media, 31
nonlinear optical crystal, 63
nonlinear susceptibility, 63
nonlinear thin optical coatings, 64
nonperiodic bounded inhomogeneous medium, 47
nonresonant, 64
nonsquare geometries, 236
normal form, 116
normal trace, 310
numerical dispersion, 153

open waveguide, 273, 274
optical data storage, 13
optical fiber, 273, 276
optical path difference, 6
optical path length, 6
optical power, 6
optical waveguides, 273
optimal design, 30
optimal design of nonlinear films, 64
optimal design problem, 54, 57, 59
optimal error estimate, 201
optoelectronics, 2
orthogonal decomposition, 279, 306
orthogonal projector, 211
Orthomin, 50
oscillations, 89
oscillatory minimizing sequences, 60

P_1-Lagrange finite-element approximation, 200
Padé approximant, 84, 85
Padé approximation, 96
Padé problem, 84
Paley–Wiener theorem, 220, 252

parameterization, 79
parameterized eigenvalue problem, 274
pattern transfer, 5
PBG waveguides, 258
perfect conductor, 75
perfectly conducting bounded obstacle, 90
perfectly conducting obstacle, 91
perfectly matched layer, 154
perimeter-to-wavelength ratio, 91
periodic (elliptic) differential operators, 216
periodic gratings, 73, 111
periodic Pauli operators, 232
periodic potential, 253
periodic radiation condition, 187
periodic structure, 55, 182
periodicity, 216
permeability, 144
permittivity, 144
permittivity function $\epsilon(x)$, 115
perturbation theory, 71
phase contour, 4
phase function, 6
phase incident wavefront, 2
phase mask lithography, 30
phase modulation, 2
phase retardation, 4
phase-coherent surface-relief gratings, 1
photolithography, 5
photonic band gap (PBG), 207
photonic crystal, 32, 207
piecewise regularity, 276
planar waveguide, 22, 276, 306
Plancherel theorem, 220
plane wave (Fourier expansion) methods, 255
plane wave spectrum (PWS) technique, 161
plane wave spectrum method, 160
plane waves, 159
PML ABCs, 156
PML matching condition, 157
PML method, 157
PMMA, 17
Poincaré's inequality, 195, 293, 296
point source, 10

INDEX

point spectrum, 216
point spread function, 10
polar coordinates, 77, 79, 292
polarization effects, 12
polarization sensitivity, 27
poles of a meromorphic function, 274
power monitor, 15
precision micromachining, 4
preconditioner, 50
principle of conservation of energy, 77
printing, 13
prohibited frequencies, 207
propagating modes, 58, 255
propagating reflected mode, 58
propagating transmitted modes, 58
pseudodifferential operator, 238
pump field intensity, 63

quadrature error, 319
quantum mechanics, 31
(quasi) characterization, 289
quasi momentum, 217, 223
quasi-periodic, 7
quasi-periodic solutions, 40
quasi-uniform mesh, 48
quasiperiodic solutions, 55

radar cross section, 90
radiating ones, 274
radiation condition, 40, 42
random impurities, 255
Rayleigh expansion, 39, 72, 113
Rayleigh hypothesis, 73, 76
Rayleigh points, 23
Rayleigh's approach, 76
Rayleigh–Sommerfeld diffraction, 169
raytracing, 6, 97
reactive ion etching, 27
real Bloch variety, 224
reciprocal, 179
reciprocal lattice, 218
reciprocity, 313
rectangular coordinate system, 145
rectangular crossed gratings, 128
rectangular gratings, 128
recursive formulas, 79, 81
reducible, 225
reduction method, 115

reference medium, 298
refractive element, 2
refractive index, 6
relaxation method, 59
relaxed formulation, 59
Rellich's theorem, 284
replication methods, 5
resolvent set, 321
resonance, 21, 22, 42, 63
resonance phenomenon, 242
resonance regime, 72
resonances, 189
resonant cavity, 30
resonant frequencies, 201
resonant waveguide gratings, 4
resonant wavelength, 24
resummation techniques, 72
rib waveguide, 273
ridge waveguide, 297
right-handedness, 179
rigorous coupled-wave approach, 112
rot_β, 278

scalar diffraction theory, 141
scaling property, 214
scanning electron microscope (SEM), 17
scattered field, 55
Schauder's lemma, 64
Schrödinger operator, 214
Schrödinger-type, 211
second harmonic field, 64
second harmonic generation, 31
second-order Mur ABC, 156
second-order nonlinear susceptibility tensor, 63
self-adjoint, 315, 321
self-adjoint operator, 210, 253, 281, 298
self-adjointness, 280
semi-infinite formulation of the BEM, 171
semi-infinite symmetric BEM (SSBEM), 170
sensors, 13
separation of variables, 77
shape interpolation basis functions, 167

Silver–Müller radiation condition, 181, 189
simple reference medium, 275
simulated annealing, 61
simulated quenching algorithms, 61
simultaneous coordinate overrelaxation, 256
single point diamond, 8
singular continuous, 226
singular frequencies, 47
singular sequence, 282, 321
singularities, 73
singularities of analytic continuation, 244
sinusoidal dielectric grating, 90
sinusoidal grating, 83
sinusoidal profile, 83
skew-symmetrical fields, 306
slab waveguide, 274
slanted volume gratings, 112
slowly varying approximation, 63
Sommerfeld radiation condition, 181, 189
sound soft media, 63
spatial phase, 6
spectral analysis of periodic differential operators, 220
spectral bands, 222
spectral theory, 275, 298
spectrum of singularities, 84
spectrum of the Dirichlet Laplacian, 234
spectrum of the Neumann Laplacian, 234
spherical coordinates, 77, 79
spurious modes, 316
spurious spectra, 257
square geometry, 233
stability, 152
stability of bounded invertibility, 47
stability result, 57
staircase phase profiles, 5
steady-state condition, 159
step-index optical fiber, 319
Strang's lemma, 319
stratified waveguides, 302
Stratton–Chu formula, 161
Stratton–Chu integral method, 160

subspace preconditioning method, 256
substrate, 40
subwavelength structures, 31
superpositions, 111
surface-relief contours, 2
surface-relief diffractive structures, 1
surface-relief grating, 112, 116
surface-relief profile, 5, 7
symmetric bilinear form, 313
symmetrical fields, 306

tangential boundary conditions, 193
tangential trace, 310
tangential transparent boundary conditions, 43
Taylor coefficients, 82
Taylor series, 80
TE eigenvalue problem, 114
TE polarization, 157, 234, 256
thin chiral coatings, 180
thin optically dense dielectric pipes, 240
threshold equation, 290, 295, 296
threshold problem, 290
tilt and decenter functions, 3
time-harmonic Maxwell equations, 39, 75, 183
TM polarization, 234, 256
Toeplitz matrix, 120
total variation seminorm, 60
total/scattered field formulation, 159
trace regularity result, 196
transfer matrix methods, 255
translations, 216
transmission conditions, 75
transmission medium, 40
transmission relations, 310
transparent boundary conditions, 41, 310, 312
transverse electric (TE), 185, 298
transverse electric (TE) polarization, 44, 112, 145
transverse electric (TE) polarized fields, 215
transverse magnetic (TM), 185
transverse magnetic (TM) polarization, 44, 112, 145

INDEX

transverse magnetic (TM) polarized fields, 215
truncated periodic structures, 63
truncation of infinite series, 121
truncation point, 171
truncations of nonlocal boundary operators, 49
tunable crystals, 258
tunable infrared (IR) filter, 16

ultranarrow bandwidth filters, 22
unbounded operator, 321
undepleted pump approximation, 63
uniform convergence, 50
unique continuation, 296
unique continuation theorem, 285
uniqueness and stability for the inverse diffraction problem, 57
unperturbed Schrödinger operator, 253
upper thresholds, 288
UV lithography, 4

vacuum wavenumber, 114
variational formulation, 41, 312
variational method, 20, 37
variational problem, 53, 313
variational techniques, 63
varying boundary, 79
vector (or edge) elements, 257
vector diffraction models, 142
velocity profile, 275
vertical-cavity surface-emitting laser (VCSEL), 13
vertices, 316
volume grating, 112, 115
volume holograms, 1

water-guided waves, 298
waveguide method, 112
waveguides, 96, 180
wavelength, 6
weak form, 193
weak formulations, 37
weak guiding assumption, 307
weak solution, 41
weak∗ L^∞ closure, 59
weak∗ compact, 60
weakly guiding optical stratified waveguides, 306
weakly guiding scalar problem, 319
Weber's result, 313
weighted L^2-space, 210
well-posed minimization problem, 60
well-posedness, 45, 49, 64, 182
wet-chemical procedures, 5
Wigner–Seitz cell, 218, 252
Wood's anomalies, 23

Yee algorithm, 152
Young's inequality, 291, 314

zero-order Fabry–Perot cavity, 29
zero-order grating, 22
zeroth-order Hankel function of the second kind, 164